OPTICAL
INSTRUMENTATION

Theory and Design

Б. Н. Бегунов, Н. П. Заказнов, С. И. Кирюшин, В. И. Кузичев

ТЕОРИЯ ОПТИЧЕСКИХ СИСТЕМ

Москва   «Машиностроение»

B. N. Begunov, N. P. Zakaznov, S. I. Kiryushin, V. I. Kuzichev

OPTICAL INSTRUMENTATION

Theory and Design

MIR Publishers Moscow

First published 1988
Revised from the 1981 Russian edition

Translated from the Russian by
M. EDELEV

На английском языке

Printed in the Union of Soviet Socialist Republics

ISBN 5-03- 000008-9

Contents

Preface *9*

Introduction *10*

1 The Fundamentals of Geometrical Optics *12*

    1.1 Fermat's Principle *12*

    1.2 The Refractive Index — Defined *14*

    1.3 The Sign Convention *14*

    1.4 The Laws of Reflection and Refraction *15*

    1.5 Total Internal Reflection *18*

    1.6 Refracting and Reflecting Surfaces *18*

2 Refraction and Reflection of Rays *20*

    2.1 Refraction of Rays at a Plane Surface *20*

    2.2 Refraction at a Spherical Surface *21*

    2.3 Reflection from a Plane Surface *23*

    2.4 Reflection from a Spherical Surface *25*

    2.5 Refraction at an Aspheric Surface *27*

    2.6 Reflection from Aspheric Surfaces *29*

3 The Perfect Optical System *31*

    3.1 The Perfect Optical System and Magnification — Defined *31*

    3.2 The Cardinal Points *31*

    3.3 Object-Image Relationships in Geometrical Optics *34*

    3.4 Angular Magnification. Nodal Points *36*

    3.5 Longitudinal Magnification *38*

    3.6 Raytracing in an Optical System Given by Its Cardinal Points *39*

    3.7 Raytracing for Tilted Object Planes *43*

    3.8 Raytracing for a Perfect System *46*

    3.9 Multielement Optical Systems *47*

4 Paraxial and 'Zero Ray' Optics *53*

    4.1 Paraxial Optics *53*

    4.2 The Optical Invariant *54*

    4.3 Numerical Raytracing with 'Zero Rays' *56*

5 Optical System Components *60*

    5.1 Materials of Optical Systems *60*

    5.2 Single Lenses *64*

    5.3 Plane-Parallel Plates *74*

    5.4 Plane, Spherical and Aspheric Mirrors *76*

    5.5 Reflecting Prisms *79*

    5.6 Refracting Prisms and Wedges *85*

    5.7 Optical Lightguides *91*

    5.8 Fresnel Lenses. Axikons *94*

6 Confining Ray Bundles in Optical Systems *101*

    6.1 Apertures *101*

    6.2 Entrance and Exit Pupils *103*

    6.3 Field of View. Vignetting. Entrance and Exit Windows *106*

6.4 Effective Aperture of Entrance Pupil *109*

7 Energy Flow in Optical Instruments   *109*
  7.1 Light Flux *112*
  7.2 Radiometry and Photometry. Symbols and Units *114*
  7.3 Polychromatic Measurements in Photometry *121*
  7.4 Radiative Transfer in Optical Systems *123*
  7.5 Energy Flow in Optical Instruments *129*
  7.6 Absorption Filters *134*
  7.7 Image Illuminance in Optical Systems *135*

8 Numerical Raytracing Through an Optical System   *140*
  8.1 Automatic Raytracing by Electronic Computer *140*
  8.2 Optical Computations with Very Narrow Astigmatic Ray Pencils *144*
  8.3 Initial Data for Numerical Raytracing *149*

9 Monochromatic Aberrations   *154*
  9.1 General *154*
  9.2 Third-Order Aberrations *156*
  9.3 Normalization of Auxiliary Rays *159*
  9.4 Spherical Aberration *162*
  9.5 Meridional Coma *167*
  9.6 The Sine Condition and Isoplanatism *168*
  9.7 Astigmatism and Field Curvature *170*
  9.8 Distortion *174*

10 Chromatic Aberrations   *176*
  10.1 Axial Chromatic Aberration *176*
  10.2 Transverse Chromatic Aberration *180*
  10.3 Spherochromatism *182*

11 The Eye as an Optical System   *185*
  11.1 The Structure of the Eye *185*
  11.2 Characteristics of the Eye *187*
  11.3 Defects of the Eye *192*

12 Illumination Systems   *195*
  12.1 Purpose and Types of Illumination Systems *195*
  12.2 Searchlight and Collimator Systems *196*
  12.3 Catadioptric Systems *201*
  12.4 Condenser Systems *205*

13 Microscopes   *208*
  13.1 The Simple Microscope *208*
  13.2 The Compound Microscope *211*
  13.3 The Resolution of a Microscope *214*
  13.4 Depth of Field *215*
  13.5 Objectives and Eyepieces *218*
  13.6 Illumination Systems *222*

14 Telescopes *225*
  14.1 The Fundamentals of the Telescope *225*
  14.2 Resolving Power and Useful Magnification *229*
  14.3 Objective Systems and Eyepieces *230*
  14.4 Focusing the Eyepiece *234*
  14.5 The Function of a Field Lens *236*
  14.6 Designing a Kepler Telescope *239*
  14.7 The Galilean Telescope *242*
  14.8 Erecting Prism Systems *244*
  14.9 Lens Erecting Telescopes *248*
  14.10 Variable Power (Zoom) Systems *251*
  14.11 Stereoscopic Telescopes *259*
  14.12 Electro-Optical Image Conversion *261*

15 Photographic Objectives *265*
  15.1 Principal Characteristics *265*
  15.2 Resolving Power and Modulation Transfer Function *269*
  15.3 Depth of Field *275*
  15.4 Light Exposure in Photography *278*
  15.5 The Principal Types of Photographic Lenses *280*

16 Optics of Television Systems *292*
  16.1 Camera and Picture Tubes *292*
  16.2 Objectives of TV Cameras *299*
  16.3 Resolution and MTF of Television Systems *301*
  16.4 Flying Spot Projection Systems *304*

17 Projection Systems *308*
  17.1 Fundamentals *308*
  17.2 Episcopes *310*
  17.3 The Projection Lantern *312*
  17.4 Size and Illumination Analysis for an Enlarger *315*

18 Optical Photoelectric Systems *319*
  18.1 Characteristics of Optical Detectors *319*
  18.2 Evaluating the Entrance Pupil Diameter *322*
  18.3 The Effect of Spectral Characteristics *323*
  18.4 Arrangements with the Detector in the Image Plane *329*
  18.5 Arrangements with the Source Image Overlapping the Detector *335*
  18.6 Arrangements with the Detector at the Exit Pupil *336*
  18.7 Some Basic Arrangements *338*

19 Optical Systems for Lasers *342*
  19.1 Properties of Laser Light *342*
  19.2 Relationships for Laser Beam Transformation *343*
  19.3 Concentrating Laser Radiation *346*
  19.4 Control of Beam Divergence *347*

19.5 The Laser Photoelectric System *350*
19.6 Optical Systems for Holography *351*
20 Anamorphic Systems   *354*
20.1 The Anamorphic Effect — Defined *354*
20.2 Cylindrical and Spherocylindrical Anamorphic Lenses *357*
20.3 Cylindrical Afocal Anamorphic Attachments *361*
21 The Design of Optical Systems   *364*
21.1 Design Techniques: General *364*
21.2 Aberration Tolerances *368*
21.3 Relating the Parameters of the Auxiliary Rays *371*
21.4 Seidel Sum Transformation for a Thin-Lens System *376*
21.5 The Basic Parameters of Thin Elements *381*
21.6 Aberrations of Aspherics *385*
21.7 Minimizing the Spherical Aberration *390*
21.8 A Cemented Doublet *394*
21.9 An Air Separated Doublet *397*
21.10 A High-Speed Separable Objective *400*
21.11 An Airspaced Triplet *403*
21.12 Reflecting Systems *406*
21.13 Catadioptric Systems *410*
21.14 Automatic Correction by Computer *415*
21.15 Summation of Aberrations *418*
21.16 Optical Specifications and Tolerances *422*
21.17 Image Evaluation *427*
21.18 Wave Aberration *429*
Appendix 1. Reflecting Prisms *432*
Appendix 2. Soviet-made Optical Materials and Products *437*
Appendix 3. Performance Specification of a Photographic Objective *442*
References *447*
Index *448*

Preface

It would be difficult to organize a reasonable course covering all the types of optical instrumentation in current use because of the immense variety of these instruments which include those designed for laser, holographic, television, and high-speed photographic applications. In arranging the material of this book we kept in mind that however sophisticated an optical system may be, its functioning can be represented as the result of the combined action of optical components and groups of elements which differ from one another in the location of the object and image. This approach is consistent with the classical division of optical systems into four types, namely, microscopes, telescopes, photographic lenses, and projection systems.

The text provides a brief and systematic outline of the theoretical principles of optical system design and also the basic principles of automatic design.

After studying this textbook the student should be able
— to correctly evaluate the principal characteristics of the instrument,
— to perform light transmission analysis and layout of optical systems,
— to select suitable optical components that fit the desired optical system, choose the source and detector, if needed, and verify this choice for consistency in providing the desired image quality,
— to design individual components of the optical system so as to achieve the desired focal length or system size,
— to design the structural elements of the system or the system components in order to arrive at a desired image quality (in terms of residual aberrations).

The outlined theory of optical systems relies on geometrical concepts, therefore it can be referred to as geometrical optics. The presentation of geometrical optics in this text can be conditionally divided into three parts: (i) geometrical optics proper, which is devoted to the fundamental concepts and laws of optics, including analysis of aberrations, (ii) the optical instrumentation theory, which draws on the material of the first part to build a geometric theory of various optical systems; and (iii) the evaluation and control of aberrations which completes the course.

Introduction

The concept of *optical system* which will be extensively used in this text embodies an arrangement of optical components, such as lenses, prisms, mirrors, plane parallel plates, and wedges, which are designed to form a desired light beam. The theory of optical systems to be presented below includes both the theory for optical components constituting the system and the performance analysis of these components arranged in various optical instruments.

Optical components appear in many diverse devices for monitoring, measurement, control, computing, recording, quality control, safety provisions, and locking, to name but a few applications. It is common to refer to instrumentation that depends for its function on optical phenomena or properties of light as *optical* instruments. Examples of such instruments may be encountered in various fields of human activity. These are microscopes, telescopes, photographic cameras, film projectors, interferometers, spectrometers, holographic setups, and the like. Optical systems harnessing laser radiation are widely used in modern technology (for welding, heat treatment, and making holes), in medicine, in physical experimentation on thermonuclear projects, and for ranging, atmospheric monitoring, and communication, to name just a few.

According to the position of the object and its image, the optical systems are conventionally divided into four classes as follows:

(i) the microscope, in which the object is at a finite distance from the system while the image is at infinity;

(ii) the telescope, where both the object viewed and its image are at infinity;

(iii) the objective lens, in which the object is at infinity while the image is at a finite distance; and

(iv) the projection system in which the object and its image are at a finite distance from the optical system.

In what follows we refer to the domain where the image is formed as the *image space* while the domain where the object lies will be referred to as the *object space*. Both the image and object spaces extent over the entire three-dimensional physical space.

It should be noted that there exist optical systems that do not form an image similar to the object in form, but rather their aim is to redistribute the rays from the object so as, for example, to produce uniform illumination of an area, which in fact will be the integral image of the light source. Such systems include the optical arrangements of illumination, photometric, and some photoelectrical instruments. These systems, however, are constituted by the aforementioned standard components such as objectives and compound microscopes.

The widespread application of diverse optical instruments and optic-based devices is accompanied by an improvement in their basic characteristics such as power, angular field, and imaging scale ensured on the basis of high-quality imagery. An important contribution to optical engineering has been made by automatic, computer-aided, optical system design.

Many fundamental physical discoveries have been harnessed by optical instrumentation and gave birth to a variety of novel devices. The discovery of the conditions that give rise to the photoelectric effect by A. G. Stoletov of Russia and H. Hertz of Germany as far back as the 19th century and later by A. Einstein have led to the development of a large family of electro-optical devices. These are extensively applied in automatic control, computer hardware, for spacecraft orientation and navigation by stars, in TV instrumentation, and in many other fields. Comprehensive investigations by N. G. Basov and A. M. Prokhorov of the USSR, and C. Townes of the USA have led to the development of lasers that today have many applications, including in the most sophisticated field of holography.

New fields of human endeavour very often stimulate research and development of optical systems for novel use. Wide-angle cinematographic projection cameras have stimulated the development of various anamorphotes. Space applications called for new improved photographic objectives. Zoom lenses have been developed which expand the possible applications in television and cinema. The use of nonspherical surfaces in the optical components of fibre and raster optics show promise for applications in optical communications.

To conclude, some of the more important Soviet research efforts in the field should be mentioned. These include the investigations by M. M. Rusinov [18] on superwide-angle objectives, and by D. S. Volosov [2] on zoom and high-speed lenses, which formed the basis for the development of new types of objectives. The investigations by D. D. Maksutov [13] on catadioptric meniscus objectives, by L. L. Slyusarev [20, 26] on aberration control of optical systems, by V. N. Churilovsky on aberration theory and catadioptric systems, and investigations by I. A. Turygin are well known in the optical engineering community.

THE FUNDAMENTALS OF GEOMETRICAL OPTICS

1.1 Fermat's Principle

We use the collective name *light* to indicate electromagnetic radiation of wavelengths 1 nm to 1 mm (these limits are fairly conditional). In this part of the text we shall be concerned with wave-like propagation of light.

If the linear dimensions of the wavefront of a light wave appreciably exceeds the wavelength (which is usually the case), the propagation of this light wave in a homogeneous medium may be regarded as rectilinear. Fig. 1.1 shows a light wave emitted by a point source A in a uniform medium. The spherical front of the wave travels to a point B through the shortest path, that is in the straight line from A to B.

When the light wave meets a medium with other properties, its wavefront undergoes deformation depending on these properties and the shape of the interface between the media. To obtain a sharp image of a point object — say of a point source emitting a divergent wave as shown in Fig. 1.1 — we trace rays from object to image through the system recalling that the rays are defined as normals to the wavefront along which light energy propagates. From geometrical considerations it is quite obvious that we have to reshape the diverging bundle of rays so that they converge in one point, that is to form a converging light wave whose centre will be the desired sharp image of the point object. Fig. 1.2 illustrates how this condition can be satisfied. The point A' will be the sharp image of A provided all the rays emanated from A meet A' simultaneously. The time these rays take to travel from A to A' must be the same, i. e. minimal as one of the rays emitted from A traverses into A' in a straight line.

To generalize this situation for the case of several homogeneous media traversed by the ray we say that if these media are labelled by 1, 2, ..., p, $p + 1$, and the clear cut interfaces dividing these media by 1, 2, ..., p, then for every ray imaging a point object into a point image we have

$$T_{min} = t_1 + t_2 + t_3 + ... + t_k + ... + t_{p+1} = \text{constant} \qquad (1.1)$$

where T_{min} is the minimal object-image travel time, $t_k = d_k/v_k$ is the time the ray takes to traverse the medium k, d_k is the ray path length in medium

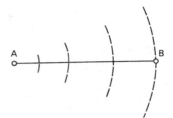

Fig. 1.1. Wavefronts of a light wave propagating in a homogeneous medium from a point source

k, v_k is the velocity of light in this medium, and p is the number of interfaces between the media.

Equation (1.1) can be rewritten as

$$\frac{d_1}{v_1} + \frac{d_2}{v_2} + \ldots + \frac{d_k}{v_k} + \ldots + \frac{d_{p+1}}{v_{p+1}} = \text{constant}$$

or

$$\frac{1}{v_1}\left(d_1\frac{v_1}{v_1} + d_2\frac{v_1}{v_2} + \ldots + d_k\frac{v_1}{v_2} + \ldots + d_{p+1}\frac{v_1}{v_{p+1}}\right) = \text{constant}$$

The ratio of the velocity at which a monochromatic light ray, i. e. a ray characterized by a definite wavelength, travels in one medium to its velocity in another medium is known as the *refractive index*, or the *index of refraction*, denoted by n. Accordingly we may rewrite the above statement as

$$d_1n_{11} + d_2n_{12} + \ldots + d_kn_{1k} + \ldots + d_{p+1}n_{p+1} = \text{constant}$$

or, with a simplifying convention for the subscripts, as

$$\sum_{k=1}^{k=p+1} d_k n_k = \text{constant} \tag{1.2}$$

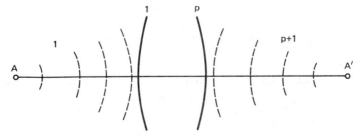

Fig. 1.2. Illustrating the condition for sharp imaging of a point

This sum is called the *optical path* of the ray. It is thus the sum of the products of the distances the ray traverses in various media by the refractive indices of these media.

From the geometric optical standpoint, an object point is imaged as a point in image space if the optical paths of all object-image rays are equal and minimal. This condition, known as *Fermat's principle*, forms the basis for many laws and relationships of geometrical optics.

1.2 The Refractive Index — Defined

The velocity of light in vacuum is independent of the wavelength and equals $299\ 792\ 458.7 \pm 1.1$ m s^{-1}. In gases, liquids and solids this velocity is less than in vacuum and depends both on the wavelength and on the status of the medium. It is convenient therefore to define the refractive index of a medium with respect to vacuum for which n is taken equal to unity. Accordingly the refractive index of a given medium relative to vacuum is the ratio of the light velocity in vacuum to that of light of certain wavelength in this medium.

The index of refraction of air at 15 °C and 101 325 Pa equals 1.00029 so that in many applications this quantity may be assumed equal to unity and independent of wavelength.

For the majority of optical media the reference refractive index has been assumed for the wavelength $\lambda = 0.58929$ μm, the yellow-orange sodium line — corresponding to Fraunhofer's D line, and is sometimes denoted by n_D. A number of countries take the reference refractive index to be measured for $\lambda = 0.58756$ μm, the yellow-orange helium line — corresponding to the d line, denoted by n_d. In the optical glass catalogue developed in co-operation by Soviet and East German experts, the reference refractive index is taken for $\lambda = 0.54607$ μm (green line of mercury — e line) and denoted accordingly by n_e.

1.3 The Sign Convention

In geometrical optics the propagation of light from left to right is assumed to be positive and the following sign convention is adopted in measurements of lines and angles.

For axially symmetric optical systems, the axis of symmetry, i. e. optical axis, is taken to be the OZ axis, and the meridional or tangential plane, i. e. the one coinciding with the page plane in optical diagrams and containing the optical axis, is assumed to be the plane *YOZ* in the right-hand Cartesian coordinate system.

Straight line segments measured in the direction of light propagation are positive as coinciding with the positive direction of the *OZ* axis. If these segments are measured in the direction opposite to that of light propagation, they receive a negative sign.

Radii of curvature of surfaces surrounding or separating optical media are assumed to be positive if the centres of curvature are on the right from these surfaces, and negative otherwise.

Because the positive direction of the *OY* axis is upwards, the line sections perpendicular to the optical axis are positive if they are above the optical axis. These relate to the heights of points where rays meet optical surfaces, and vertical line sections in object and image representations.

An angle is regarded as positive if the axis from which the measurement is made has to be turned around the angle vertex clockwise to cover the sector between the angle sides and is regarded as negative otherwise.

1.4 The Laws of Reflection and Refraction

With reference to Fig. 1.3 consider a bundle of parallel rays travelling through a plane interface between two homogeneous media of refractive indices n and n'. Assume $n < n'$, that is the light comes from the medium of lesser optical density to one of a greater optical density.

According to Fermat's principle (see eq. (1.2)), the plane incident wavefront remains plane upon refraction on the condition that

$$CAn = BDn' \tag{1.3}$$

From Fig. 1.3

$$CA = AB\sin\varepsilon \quad \text{and} \quad BD = AB\sin\varepsilon' \tag{1.4}$$

Equations (1.3) and (1.4) lead us to *Snell's law*

$$n\sin\varepsilon = n'\sin\varepsilon' \tag{1.5}$$

where ε is the angle of incidence, i. e. the angle between the normal to the surface at the point of incidence and the incident ray, and ε' is the angle of refraction which is the angle between the same normal and the refracted ray. In addition to this mathematical expression, *the law of refraction* states that the incident and refracted rays lay in the same plane with the normal to the surface at the point of incidence.

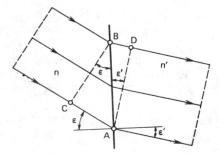

Fig. 1.3. Refraction of a parallel ray pencil at a plane interface

Consider now the ray incident on a plane reflecting surface, as shown in Fig. 1.4, and on reflection redirected remaining in the same medium of refraction index n. In view of the sign convention, the refraction law (1.5) yields

$$\sin\varepsilon = \sin(-\varepsilon') \qquad (1.6)$$

which is achievable on the condition that $n = -n'$. This change in sign of the refractive index is due to the reversal of the ray propagation direction on reflection. The condition (1.6) makes us conclude that $\varepsilon' = -\varepsilon$, that is the angle of reflection ε' equals the angle of incidence ε (in absolute value). This is the *reflection law* usually augmented by the notion that the incident and reflected rays lay in one plane with the normal to the reflecting surface at the point of incidence.

The laws of refraction and reflection remain valid on reversing the direction of rays for the opposite, i. e. there holds the principle of *reversibility* of light paths according to which the refracted ray may be changed for the incident ray, and the incident ray for the refracted, the reflected ray may be changed for the incident and the incident for the reflected.

Another notion of great use in geometrical optics is the assumption of *mutual independence of light paths* — it states that rays propagate independently of one another. More often than not these laws and notions are exploited in conjunction.

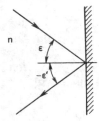

Fig. 1.4. The idea of specular reflection

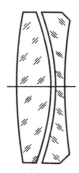

$$r_1 = 154.88 \qquad d_1 = 12 \qquad n_1 = n_3 = n_5 = 1$$
$$r_2 = -74.82 \qquad d_2 = 1.2 \qquad n_2 = 1.5713$$
$$r_3 = -74.43 \qquad d_3 = 2.5 \qquad n_4 = 1.7232$$
$$r_4 = 304.8$$

Fig. 1.5. A lens system

$$r_1 = -125.32 \quad n_1 = n_3 = 1$$
$$r_2 = -59.43 \quad d = -42.80 \qquad n_2 = -1$$

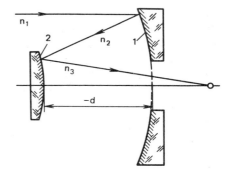

Fig. 1.6. A mirror reflection system

$$r_1 = -88.31 \qquad\qquad n_1 = 1$$
$$\qquad\qquad\quad d_1 = 4$$
$$r_2 = -90.78 \qquad\qquad n_2 = 1.5183$$
$$\qquad\qquad\quad d_2 = 56$$
$$r_3 = -264.2 \qquad\qquad n_3 = 1$$
$$\qquad\qquad\quad d_3 = -52$$
$$r_4 = -83.18 \qquad\qquad n_4 = -1$$
$$\qquad\qquad\quad d_4 = 15$$
$$r_5 = 84.72 \qquad\qquad n_5 = 1$$
$$\qquad\qquad\quad d_5 = 6$$
$$r_6 = -90.36 \qquad\qquad n_6 = 1.5713$$

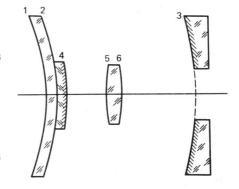

Fig. 1.7. A mirror-lens (catadioptric) system

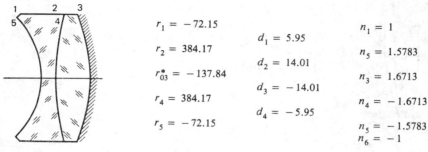

$$r_1 = -72.15$$
$$d_1 = 5.95$$
$$r_2 = 384.17$$
$$d_2 = 14.01$$
$$r_{03}^* = -137.84$$
$$d_3 = -14.01$$
$$r_4 = 384.17$$
$$d_4 = -5.95$$
$$r_5 = -72.15$$

$$n_1 = 1$$
$$n_5 = 1.5783$$
$$n_3 = 1.6713$$
$$n_4 = -1.6713$$
$$n_5 = -1.5783$$
$$n_6 = -1$$

Fig. 1.8. A mirror-lens system
with an aspheric surface

---

* This is the radius of curvature at the vertex of the aspheric surface
$y^2 + x^2 + 275.68z = 0$

1.5 Total Internal Reflection

The Snell law (1.5) indicates that $|\varepsilon'| > |\varepsilon|$ for $n > n'$. As the angle of incidence ε increases in absolute value, the angle of refraction ε' also increases to reach 90° at some ε_{cr}, i. e. on refraction the incident ray will graze the surface. Accordingly, for $\varepsilon' = 90°$

$$\sin \varepsilon_{cr} = n'/n \qquad (1.7)$$

A further increase in the angle of incidence causes reflection of the ray in agreement with the law of reflection. This phenomenon is known as *total (internal) reflection*, and the angle ε_{cr} defined by (1.7) is referred to as the *critical angle*.

To illustrate the order of magnitude, for glass of $n_e = 1.5183$ and $n' = 1$ (air) we have $\varepsilon_{cr} \approx 41°12'$.

1.6 Refracting and Reflecting Surfaces

As will be recalled, the optical system is a setup of optical components designed to form in a desired way the light beams confined in a certain solid angle.

Optical components are solids confined between some sort of surfaces which can be flat, spherical, and aspheric surfaces: cylindrical, axially symmetric surfaces of second and higher orders, to name but a few. These surfaces may be refracting and reflecting.

An optical system is called centred if the centres of spherical surfaces and axes of symmetry of other surfaces lie in one line called the *optical axis*.

The sign convention adopted above allows any optical system constituted by components with refracting and reflecting surfaces to be represented by sets of figures without recourse to graphics. Figs. 1.5 to 1.8 give some examples of such representation.

2

REFRACTION AND REFLECTION OF RAYS

2.1 Refraction of Rays at a Plane Surface

Optical components involving plane surfaces include planoconvex and planoconcave lenses, wedges, prisms, plane-parallel plates (protective glasses, second surface mirrors, filters, etc.).

With reference to Fig. 2.1 consider a ray incident on a plane interface between two media of refractive indices n_1 and n_2 such that $n_1 < n_2$. The ray coming from a point A on the system axis meets the interface at a point M distanced h above the optical axis. According to Snell's law

$$\sin \varepsilon' = \frac{n_1}{n_2} \sin \varepsilon$$

It will be seen from the figure that $\varepsilon = \sigma$ and $\varepsilon' = \sigma'$ therefore

$$\sin \sigma' = \frac{n_1}{n_2} \sin \sigma$$

Given the point A is at a distance s from the interface, $h = s \tan \sigma$. Since from the diagram $h = s' \tan \sigma'$ we have $s' = s \tan \sigma / \tan \sigma'$. Assuming A represents a point object, the section s' and angle σ' define the position of the respective virtual image at A'.

At $\sigma = 0$, the angle $\sigma' = 0$ and the rays normal to the plane surface survive on refraction without any change in their direction.

Rays emanating from one point or meeting at one point form a bundle referred to as *homocentric* in this text. One of the objectives in optical system design is to retain the homocentric property of ray bundles on passage through the system (see Section 1.1).

Let us see what happens with a homocentric ray bundle on refraction at a plane surface. We consider for this purpose how s' varies with h, i. e. for various angles σ.

Referring to Fig. 2.1 we have

$$\sin \sigma = \sin \varepsilon = h / \sqrt{h^2 + s^2}$$

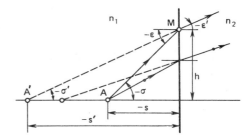

Fig. 2.1. Refraction of a ray at a
plane interface between two media

and

$$\sin\sigma' = \sin\varepsilon' = h/\sqrt{h^2 + s'^2}$$

Substituting these values for the sines in Snell's law (1.5) yields

$$s' = \frac{n_2}{n_1}\sqrt{s^2 + (1 - n_1^2/n_2^2)h^2} \qquad (2.1)$$

This expression suggests that the homocentricity of the ray bundle is not conserved as the position of the image, defined by s', is a nonlinear function of the quantity h, that is the height where rays meet the refracting surface. Accordingly, for greater h, by absolute value, there corresponds greater s' defining, we recall, the spacing of the image produced by the specific ray from the interface.

Thus the image of a point object produced by the ray bundle refracted at the plane surface will be blurred as the image of this point will consist of many point images shifted with respect to one another.

2.2 Refraction at a Spherical Surface

Consider now a ray travelling in a medium with refractive index n_1 and striking a spherical interface of radius r with another medium of refractive index n_2 such that $n_2 > n_1$. The notation for this situation is obvious from Fig. 2.2, showing that the ray emanated from A hits the interface at a point M. For the sake of consideration we construct two spheres of radii rn_1/n_2 and rn_2/n_1 concentric with the interface.

To obtain the direction of the refracted ray, the incident ray is continued to intersect the sphere rn_2/n_1 at B, and B is connected with the centre C. The point D where BC intersects the sphere rn_1/n_2 defines the direction MD of the refracted ray. We shall illustrate this statement. By construction

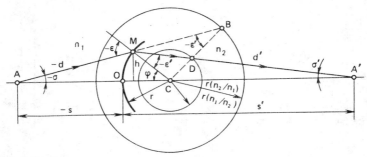

Fig. 2.2. Refraction at a spherical interface

we have

$$\frac{CD}{CM} = \frac{r(n_1/n_2)}{r} = \frac{n_1}{n_2} \quad \text{and} \quad \frac{CM}{CB} = \frac{r}{r(n_2/n_1)} = \frac{n_1}{n_2}$$

Consequently, $\triangle MCD \backsim \triangle BCM$, whence $\angle MBC = \angle CMD = -\varepsilon'$. By construction $\angle CMB = -\varepsilon$, and using the law of sines we have from $\triangle BCM$

$$\frac{CM}{CB} = \frac{\sin \varepsilon'}{\sin \varepsilon} = \frac{n_1}{n_2}$$

The above geometric construction enables one to obtain the image A' of an object point A without having to recur to the height h of incidence of the ray on the spherical interface. To demonstrate, with the notation of Fig. 2.2 we get from $\triangle AMC$

$$\sin \varepsilon = \frac{r-s}{r} \sin \sigma \qquad (2.2)$$

By the law of refraction

$$\sin \varepsilon' = \frac{n_1}{n_2} \sin \varepsilon \qquad (2.3)$$

From triangles AMC and CMA' we have $-\varepsilon = \varphi - \sigma$ and $\varphi = \sigma' - \varepsilon'$, whence

$$\sigma' = \sigma - \varepsilon + \varepsilon' \qquad (2.4)$$

By the law of sines we get from $\triangle CMA'$ $r - s' = r \sin \varepsilon'/\sin \sigma'$ or

$$s' = r(1 - \sin \varepsilon'/\sin \sigma') \qquad (2.5)$$

Formulae (2.2)-(2.5) define the position of the image for an axial object

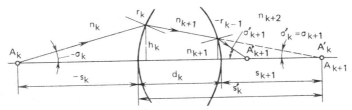

Fig. 2.3. Refraction of a ray by two spherical surfaces

point in a system with a spherical refracting surface. There will be a host of images for one object point, of course, as the quantity s' is a nonlinear function of angle σ defining the position of a ray emanating from a point on the axis. Hence, an axial point will be imaged into a multitude of points lying on the optical axis. In other words, the homocentricity of a ray bundle is not conserved in a system with a spherical refracting surface.

As a rule, an optical system involves a number of refracting surfaces. To assess the effect of these surfaces, one should apply Eqs (2.2)-(2.5) to each refracting surface in succession observing that the image obtained on refraction at the first surface is the object for the next refracting surface, and so on. Therefore, $\sigma_2 = \sigma_1'$, $\sigma_3 = \sigma_2'$, etc.

With reference to Fig. 2.3 exemplifying a system with two refracting surfaces of opposite curvature we have

$$s_{k+1} = s_k' - d_k \qquad (2.6)$$

where d_k is the vertex to vertex distance of surfaces k and $k + 1$.

One of the requirements causing rather sophisticated system arrangements is due to a homocentric ray bundle at the exit from the system.

2.3 Reflection from a Plane Surface

Figure 2.4 illustrates imaging in a plane mirror. The image A' of a point A, derived in accordance with the law of reflection ($\varepsilon = -\varepsilon'$), is *virtual*, i. e. is constructed as the intersection of the reflected rays continued behind the surface. It is seen to lie a distance $s' = -s$ from the surface.

The image $A'B'$ of a line segment AB can be traced in a similar manner. The plane mirror is seen to produce ideal images equal in size to the objects being reflected.

All laws related to refraction are valid for reflection as well, because the law of reflection may be regarded as a particular case of the refraction law at $n_2 = -n_1$.

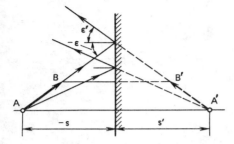

Fig. 2.4. Image formation by a plane mirror

Figure 2.5 illustrates that to deflect the reflected ray through a certain angle ψ the reflecting surface (mirror) should be rotated from position I to position II through an angle φ, the angles being related as $\psi = 2\varphi$, i. e., the angle of deflection of the reflected ray is twice the angle through which the mirror has been rotated.

In a system of two plane mirrors, labelled *1* and *2* in Fig. 2.6, placed at an angle γ to one another, the incident ray turns after double reflection through an angle ω which is twice the included angle between the mirrors, namely,

$$\omega = 2\gamma \qquad (2.7)$$

This angle is independent of the direction at which the incident ray strikes the system, therefore rotation of a system about the line at which the mirrors intersect leaves the position of the image unaltered.

One plane reflecting surface produces a mirror image. One more mirror is needed to regain the original erected orientation. A third mirror added to the system of two again produces a mirror image, i. e., such a system is equivalent to a single mirror in its effect. An obvious conclusion is that a system of odd number of mirrors fails to produce an erect image.

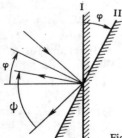

Fig. 2.5. Deflection of a ray by turning a plane mirror

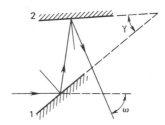

Fig. 2.6. A system of two plane mirrors with included angle γ

2.4 Reflection from a Spherical Surface

Figure 2.7 presents the notation for the reflection of rays from a concave spherical surface of radius r. Given that the incident ray is emitted at an angle σ to the axis from a point A a distance s from the vertex O of the surface, we are to locate the reflected ray, i. e. to derive angle σ' and distance s'. We see also that the ray meets the surface at a point M elevated h above the axis, and the centre of curvature of the sphere is at C.

From $\triangle AMC$ we have

$$\sin \varepsilon = \frac{q}{r}\sin \sigma \qquad (2.8)$$

where $q = r - s$.

By the reflection law $\varepsilon' = -\varepsilon$. From the triangle AMA'

$$\sigma' = \sigma + 2\varepsilon' \qquad (2.9)$$

From the triangle CMA'

$$q' = r\sin \varepsilon'/\sin \sigma' \qquad (2.10)$$

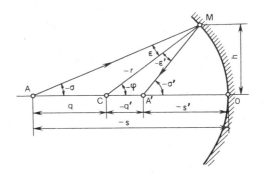

Fig. 2.7. Reflection from a concave spherical surface

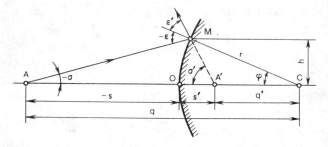

Fig. 2.8. Reflection from a convex spherical surface

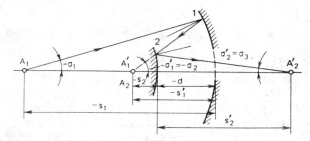

Fig. 2.9. Reflection in a system of two mirrors

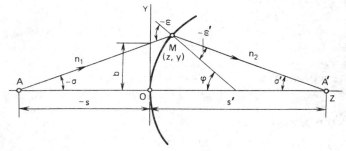

Fig. 2.10. Refraction at a second-order surface

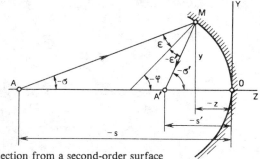

Fig. 2.11. Reflection from a second-order surface

To compute the distance s' locating the image A', we note that

$$s' = r - q' \qquad (2.11)$$

Consideration of expressions (2.8) through (2.11) leads us to conclude that s' is a nonlinear function of σ. Therefore, a spherical reflecting surface cannot retain the homocentricity of ray bundle on reflection.

Retaining the notation of the previous figure, Fig. 2.8 illustrates the reflection of a ray from a convex spherical surface.

The values of σ' and s' defined by expressions (2.8) through (2.11) provide the initial basis in the computing of ray travel through a subsequent reflecting or refracting surface. Fig. 2.9 gives an example of such situation. Here, the obvious change of notation is as follows: $\sigma_2 = \sigma_1', s_2 = s_1' - d$. To calculate s_2, the distance d must be taken with a minus sign.

The height of the point M in Fig. 2.7 or Fig. 2.8 can be calculated as $h = r\sin\varphi = r\sin(\sigma' - \varepsilon')$.

2.5 Refraction at an Aspheric Surface

Aspheric surfaces have some intrinsic advantages for optical systems as compared with ordinary spherical optics. However, aspherics are much more difficult to manufacture and control and therefore are encountered in specific applications only. Fortunately, there are indications that the technical difficulties barring extensive application of aspheric components are being overcome.

The surfaces simplest to manufacture and therefore often involved in optical systems are those of second order, that is, paraboloids, ellipsoids, and hyperboloids. We consider refraction at a second order surface with reference to the nomenclature presented in Fig. 2.10.

The raytracing in this case is based on solving a system of two equations [3, 18], one for the profile of the surface, the other being the equation of the ray incident on the surface.

The equation of the surface with its vertex at the origin is as follows

$$y^2 = a_1 z + a_2 z^2 \qquad (2.12)$$

This is an ellipse for $a_2 < 0$, a hyperbola for $a_2 > 0$, and a parabola at $a_2 = 0$.

The equation of the ray is written as

$$y = az + b \qquad (2.13)$$

Eliminating the ordinate y between (2.12) and (2.13) yields

$$a^2 z^2 + 2abz + b^2 = a_1 z + a_2 z^2$$

while elimination of the abscissa z between these equations yields

$$y^2 = a_1 \frac{y - b}{a} + a_2 \left(\frac{y - b}{a} \right)^2 \tag{2.14}$$

Because the abscissa z of the point of incidence is usually small in absolute value, for better accuracy of computations it is desirable to determine y from Eq. (2.14). This leads to

$$y = \frac{a_1 a - 2a_2 b \pm a \sqrt{a_1^2 - 4b (a_1 a - a_2 b)}}{2(a^2 - a_2)} \tag{2.15}$$

The coefficients a and b in Eq. (2.13) are as follows

$$a = - \tan \sigma$$
$$b = s \tan \sigma \tag{2.16}$$

The coefficient a_1 in Eq. (2.12) is twice the radius of the meridional section at the vertex

$$a_1 = 2r_0 \tag{2.17}$$

With account of (2.16) and (2.17), Eq. (2.15) rewrites as follows:

$$y = \frac{1 + \dfrac{s}{r_0} a_2 \pm \sqrt{1 + \dfrac{s}{r_0} \left(2 + \dfrac{s}{r_0} a_2 \right) \tan^2 \sigma}}{\dfrac{a_2}{\tan \sigma} - \tan \sigma} r_0 \tag{2.18}$$

The abscissa z of the incidence point is defined by the equation of the incident ray (2.13).

If the meridional section of the surface is a parabola, then Eq. (2.12) has $a_2 = 0$ and the solution (2.18) is simplified to the form

$$y = - \frac{r_0}{\tan \sigma} \left(1 \pm \sqrt{1 + 2 \frac{s}{r_0} \tan^2 \sigma} \right)$$

Given the coordinates z and y of the incidence point (point M in Fig. 2.10), the angle φ between the normal at this point and the OZ axis is obtained from $\tan \varphi = \partial z / \partial y$ which in this case has the form

$$\tan \varphi = \frac{y}{a_1/2 + a_2 z} = \frac{y}{r_0 + a_2 z} \tag{2.19}$$

(for the case of a parabola, $a_2 = 0$).

With reference to Fig. 2.10 (see also Fig. 2.2) we have

$$\varepsilon = \sigma - \varphi \qquad (2.20)$$

The angle of refraction ε' is determined from Snell's law as

$$\sin \varepsilon' = (n_1/n_2)\sin \varepsilon \qquad (2.21)$$

The angle which the refracted ray makes with the optical axes is obtained as

$$\sigma' = \varphi + \varepsilon' \qquad (2.22)$$

The coordinate s' of the image A' is given as

$$s' = y/\tan \sigma' + z \qquad (2.23)$$

These formulae applied in succession provide a numerical raytracing for the refraction at a surface of second order. This technique is applicable to surfaces of higher orders. For a number of surfaces, the numerical raytracing for each next surface is initiated with a set of data including the values of σ_k' and s_k', so that $\sigma_{k+1} = \sigma_k'$ and $s_{k+1} = s_k' - d_k$.

2.6 Reflection from Aspheric Surfaces

In the numerical raytracing for the case of an aspheric reflecting surface, the first thing to determine is the coordinates of the incidence point, for example, by the approach outlined in Section 2.5. Then the formulae (2.19) through (2.23) are applied, using the equality $\varepsilon' = -\varepsilon$ in plane of Eq. (2.21), related to the case of refraction.

By way of illustration we consider the reflection of a ray from the second order surface for which the meridional section is shown in Fig. 2.11. The equation of this surface is given by Eq. (2.12).

The ray emanates from the point A and meets the surface at a point M with coordinates z and y. On reflection this ray intersects the z axis at a point A', the image of A.

The angle between the normal at M to the surface and the z axis is computed with Eq. (2.19) as follows

$$\tan\varphi = y/(0.5a_1 + a_2 z)$$

The angle of incidence $\varepsilon = \sigma - \varphi$, the angle of reflection $\varepsilon' = -\varepsilon$, and the angle between the reflected ray and the optical axis $\sigma' = \varphi + \varepsilon'$. The image point A' is a distance s' from the vertex of the reflecting surface O:

$$s' = y/\tan\sigma' + z \qquad (2.24)$$

This equation is outwardly similar to (2.23), the difference being that in this case tan σ' relates to the reflected ray. The coordinates z and y for this equation are to be determined with Eqs (2.13) and (2.18).

3

THE PERFECT OPTICAL SYSTEM

3.1 The Perfect Optical System and Magnification — Defined

In this context we shall refer to an optical system as perfect if it images each point of the object as a point in image space and preserves the given scale of imaging.

Actual optical systems as a rule cannot form an absolutely sharp image completely corresponding to the object even if we assume for a moment that they are free from diffraction effects. Nevertheless the idea of perfect imagery produced by a perfect optical system is used in designing optical systems with tolerable deviations from the perfect counterpart. The following conditions should be met in order that such a system convert a homocentric ray pencil in object space into a homocentric ray pencil in image space:

(i) to every point in object space there corresponds a point in image space, and

(ii) to each straight line in object space there corresponds a straight line in image space.

These object-image corresponding points and lines, including rays, are called *conjugates*. It will be recalled that both object space and image space extend over the entire physical space.

The *transverse* (or *lateral*) *magnification* of an optical system, denoted by β in this text, is the ratio of a line segment perpendicular to the optical axis to the respective image size also taken perpendicular to the axis, in the designations of this book $\beta = y'/y$.

For perfect optical systems with circular symmetry, the transverse magnification is the same over the entire image field. In optical systems of double symmetry, transverse magnification is different in two mutually perpendicular directions of the same image plane.

3.2 The Cardinal Points

Among the multitude of points of object space there are infinitely distant points. Every luminous point at infinity gives rise to a parallel pencil

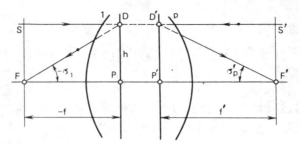

Fig. 3.1. The principal foci and principal points of an optical system

of light rays. Consider such a parallel pencil of rays incident on a refracting surface of an axially symmetric optical system. This system, if it is a perfect one, makes this pencil converge into an axial image point at F' (Fig. 3.1) called the *second* or *image side principal focus*. This point and the object point at infinity are conjugates. The plane through the second focal point and perpendicular to the axis is called the *second* or *image side principal focal plane* of the optical system.

If we now imagine a parallel pencil of rays incoming from the right on the rightmost surface of our perfect multicomponent system — this may be the pth refracting surface — the system will focus this pencil at a point F called the *first* or *object side principal focus* (any point where rays meet is a focus). The plane through this point and perpendicular to the optical axis is the *object side focal plane* of the system. Obviously, the point at infinity having launched the parallel ray pencil and the object side focal point are object and image conjugates.

For the sake of raytracing the effect of such a multicomponent optical system, consisting of several more or less thin lenses — the type used in a camera, say, may be represented with the help of a pair of conditional conjugate planes with magnification unity between them. In most practical systems these will be virtual conjugates, i. e., lying inside the system as in Fig. 3.1, where their intersections with the axis, known as the *principal points*, are denoted P and P'. These planes are called the *principal planes*, first and second for the axial points P and P', respectively.

Figure 3.1 indicates that the principal planes can be located at the intersection of rays parallel to the optical axes and incoming to the system from left and right and the continuation of the rays passing through the principal focal points F and F'.

The distance f from the first principal plane to the object side focal point is called the *first*, or *object side*, *focal length*. Similarly, the *second focal length* is the distance from the second principal plane to the second principal focus.

The three pairs of points — focal, principal, and nodal (of which we shall learn below) — are called the *cardinal points* of the system.

Figure 3.1 shows two pairs of conjugate rays: SD and $D'F'$, and FD and $D'S'$. Consequently, points D and D' lying in the principal planes at the same height h above the optical axis and having been derived as the intersections of a pair of rays conjugate to the other pair are themselves conjugate.

From Fig. 3.1 we obtain for the rays incoming on the system at a height h the following formulae for the focal lengths

$$f' = h/\tan \sigma'_p \quad \text{and} \quad f = h/\tan \sigma_1$$

where σ'_p is the angle at which the ray emerging from the system crosses the optical axes in image space, and σ_1 is the angle at which the ray traversed the system in the other direction crosses the axis in object space.

For small heights h of incidence of rays travelling close to the optical axis — i. e. for *paraxial rays* — the focal lengths may be defined as follows

$$f' = h/\sigma'_p \tag{3.1}$$

$$f = h/\sigma_1 \tag{3.2}$$

A perfect optical system may be represented as being infinitely thin. Then the first and second principal planes coalesce. Referring to Fig. 3.2 showing that this system, given by its focal lengths f and f', separates media with refractive indices n_1 and n_{p+1} imagine that a pencil of parallel rays strikes the system at a small angle σ. This ray pencil emanates from an infinitely distant point B in object space. The ray labelled 1 passes through the first focal point F and meets the system at a point K. It leaves the system parallel to the optical axes and meets the second focal plane at B' being the image of the point B at infinity.

The ray labelled 2 passes through the coinciding principal points P making with the axis an angle of incidence ε equal to σ. It is worth noting

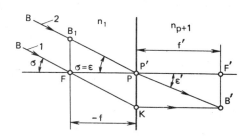

Fig. 3.2. Passage of a parallel ray pencil through a perfect optical system

that the optical axis is normal to all the surfaces of a centred optical system, whose axial points coincide at point P in the case of indefinitely thin system. Ray 2 meets the system at an angle ε' and passes through the point B'.

From the construction of Fig. 3.2 we conclude that $|FB_1| = |PK| = |F'B|$, consequently $-f \tan \varepsilon = f' \tan \varepsilon'$. As $\varepsilon \to 0$, ε' also tends to zero, therefore for small ε and ε' we have $-f \sin \varepsilon = f' \sin \varepsilon'$. Observing that by the refraction law $\sin \varepsilon / \sin \varepsilon' = n_{p+1}/n_1$ we get

$$-f/f' = n_1/n_{p+1} \tag{3.3}$$

Thus in the paraxial ray approximation we arrived at the following rule: the ratio of the focal lengths in a perfect optical system with refracting surfaces is equal to the ratio of the refractive indices of the media on either side of the optical system taken with the opposite sign. The minus sign indicates that the principal foci F and F' lie on either side of the coinciding principal planes of this system.

When the optical system under examination is interposed in a homogeneous medium, say in air of $n_1 = n_{p+1} = 1$, we have $f' = -f$, i. e. the first and second focal lengths are numerically equal.

3.3 Object-Image Relationships in Geometrical Optics

In this section we derive the raytracing equations connecting object and image points with reference to the construction in Fig. 3.3. The system is given by its cardinal points, therefore in graphical construction it is suffi-

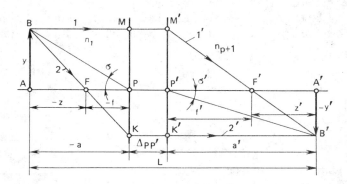

Fig. 3.3. Derivation on Newton's formula and a conjugate distance equation

cient to represent it by a skeleton of its principal planes. The object is a line segment AB of length y perpendicular to the optical axis at A on the object side. The system images point B into a point B' which occurs as the intersection of two rays in image space conjugate to the two rays emanating from B in object space.

The ray labelled 1 travels parallel to the optical axis and on changing direction at M' on the second principal plane and becomes the conjugate ray $1'$ passing through the second focus F'. The ray labelled 2 travels from B in object space through the first focus F. It changes direction at point K and in image space its conjugate will be ray $2'$ parallel to the optical axis. These two rays $1'$ and $2'$ unite in B', the image of B. Notice the practice of convenient raytracing in graphical construction by launching rays parallel to the optical axis and through a focal point. These rays are selected because their way of travel behind the system is well known.

Denote the distance from the first focus to A by $-z$ and the distance from the second focus F' to A' by z'. From two pairs of similar triangles we then obtain

$$-y'/y = -f/-z = z'/f'$$

whence

$$zz' = ff' \tag{3.4}$$

This is *Newton's conjugate distance equation* relating the conjugates measured from the foci rather than the principal planes.

If the optical system is immersed in a uniform medium, then $f' = -f$ and Eq. (3.4) becomes

$$zz' = -f'^2 \tag{3.5}$$

Let the distances of A and A' from the principal planes be a and a' respectively. Then from Fig. 3.3, $z = a - f$ and $z' = a' - f'$. Substituting these expressions for z and z' into (3.4) results in a conjugate distance equation for axial points measured from the principal planes, namely,

$$\frac{f'}{a'} + \frac{f}{a} = 1 \tag{3.6}$$

For $f = -f'$ this equation rewrites as

$$\frac{1}{a'} - \frac{1}{a} = \frac{1}{f'} \tag{3.7}$$

The same problem of determining the position of axial image points given a set of the relevant initial data is solvable with the help of the

3*

transverse or lateral (*linear*) *magnification* β, being the ratio of image to object heights. To demonstrate, from Fig. 3.3 we have

$$\beta = y'/y = -f/z = -z'/f' \tag{3.8}$$

Substitution of $a - f$ for z and $a' - f'$ for z' carries these equalities to

$$a = \frac{(\beta - 1)f}{\beta} = \frac{n_1}{n_{p+1}} \frac{1 - \beta}{\beta} f' \tag{3.9}$$

$$a' = (1 - \beta)f' \tag{3.10}$$

For $n_1 = n_{p+1}$

$$a = \frac{1 - \beta}{\beta} f' \tag{3.11}$$

When the position of the object , say a line segment y perpendicular to the optical axis, is specified by a distance a, then Eq. (3.9) or Eq. (3.11) yields the value of the transverse magnification β, and Eq. (3.8) gives the size of the image, y'.

Denote the distance between the conjugate axial points of the object and image (A and A') by L, and the spacing between the principal points along the axis by $\Delta_{PP'}$. Assuming the values of L, $\Delta_{PP'}$, and $\beta = y'/y$ are known we get for $n_1 = n_{p+1}$

$$f' = -\frac{(L - \Delta_{PP'})\beta}{(1 - \beta)^2} \tag{3.12}$$

$$a' = -\frac{(L - \Delta_{PP'})\beta}{1 - \beta} \tag{3.13}$$

$$a = -\frac{L - \Delta_{PP'}}{1 - \beta} \tag{3.14}$$

3.4 Angular Magnification. Nodal Points

The angular magnification γ of an optical system is defined as the ratio of the tangents of angles the ray makes with the optical axis on the image and object side:

$$\gamma = \tan \sigma_p'/\tan \sigma_1 \tag{3.15}$$

For an infinitely thin perfect optical system schematized in Fig. 3.4, $\gamma = a/a'$, and upon incorporation of a and a' from Eqs. (3.9) and (3.10)

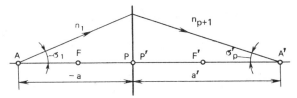

Fig. 3.4. Construction to deduce the angular magnification

$$\gamma = \frac{n_1}{n_{p+1}} \frac{1}{\beta} \qquad (3.16)$$

The conjugate axial points of unit angular magnification are called the *nodal points* of the system. These points are usually denoted by N and N'. Geometrically, rays entering the system directed towards the axial point N leave it as though from the axial point N' and make the same angle with the axis. Equation (3.16) suggests that with the same medium on both sides of the system the nodal points coincide with the principal points for which $\beta = 1$.

If the optical system separates media with different refractive indices n_1 and n_{p+1}, then for $\beta = 1$, i. e. in the principal planes (see Fig. 3.3), and for small σ and σ' we obtain

$$\gamma_{PP'} = \tan \sigma' / \tan \sigma \approx n_1/n_{p+1} = -f/f' \qquad (3.17)$$

With reference to Fig. 3.5 we locate the nodal points for a system given by its principal foci and points for the case of $n_{11} \neq n_{p+1}$. Let the distances of the first and second nodal points from the respective foci be denoted by z_N and $z_{N'}$, then from Eq. (3.16) with account of $\gamma = 1$ we get $\beta_{NN'} = -f/f'$, and Eq. (3.8) yields $\beta_{NN'} = -z'_{N'}/f' = -f/z_N$. Accordingly, $z'_{N'} = f$ and $z_N = f'$.

The spacing $\Delta_{NN'}$ between the nodal points is determined from

$$-f + \Delta_{PP'} + f' = z_N + \Delta_{NN'} - z'_{N'}$$

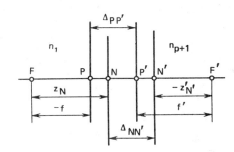

Fig. 3.5. The nodal points of an optical system

where $\Delta_{PP'}$ is the spacings between the principal points. Since $z_N = f'$ and $z'_{N'} = f$, we have $\Delta_{NN'} = \Delta_{PP'}$, that is, the spacings between the nodal points and principal points in a system are identical.

3.5 Longitudinal Magnification

The *longitudinal magnification* α of an optical system is defined as the ratio of an infinitesimally small axial line segment in image space to the conjugate line segment in object space, viz.,

$$\alpha = \frac{\partial z'}{\partial z}$$

Figure 3.6 shows the conjugate line segments $\Delta z'$ and Δz whose ratio in the limit of $\Delta z \to 0$ gives the longitudinal magnification α.

We obtain the ratio $\partial z'/\partial z$ by differentiating Newton's conjugate distance equation $zz' = ff'$ with respect to z' and z. This yields $z\partial z' + z'\partial z = 0$, whence

$$\alpha = \frac{\partial z'}{\partial z} = -\frac{z'}{z}$$

Since $z' = ff'/z$, then

$$\alpha = -\frac{ff'}{z^2}$$

which with account of

$$-\frac{f}{z} = \beta \quad \text{and} \quad \frac{f'}{z} = -\frac{n_{p+1}}{n_1}\frac{f}{z} = \frac{n_{p+1}}{n_1}\beta$$

becomes

$$\alpha = \frac{n_{p+1}}{n_1}\beta^2 \tag{3.18}$$

Observing the relationship (3.16) between the transverse and angular magnification leads to

$$\alpha\gamma = \beta$$

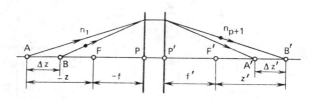

Fig. 3.6. Construction to find the longitudinal magnification

With the same medium on both sides of the optical system Eq. (3.18) reduces to $\alpha = \beta^2$.

3.6 Raytracing in an Optical System Given by Its Cardinal Points

This section will be devoted to graphical raytracing analysis in the imaging of point and line objects by a perfect optical system specified by its cardinal points. The numerical substantiation of this analysis will be drawn on Eqs. (3.4) through (3.14) derived in Section 3.4. For simplicity we assume that there is the same medium on both sides of the system.

The aim of this treatment is to illustrate that the graphical solution of a raytracing problem provides an easily obtainable and readily tractable solution for many cases of practical interest. We wish to display a few constructions for later convenience. Before we begin we wish to note that for a system having the same medium on both sides Eq. (3.17) suggests identical angles σ and σ' made in Fig. 3.3 by the rays BP and $P'B'$ with the optical axis.

Consider the construction of the image for a line segment perpendicular to the optical axis of the system with reference to Fig. 3.7. For the point B, the image B' is determined as the intersection in image space of two rays emanating from B in object space. These may be rays labelled 1 and 2 and the corresponding rays $1'$ and $2'$ in image space. This choice of rays has no limitations connected with the equality of refractive indices on either side of the system.

Another choice of rays can be rays 1 and 3 in object space and the conjugate rays $1'$ and $3'$ in image space, the rays 3 and $3'$ being parallel as the angular magnification at the principal points, coinciding with the nodal points in this case, equals unity, i. e. $\sigma = \sigma'$.

A third construction involves rays 2, $2'$ and 3, $3'$, and is seen to be similar to that of choice two.

The image of the axial point A, being the origin of the perpendicular

Fig. 3.7. Three alternative constructions for point imagery by an optical system with the same medium on either side

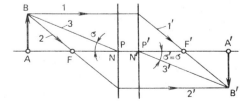

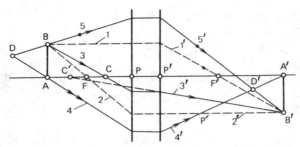

Fig. 3.8. Constructing the image of an off-axis point on the base of the construction for another extra-axial point

through B to the optical axis, can be traced as such — by dropping the perpendicular from B' to the axis.

Once the images of such two points as A and B has been constructed we can trace out the image of any point from object space. To demonstrate this statement we refer the reader to Fig. 3.8 which illustrates the construction of the images A' and B' by the rays 1 and 3 parallel to the optical axis and passing through the respective focus for the system with $f' \neq -f$. As before the axial points A and A' are the bases of perpendiculars through B and B', respectively.

To construct the image of an axial point C we draw ray 3 emanating from B through this point. Obviously, the ray $3'$ conjugate to the ray 3 in image space is to pass through B'. The second ray launched from C along the optical axis passes the system undeflected. The image, this time virtual, is produced where the continuation of ray $3'$ intersects the optical axis — point C' on the object side.

The image of an off-axis point D can be traced by launching two rays labelled 4 and 5 through A and B in object space. Beyond the system these rays are to pass through the respective image points. The intersection of rays $4'$ and $5'$, conjugate to rays 4 and 5, produces a point D', the image of D.

Whereas the second focus F' is the image of an infinitely distant point emitting a pencil of rays parallel to the optical axis, the second focal plane is the multitude of images of object points at infinity in object space, i. e. the image of an infinitely distant plane. The first focal plane is defined in a similar manner by reverting the direction of rays.

Figure 3.9 illustrates still another method of construction in imaging a line segment AB perpendicular to the optical axis on the object side. We let parallel rays labelled 1 and 2 emanating from an object point G at infinity pass through points A and B. The intersection of the conjugate rays $1'$ and

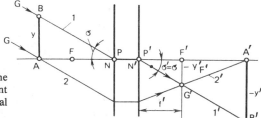

Fig. 3.9. Constructing the image of a line segment perpendicular to the optical axis

$2'$ produces point G' in the second focal plane — the image of the object point G. We have used in this construction the fact that $f' = -f$, that is the principal and nodal points in this system coincide and $\sigma' = \sigma$, that is the ray $1'$ emergent from the system is parallel to ray 1. We tacitly imply also that there is a ray from the point A along the optical axis which survives the system undeviated, so that the image of A is found as the intersection of ray $2'$ with the ray along the optical axis at A'. The point B' lies on the perpendicular to the optical axis at A' and is located by crossing with ray $1'$ conjugate to ray 1.

With account of the fact that $\sigma' = \sigma$ the spacing $y'_{F'}$ of G' from the optical axis is given as

$$|y'_{F'}| = f' \tan \sigma$$

Figure 3.10 presents the graphical imaging of an erect line segment AB achieved in the traditional way by two rays — parallel to the optical axis and passing through the focus. In the situation at hand $n_{p+1} \neq n_1$, therefore $\sigma \neq \sigma'$ and no other choice of construction rays will suit the purpose of imaging an off-axis point. We use this construction to evaluate some useful relationships. To this end we launch another ray from B to the first principal point P. After refraction by the system this ray emerges as $P'B'$. We also mark off segments FG and $F'R$ on the perpendiculars from the principal foci as shown. With reference to the figure we have

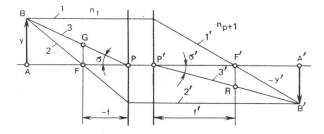

Fig. 3.10. Construction to find the image of an off-axis point

$FG = -f\tan\sigma$ and $F'R = f'\tan\sigma'$, so that

$$\frac{FG}{F'R} = -\frac{f}{f'}\frac{\tan\sigma}{\tan\sigma'}$$

From Eq. (3.17) $\tan\sigma/\tan\sigma' = -f'/f$, hence $|F'R| = |FG|$. This leads us to another method of graphical construction of images for off-axis points or line segments perpendicular to the optical axis. In addition to one of the traditional rays, *1* or *2* in object space, there should be drawn ray *3* to the first principal point. The conjugate of this ray, *3'*, pierces the second focal plane as far below the focal point as ray *3* pierces the first focal plane above the first focal point.

An obvious conclusion from this construction is that the course of a ray passing through a principal point is always known in an optical system given by its cardinal points.

For many simple optical systems as well as one-component systems, such as a simple lens or a few cemented lenses, the pole-to-pole dimension along the optical axis is usually small compared with the radii of refracting surfaces. Therefore, graphical raytracing for such systems is common to show as for a system with coinciding principal planes. A correction for the distance between the object and image is accounted for as the distance between the principal planes.

Figure 3.11 illustrates four choices of construction rays in imaging a line segment y perpendicular to the optical axis by a positive ($f' > 0$) thin optical system for the case of $n_{p+1} = n_1$ and the object lying ahead of the first focus.

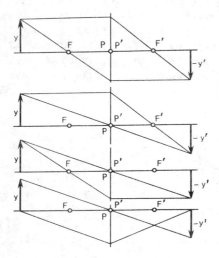

Fig. 3.11. Alternative ray pairs in graphic raytracing

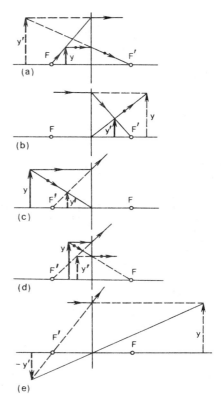

Fig. 3.12. Examples of graphic imagery   (e)

Figure 3.12 exemplifies the construction of images for a line segment y by positive, at (a) and (b), and negative ($f' < 0$), at (c) through (e), thin optical systems for various placements of y. Diagrams (b) and (e) illustrate the situations with virtual objects. These situations arise when the object is an image produced by the preceding optical system not shown in these diagrams.

3.7 Raytracing for Tilted Object Planes

Consider raytracing techniques for the case when the object plane or line is no longer perpendicular to the optical axis. An example of practical interest may be drawn from aerial surveying where the optical axis of the camera is usually other than vertical. As a result the aerophotograph has a

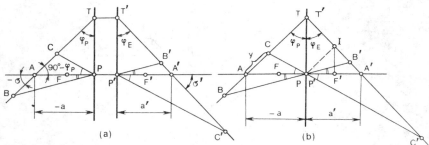

Fig. 3.13. Oblique object and image planes

scale variable over the image. In order to ensure a constant scale over the entire field being surveyed, in printing the optical axis of the projector lens is set obliquely to the plane of the aerial negative being the object in this imaging.

Figure 3.13 shows raytracing procedures for a tilted object plane represented by its meridional section BC which makes with the optical axis an angle of $90° - \varphi_P = -\sigma$. The angle φ_P is the dihedral angle between the object plane and the principal planes of the system given by its principal points and focal lengths. We wish to construct the image of this oblique plane. We note that as usually A denotes the object axial point whose image A' can be easily computed and constructed.

For the sake of construction we continue the object plane until it intersects with the first principal plane. The trace of this intersection in the meridional section is point T. This point is imaged onto the second principal plane at T' with unit magnification.

Because points A' and T' are the images of points belonging to the meridional section of the object plane, the line $A'T'$ will represent the meridional section of the image plane.

The dihedral angle $\varphi_E = 90° - \sigma'$ between the image plane and the second principal plane can be determined as

$$\tan \varphi_E = -\frac{a'}{a} \tan \varphi_P \quad \text{or} \quad \tan \varphi_E = -\beta \tan \varphi_P$$

where β is the transverse magnification of the system.

The images of points B, C and any others from the tilted plane can be obtained by tracing rays through these points and the first nodal point (coinciding with the principal point in this case). These rays emerge from the system at the same angles with the optical axis they make on entering the system.

The meridional section of oblique planes of the object and image is

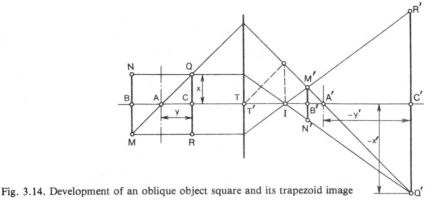

Fig. 3.14. Development of an oblique object square and its trapezoid image

represented by pairs of conjugate rays in object and image spaces (Czapski condition). A particular case of Czapski's condition is that the planes of objects and images perpendicular to the optical axis are conjugates.

All the reasoning we have used in the above construction remains valid after transition to the infinitely thin optical system. The respective construction is illustrated in Fig. 3.13b. If we set up in the oblique object plane a rectangular system of coordinates (x, y) with y being along the edgewise view in the figure and the origin at A, and a similar system of coordinates (x', y') tied up with the conjugate image plane, then given a point in the object plane (x, y) its image can be computed by

$$y' = \frac{a'y\cos \varphi_P}{\cos \varphi_E(a + y \sin \varphi_P - y \cos \varphi_P \tan \varphi_E)}$$

$$x' = \frac{y'x \cos \varphi_E}{y \cos \varphi_P}$$

These expressions can be used to compute the transverse magnification for any conjugate line segments drawn perpendicular to the optical axis through the conjugate points in the meridional section of the oblique plane under examination.

Figure 3.14 shows the development of object and image oblique planes. A square $MNQR$ in the object plane is imaged by the optical system to the trapezoid $M'N'Q'R'$. In constructing this view we used the vanishing point I (see Fig. 3.13b) lying in the second focal plane and located as the intersection of the ray $T'A'$ with this plane.

Within the segment BC of the oblique plane in Fig. 3.13 we have

$$|\beta|_{\min} = B'M'/BM = x'_B/x_B$$

$$|\beta|_{\max} = C'R'/CR = x'_C/x_C$$

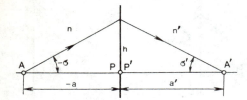

Fig. 3.15. Ray travel through a perfect optical system

3.8 Raytracing for a Perfect System

A ray emergent from an axial point and striking the optical system at a height h above the optical axis is defined by the angle σ it makes with the axis. This statement is illustrated in Fig. 3.15. Let us determine the angle σ' between the conjugate ray, producing the image point at A', and the optical axis of the perfect optical system with coinciding principal planes given by its focal lengths.

With reference to Fig. 3.15 we have $a = h/\tan \sigma$ and $a' = h/\tan \sigma'$. Incorporating these expressions into (3.6) carries it to

$$\frac{f' \tan \sigma'}{h} + \frac{f \tan \sigma}{h} = 1$$

whence

$$\tan \sigma' = -\frac{f}{f'} \tan \sigma + \frac{h}{f'}$$

or

$$\tan \sigma' = -\frac{f}{f'} \tan \sigma + \frac{h}{n'}$$

where $\phi = n'/f'$ is the *power* of the optical system.

In the general form for a multielement system this formula, sometimes called the angle formula, is as follows

$$\tan \sigma_{k+1} = -\frac{f_k}{f_k'} \tan \sigma_k + \frac{h_k \phi_k}{n_{k+1}} \qquad (3.19)$$

Replacing the ratio of the focal lengths with that of refractive indices (Eq. (3.3)) carries the angle formula to

$$\tan \sigma_{k+1} = \frac{n_k}{n_{k+1}} \tan \sigma_k + \frac{h_k \phi_k}{n_{k+1}} \qquad (3.20)$$

For situations with the system immersed in air we have

$$\tan \sigma_{k+1} = \tan \sigma_k + h_k \phi_k \qquad (3.21)$$

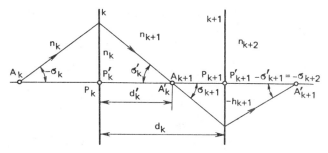

Fig. 3.16. Construction for numerical raytracing

To compute the incidence heights h_k we refer to Fig. 3.16. From similar triangles having a common vertex at A_k' (A_{k+1}) we have

$$\frac{h_k}{a_k'} = -\frac{h_{k+1}}{d_k - a_k'}$$

or $h_{k+1} = h_k - h_k d_k / a_k'$, where d_k is the distance between components k and $k+1$. Observing that $h_k/a_k' = \tan \sigma_{k+1}$ we finally get

$$h_{k+1} = h_k - d_k \tan \sigma_{k+1} \qquad (3.22)$$

Thus we arrived at the incidence height formula.

With the derived formulae for angles and incidence heights a numerical raytracing can be carried out for a perfect optical system of any complexity, or the powers of the system components can be determined for a given ray travel.

3.9 Multielement Optical Systems

In this section we wish to determine the power of an optical system involving a few elements specified by their powers K and interelement spacings d. The power of the system equivalent to a system of p components is given as

$$\phi_{eq} = \frac{n_{p+1} \tan \sigma_{p+1}}{h_1} \qquad (3.23)$$

where n_{p+1} is the refractive index on the image side of the system, σ_{p+1} the angle which the ray leaving the system makes with the optical axis, and h_1 is the height at which the ray parallel to the optical axis is incident on the first (front) component of the system.

The power of the system is determined with the use of the angle formula (3.20) as follows

$$\tan \sigma_1 = 0$$

$$\tan \sigma_2 = h_1 \frac{\phi_1}{n_2}$$

$$\tan \sigma_3 = \frac{n_2}{n_3} \tan \sigma_2 + h_2 \frac{\phi_2}{n_3}$$

. . . .

These equations lead to the following expression

$$\tan \sigma_{p+1} = \frac{1}{n_{p+1}} (h_1 \phi_1 + h_2 \phi_2 + \ldots + h_p \phi_p)$$

Reverting to (3.22) we now obtain

$$\phi_{eq} = \frac{1}{h_1} \sum_{k=1}^{k=p} h_k \phi_k \tag{3.24}$$

Let us determine the power of a system consisting of two components labelled *1* and *2* in Fig. 3.17. The incidence height of the ray shown at component *2* is defined by (3.22) in view of (3.20), namely,

$$\tan \sigma_1 = 0$$
$$\tan \sigma_2 = h_1 \phi_1 / n_2$$
$$h_2 = h_1 (1 - \phi_1 d / n_2)$$
$$\tan \sigma_3 = h_1 [\phi_1 / n_3 + (1 - \phi_1 d / n_2) \phi_2 / n_3]$$

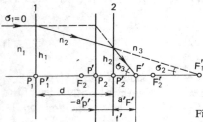

Fig. 3.17. Construction for a two-element system

Substituting the expression for $\tan \sigma_3$ into (3.23) we get

$$\phi = \phi_1 + \phi_2 - \phi_1\phi_2 d/n_2 \qquad (3.25)$$

The distance from component *2* to the equivalent second focus of the system is $a'_{F'} = h_2/\tan \sigma_3$ or

$$a'_{F'} = f'\left(1 - \phi_1\frac{d}{n_2}\right) \qquad (3.26)$$

while the distance of this component to the second principal plane of the system is $a'_{H'} = a'_{F'} - f'$.

The first focal length, and the position of the first principal focus and the first principal plane of the equivalent optical system is defined by reversing the direction of rays. Then in view of (3.26) we obtain the distance from component *1* to the first equivalent focus

$$a_F = f\left(1 - \phi_2\frac{d}{n_2}\right)$$

The first principal plane of the equivalent system is located at a distance of $a_H = a_F - f$ from the first component.

If both components of the system are immersed in air, then the power of the system is

$$\phi = \phi_1 + \phi_2 - \phi_1\phi_2 d \qquad (3.27)$$

and the distance from component *2* to the equivalent second focus is

$$a'_{F'} = f'(1 - \phi_1 d) \qquad (3.28)$$

There are situations where the space between the components is filled with air (air-spaced lenses) and the media beyond the system are other than air, i. e. $n_1 \neq 1$, $n_2 = 1$, and $n_3 \neq 1$. Then the power of the system is determined with the formula (3.27) where $\phi_1 = 1/f'_1$ and $\phi_2 = n_3/f'_2$.

In some two-element systems, called doublets, the second focus of the front component is made to coincide with the first focus of the back component. In this case the spacing between the components surrounded by air is equal to the sum of their back focal lengths and the power of the system is zero according to (3.27). This type of optical system is called *telescopic*. It has an infinite focal length, $f' = \infty$.

Now we determine the power of a triplet, i. e. a system consisting of three infinitely thin lenses, as shown in Fig. 3.18. We again invoke the formulae for angles and incidence heights to derive the value of $\tan \sigma_4$ for $\sigma_1 = 0$, to get

$$\phi = \frac{n_4}{f'} = \frac{n_4\tan \sigma_4}{h_1}$$

or
$$\phi = \phi_1 + \phi_2 + \phi_3 - \frac{\phi_1}{n_2} d_1 (\phi_2 + \phi_3)$$

$$- \frac{\phi_3}{n_3} d_2(\phi_1 + \phi_2 - \frac{\phi_1\phi_2}{n_2} d_1) \tag{3.29}$$

The distance $a'_{F'}$ of the second focal point F' from the back component is determined as

$$a'_{F'} = \frac{h_3}{\tan \sigma_4} = \frac{n_4 h_3}{\phi h_1}$$

$$= \frac{n_4}{\phi} \left(1 - \frac{\phi_1}{n_2} d_1 - \frac{\phi_1}{n_3} d_2 - \frac{\phi_2}{n_3} d_2 + \frac{\phi_1\phi_2 d_1 d_2}{n_2 n_3} \right)$$

or

$$a'_{F'} = \frac{n_4}{\phi} \left[1 - \phi_1 \left(\frac{d_1}{n_2} + \frac{d_2}{n_3} \right) - \frac{\phi_2}{n_3} d_2 \left(1 - \frac{\phi_1}{n_2} d_1 \right) \right] \tag{3.30}$$

For a system immersed in air with all its components, Eqs. (3.29) and (3.30) become

$$\phi = \phi_1 + \phi_2 + \phi_3 - \phi_1 d_1(\phi_2 + \phi_3) - \phi_3 d_2(\phi_1 + \phi_2 - \phi_1\phi_2 d_1) \tag{3.31}$$

$$a'_{F'} = [1 - \phi_1(d_1 + d_2) - \phi_2 d_2(1 - \phi_1 d_1)]/\phi$$

For an optical system constituted by three infinitely thin contacting components so that $d_1 = d_2 = 0$, the power is $\phi = \phi_1 + \phi_2 + \phi_3$, and according to (3.31) $a'_{F'} = f'$.

For an infinitely thin system of p infinitely thin components in contact we obtain (see Eq. (3.24))

$$\phi = \sum_{k=1}^{k=p} \phi_k \tag{3.32}$$

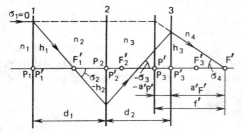

Fig. 3.18. Notation for a three-element system

that is, the power of an infinitely thin system equals the sum of the powers of the infinitely thin components.

The expressions derived in this section are valid for components with separate principal planes, i. e. with $\Delta_{PP'} \neq 0$. The respective spacings d_k are then measured from the second principal point of the preceding component to the first principal point of the next.

The problem of parameter evaluation for the equivalent system can be solved graphically. The respective construction will be carried out in steps, i. e., the image formed by the first component will be the object for the next component and so on.

4

PARAXIAL AND ZERO RAY OPTICS

4.1 Paraxial Optics

In a centred system, i. e. one consisting of a number of spherical refracting (dioptric) and reflecting (catadioptric) surfaces having their centres on a common axis, numerical raytracing can be carried out with the expressions (2.2)-(2.5) and (2.6).

Equations (2.2)-(2.5) as written for one refracting spherical surface yield an invariant of the form

$$n_1(s - r)\sin \sigma = n_2(s' - r)\sin \sigma' \qquad (4.1)$$

where n_1 and n_2 are the refractive indices of media separated by the sphere of radius r, s and s' are the object and image distances measured along the axis from the vertex (pole) of the sphere, and σ and σ' are the slope angles made by the ray with the axis at the object and image axial points. It will be recalled that in the general case Eqs. (2.2)-(2.5) give s' as a function of both s and σ, i. e. for one and the same s the distance s' will depend on σ and the convergent bundle of rays will no longer be homocentric.

The homocentricity of a ray bundle is conserved if

$$\frac{\sin \sigma}{\sin \sigma'} = \text{constant} \qquad (4.2)$$

Beyond some specific cases this condition is satisfied at small (in absolute value) angles σ and σ'.

Figure 4.1 shows a ray making with the optical axis a small angle σ and meeting the refracting surface at a small height h. These conditions imply small values of the angles φ and σ'. For these angles we may safely replace their tangents and sines with angle radian measures and write $s\sigma = s'\sigma'$, so the condition (4.2) becomes

$$\frac{\sin \sigma}{\sin \sigma'} \approx \frac{\sigma}{\sigma'} = \frac{s'}{s} = \text{constant} \qquad (4.3)$$

Substituting the thus derived sine ratio into (4.1) yields for the distance of the axial image point from the refracting surface vertex

$$s' = \frac{n_2 rs}{(n_2 - n_1)s + n_1 r} \qquad (4.4)$$

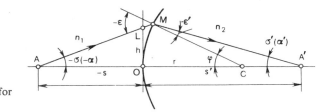

Fig. 4.1. Notation for
paraxial raytracing

For a given s this quantity is constant and independent of σ. Consequently, the homocentricity of the ray bundle emergent from the spherical refracting surface is conserved subject to the condition (4.3). Thus, for an axial ray pencil making small angles with the optical axis and incident on the optical system at a small height, the system consisting of centred surfaces behaves as a perfect system. These rays are called *Gaussian* or *paraxial* and the treatment involving such rays is referred to as the *paraxial* or *Gaussian approximation*.

We denote the angles a paraxial ray makes with the optical axis by α. Values of α must be such that

$$\sin \alpha \approx \tan \alpha \approx \alpha, \quad \text{and} \quad \cos \alpha \approx 1 \tag{4.5}$$

With such angles the law of refraction becomes

$$n_1 \varepsilon = n_2 \varepsilon' \tag{4.6}$$

because once the condition (4.5) is satisfied, $\sin \varepsilon \approx \varepsilon$ and $\sin \varepsilon' \approx \varepsilon'$.

With reference to Fig. 4.1, the paraxial ray meets the refracting axially symmetric surface at a point M lying at a small height h, therefore for both spherical and aspheric surfaces this point may be regarded as coinciding with the point L where this ray pierces the plane tangent to the surface at its axial point O. Accordingly, for paraxial rays the refracting surfaces of the system may be replaced by planes tangent to these surfaces at their axial points.

After some simple algebra (4.4) becomes

$$n_1 \left(\frac{1}{s} - \frac{1}{r} \right) = n_2 \left(\frac{1}{s'} - \frac{1}{r} \right) \tag{4.7}$$

This product is a refraction invariant valid for paraxial rays. This relation, its derivative equation

$$\frac{n_2}{s'} - \frac{n_1}{s} = \frac{n_2 - n_1}{r} \tag{4.8}$$

and eq. (4.4) relate the axial intercepts s and s' on object and image measured from the vertex of the refracting surface in paraxial raytracing.

For a reflecting surface ($n_2 = - n_1$), Eq. (4.8) takes the form

$$\frac{1}{s'} + \frac{1}{s} = \frac{2}{r} \tag{4.9}$$

Equations (4.4) and (4.7)-(4.9) can be employed to solve for axial image intercepts s_k' in tracing the paraxial ray through each surface in the system. Having derived s_1' related to the first surface, Eq. (2.6), $s_2 = s_1' - d_1$ must be invoked to step up to the next surface, where d_1 stands for the spacing between the vertices of the first and second surfaces.

4.2 The Optical Invariant

Consider the imagery of an off-axis point B into B' by a spherical refracting surface of radius r. The respective construction is presented in Fig. 4.2. The point B lies in the plane perpendicular to the optical axis and intersecting this axis at A.

We construct the image A' of point A by tracing a paraxial ray which makes with the optical axis an angle α. This ray makes with the normal to the surface at the point of incidence an angle ε. Substituting this angle along with n_1 and n_2 into (4.6) yields the angle of refraction ε'. Laying off this angle locates A' at intersection with the optical axis. Now we trace a ray through B and the centre of the refracting sphere C, and lay off on this ray a point A_1 as distant from C as point A. Points A and A_1 are essentially identical for the spherical refracting surface. Therefore the image A_1' of A_1 will be located at the same distance $- R'$ from C as the image A' of point A.

Hence, for one spherical refracting surface, conjugate elements loci make up spheres concentric with the refracting one. The radii of the object sphere and image sphere, R, and R', are related by a dependence derivable, for example, from (4.6),

$$R' = \frac{n_1}{n_2} \frac{\alpha}{\alpha'} R \tag{4.10}$$

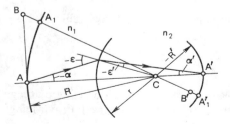

Fig. 4.2. Image formation by a spherical surface

where according to (4.3) $\alpha/\alpha' = s'/s$. Upon differentiation it gives

$$dR' = \frac{n_1}{n_2}\,\frac{\alpha}{\alpha'}\,dR$$

Consequently, as R increases (see Fig. 4.2), R' decreases in absolute value (R' is negative valued). Thus the image B' of point B is a distance $B'C < |R'|$ from the centre C. A spherical refracting surface, therefore, cannot image a line segment perpendicular to the optical axis into a perpendicular line in image space. Thus two planes perpendicular to the optical axis will be conjugate elements in object and image space on the condition that their size does not exceed the paraxial region.

Figure 4.3 shows the construction of the image y' for a small line segment y perpendicular to the optical axis on refraction at a spherical surface centred on C and separating two media of refractive indices n_1 and n_2 such that $n_1 < n_2$. The ray from the top of y to the centre C emerges from the surface undeviated and cuts off on the perpendicular erected a distance s' from O a line segment $-y'$ which is the image of y.

The image y' can be constructed also by tracing a ray through the vertex of the refracting surface O. Because for paraxial rays $n_1\varepsilon = n_2\varepsilon'$, the ray incident at an angle ε to the normal, which in this case coincides with the axis, emerges from the surface at an angle ε' to the axis. With reference to Fig. 4.3,

$$\frac{-y'}{y} = \frac{s'\varepsilon'}{-s\varepsilon}$$

which on invoking the refraction law becomes

$$\frac{-y'}{y} = \frac{s'n_1}{-sn_2} = \frac{n_1\alpha}{-n_2\alpha'}$$

whence

$$n_1 y \alpha = n_2 y' \alpha' \tag{4.11}$$

This equation, known as the Lagrange Law or the Smith-Helmholtz equation, was independently discovered by several people. The quantity standing on either side is called the Lagrange invariant, or simply the optical invariant. Since $n_1/n_2 = -f/f'$ (see Eq. (3.3)), Eq. (4.11) rewrites

$$fy\,\alpha = -f'y'\alpha' \tag{4.12}$$

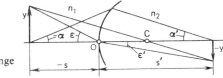

Fig. 4.3. Deduction of the Lagrange invariant

Equalities (4.11) and (4.12) can be extended to any number of refracting spherical and aspheric surfaces in the Gaussian, i. e. paraxial, approximation. For a dioptric system consisting of p surfaces we may write

$$n_1 y_1 \alpha_1 = n_2 y_1' \alpha_1' = \ldots = n_{p+1} y_p' \alpha_p'$$

where the subscript 1 refers to the object space of the first surface, and the subscript $p + 1$ to the image space of the back surface.

For a reflecting surface $n_2 = -n_1$ the Lagrange invariant reduces to the two-factor product

$$y\alpha = -y'\alpha'$$

4.3 Numerical Raytracing with 'Zero Rays'

The paraxial approximation incurs many inconveniences in focal length computations due to infinitesimal values of incidence heights and angles appearing in Eqs. (3.1) and (3.2). These inconveniences can be avoided by introducing the concept of 'zero rays'. A 'zero ray' is a fictitious ray refracting (reflecting) in the same manner as a *paraxial ray* and meeting the optical axis at the same distances as the paraxial ray, but traversing the principal planes at actual, i. e. non-Gaussian, heights above the axis. The raytracing in Chapter 3 in fact relies on this 'zero-ray' concept. The equations for slope angles (3.20) and incidence heights (3.22) form a base for zero-ray tracing including the evaluation of focal lengths of the system with known powers of the surfaces or components involved.

We note that the slope angles σ_k and σ_{k+1}, and incidence heights h_k derived in the 'zero-ray' approximation, are close in value to the angles and heights made by real rays traversing the system.

We replace the power of the kth surface in (3.20) by its expression through the structural parameters related to this surface. We also let $s = -\infty$ in (4.7), i. e. put the object point at infinity, then the distance from the vertex of the refracting surface to the image of this point is $s' = f'$. Under the circumstances the power of the kth surface is expressed through the structural parameters of the spherical surface, namely,

$$\phi_k = \frac{n_{k+1}}{f_k'} = -\frac{n_k}{f_k} = \frac{n_{k+1} - n_k}{r_k} \qquad (4.13)$$

Substituting this expression for power into Eq. (3.20) yields the 'zero-ray' equation for slope angles

$$\tan \sigma_{k+1} = \frac{n_k}{n_{k+1}} \tan \sigma_k + h_k \frac{n_{k+1} - n_k}{n_{k+1} r_k} \qquad (4.14)$$

The equation for incidence height of a 'zero ray' retains its form (3.22), viz.,

$$h_{k+1} = h_k - d_k \tan \sigma_{k+1} \qquad (4.15)$$

Equation (4.14) can be used to derive an expression for radius, which for given incidence height and slope angles yields the radii of the spherical surfaces constituting the optical system, viz.,

$$r_k = \frac{h_k(n_{k+1} - n_k)}{n_{k+1}\tan \sigma_{k+1} - n_k\tan \sigma_k} \qquad (4.16)$$

For a reflecting surface, where $n_{k+1} = -n_k$, the expression for radius takes the form

$$r_k = \frac{2h_k}{\tan \sigma_{k+1} + \tan \sigma_k}$$

Applying equations (4.14) and (4.15) step by step enables the passage of a zero ray to be computed for a catadioptric system of a few refracting and reflecting surfaces.

The tracing of a 'zero ray' is employed to determine the second focal length f' and the distance $s'_{F'}$, sometimes called *back focal length* of the system, which is the distance from the vertex of the rear surface to the second focal point of the system. Assuming the angle σ_1 being zero yields the first and subsequent equations for zero-ray angles in the form

$$\tan \sigma_2 = h_1(n_2 - n_1)/n_2 r_1$$
$$\tan \sigma_3 = (n_2/n_3)\tan \sigma_2 + h_2(n_3 - n_2)/n_3 r_2$$
$$\tan \sigma_4 = (n_3/n_4)\tan \sigma_3 + h_3(n_4 - n_3)/n_4 r_3$$
$$\cdots\cdots\cdots\cdots\cdots\cdots\cdots\cdots\cdots\cdots\cdots\cdots\cdots$$

The respective equations for incidence heights are as follows

$$h_2 = h_1 - d_1\tan \sigma_2$$
$$h_3 = h_2 - d_2\tan \sigma_3$$
$$\cdots\cdots\cdots\cdots\cdots\cdots\cdots$$

For a system of p surfaces, schematized in Fig. 4.4, the second focal length and the back focal length are given as

$$f' = h_1/\tan \sigma_{p+1} \qquad (4.17)$$
$$s'_{F'} = h_p/\tan \sigma_{p+1} \qquad (4.18)$$

With the reversed direction of the zero ray Eqs. (4.17) and (4.18) can be used to determine the first focal length f, and the front focal length s_F of the system. The last radius of curvature in the system becomes the first one,

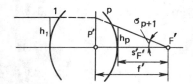

Fig. 4.4. The second focal length and the back focal length

the signs of the radii of curvature change for the opposite, and the subscripts of spacings and refractive indices are also changed. The result is taken with the opposite sign.

To compute a ray course through a plane refracting surface perpendicular to the optical axis, the radius of curvature is let be infinite in the respective equation.

When the optical system examined is a catadioptric one, i. e., involves a reflecting surface, numbered k, for example, then in Eqs. (4.14) and (4.15) related to this surface one should take into account that $n_{k+1} = -n_k$, and d_k also changes its sign because the reflected ray changes its direction. In Soviet publications the 'zero ray' equations for angles and heights assume the form somewhat different from Eqs. (4.14) and (4.15) in that the tan σ are conventionally denoted by σ

$$\sigma_{k+1} = (n_k/n_{k+1})\sigma_k + h_k(n_{k+1} - n_k)/n_{k+1}r_k \qquad (4.19)$$

$$h_{k+1} = h_k - \sigma_{k+1}d_k \qquad (4.20)$$

$$r_k = h_k(n_{k+1} - n_k)/(\sigma_{k+1}n_{k+1} - \sigma_k n_k) \qquad (4.21)$$

Equations (4.17) and (4.18) accordingly become

$$f' = h_1/\sigma_{p+1} \qquad (4.22)$$

$$s'_{F'} = h_p/\sigma_{p+1} \qquad (4.23)$$

In addition to the focal lengths and back focal length, the 'zero-ray' tracing defines the position of the image and the lateral magnification of the system for finite object distances. For simplicity the incidence height at the front surface is usually assumed equal to its radius, i. e. $h_1 = r_1$. Then for finite object distance s_1 from the system $\sigma_1 = r_1/s_1$, and the back focal length (image distance from the back surface) will be

$$s'_p = h_p/\sigma_{p+1}$$

where h_p is the incidence height at the back surface of the system, and σ_{p+1} stands for the tangence of the angle the zero ray makes with the axis in image space.

The lateral magnification of the system can be found with the formula

$$\beta = \frac{n_1}{n_{p+1}} \frac{\sigma_1}{\sigma_{p+1}} \qquad (4.24)$$

derived from (3.15) and (3.16). In this expression, n_1 and n_{p+1} are the refractive indices of the object and image media on either side of the system including p surfaces, and σ_1 and σ_{p+1} are the tangents of the zero-ray angles in object and image spaces respectively.

Of course the raytracing procedure may be programmed and relegated to a computer, however the student is recommended to perform these computations with a pocket calculator to better comprehend the essence of these computations and to have a possibility to check out each step of the procedure.

5

OPTICAL SYSTEM COMPONENTS

5.1 Materials of Optical Systems

Any substance occupying a certain volume in the optical system and transmitting visible radiation is referred to as an optical medium. This may be gaseous, say air or other naturally occurring gases, solid — for example, glass for making lenses, prisms, etc., crystalline materials for optical components, films and coatings, liquids, say, oils of various origin, and specially prepared media.

Optical components are usually manufactured of colourless or coloured optical glass, quartz, ceramic glass, crystals, plastics, and other materials.

The main material for optical component manufacture is nowadays optical colourless glass. Depending on the chemical analysis designed for particular applications, this glass must exhibit a certain set of optical constants of interest for the optical engineer. These are refractive indices for various wavelengths in the region where it is to be used and derivable quantities such as mean dispersions, V-values, and relative partial dispersions.

A large diversity of glasses of various characteristics is a necessary prerequisite for manufacturing optical instruments·of high performance.

The refractive index n_e measured for the wavelength 546.07 nm is adopted by many optical establishments as the principal index of refraction. This value is used specifically by the catalogue of optical glass compiled by Soviet and East German designers. This catalogue lists for each brand of glass refractive indices for 23 wavelengths corresponding to characteristic lines of chemical elements and for 12 laser wavelengths.

As will be recalled, dispersion is concerned with descriptions of the variation of refractive index, n, with wavelength, λ. With most transparent substances, n increases as λ decreases. There are a number of expressions describing this dependence, for example, Sellmeier's equation and Cauchy dispersion formula. In the aforementioned catalogue the respective expression is

$$n_\lambda^2 = A_1 + A_2\lambda^2 + A_3\lambda^{-2} + A_4\lambda^{-4} + A_5\lambda^{-6} + A_6\lambda^{-8}$$

where λ is the wavelength in micrometres, and the coefficients A_1 through A_6 are specified in the catalogue for each type of glass. For a wavelength range from 0.365 to 1.0139 μm this expression gives n values accurate to within $\pm 1 \times 10^{-5}$.

In the visible range, the refractive index of air at 15 °C and 101 325 Pa is $n_{air} = 1.00027$ to 1.00029. For most practical applications the index of refraction for air is assumed to be unity, i. e. equal to that of vacuum, although the former is temperature and pressure dependent.

The difference of refractive indices taken for certain wavelengths is known as the *mean dispersion*. The optical materials for use in near ultraviolet and blue ranges of the spectrum are characterized by the mean dispersion $n_i - n_g$. The visible band is characterized by the mean dispersions $n_{F'} - n_{C'}$ and $n_F - n_C$, where the subscripts refer to the respective lines of chemical elements, such as hydrogen or mercury, indicated in catalogues. Finally, in the infrared, the mean dispersions $n_r - n_{1013.9}$ and $n_{1013.9} - n_{2249.3}$ are employed.

The ratio of the type $V_3 = (n_3 - 1)/(n_1 - n_2)$ is called the *constringence* or V-value. When written in the form

$$V_e = \frac{n_e - 1}{n_{F'} - n_{C'}} \quad \text{or} \quad V_D = \frac{n_D - 1}{n_F - n_C}$$

the ratio is also termed the Abbe number. The inverse of this ratio, also used in optical texts, is known as the *dispersive power*.

For intermediate differences, *partial dispersions*, say $\Delta n = n_F - n_D$, are quoted, or *relative partial dispersions* given as

$$\frac{.\Delta n}{n_{F'} - n_{C'}} \quad \text{or} \quad \frac{\Delta n}{n_F - n_C}$$

The aforementioned catalogue, for example, quotes relative partial dispersions for 24 spectral bands when characterizing each glass type.

Glasses are somewhat arbitrarily divided into two groups, the crown glasses and the flint glasses, crowns having a V-value of 50 or more, flints having a V-value of 50 or less.

In the Soviet nomenclature colourless optical glass is divided into a number of types according to the values of refractive index n_e and constringence (V-value) V_e as follows: light crown, denoted by LK (transliterated); phosphate crown, FK; dense phosphate crown, TFK; crown, K; barium crown, BK; dense crown, TK; extra-dense crown, STK; special (anomalous dispersion) crown, OK; crown flint, K; barium flint, BF; dense barium flint, TBF; light flint, LF; flint, F; dense flint, TF; extra dense flint, STF; and special (with unusual profile of dispersion) flint.

Figure 5.1 demonstrates the distribution of the aforementioned types of glass on a n_e versus V_e diagram. Special glasses of OK and OF types may be prepared to appear at any field of the diagram belonging to crowns or flints. The crown glasses are seen to possess greater V-values and smaller refractive indices compared to that of flints. Note that the V-values are conventionally plotted in reverse, i. e. descending order.

In addition to the optical constants listed above optical system designers employ some other characteristics of optical glass. These include the temperature coefficient, $\beta_{abs, t, \lambda} = \Delta n_{abs, \lambda}/\Delta t$, representing the change in the absolute refractive index of the glass at a wavelength λ when glass temperature alters by one degree. For glasses, refractive index increases with the ambient temperature.

Another important characteristic of glass is its transmission discussed in more detail in Chapter 7.

Choosing an appropriate type of optical glass for certain operating conditions the designer must take into account the resistance of glass to moist atmospheric conditions and weakly acidic solutions, to ionizing radiation, and such physical characteristics of glass as the coefficient of linear expansion, thermal conductivity, heat capacity, density, modulus of elasticity, shear modulus, and some electrical and magnetic characteristics.

In the Soviet design classification optical glass is categorized with some figures of merit as follows: (1) deviation range of n_e, for example, a 1st category glass has $n_e = \pm 2 \times 10^{-4}$ whereas a 5th category glass has a figure of $\pm 20 \times 10^{-4}$; (2) deviation of the mean dispersion $n_{F'} - n_{C'}$; (3) uniform quality of refractive index and mean dispersion for a lot of works to be polished; (4) berefringence; (5) attenuation constant which is numerically equal to the inverse distance over which the radiation flux is attenuated owing to absorption and scattering in the glass to 1/10th of its initial value; (6) striations, in glass; (7) bubble frequency; and (8) optical homogeneity reckoned as the homogeneous refractive index over the bulk of the glass sample. For specific grades of glass, these characteristics are treated in more detail in the Soviet-GDR catalogue mentioned earlier.

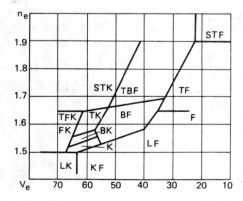

Fig. 5.1. Refractive index, n_e, versus constringence, V_e

Some Soviet glass types recommended for optical component manufacture are listed in Table A1 of the Appendix. We note that series 100 glass — those with the numerical index varying from 100 to 199 — are non-browning formulations — this browns less of all when exposed to nuclear radiation. Soviet designers of optical instrumentation look up the required types of glass in the Soviet State standards GOST 3514-76 and GOST 13659-78, and in the Soviet-GDR catalogue [28]. Elsewhere other catalogues are in use such as those of Schott and Gen. (Jena, GDR), Ohara Glass (Japan), Chance Bros., Ltd. (England), Parra Mantois et Cie (France), Baush and Lomb Co. (Rochester, New York), to list just a few manufacturers of optical glass.

Absorption filters are manufactured of coloured optical glasses produced by adding various dye stuffs to optical glass mixes. The principal characteristic of coloured glasses is that their transmission is a function of wavelength λ.

Ground glass and opal glass are used to manufacture optical components which scatter the incident radiation in a diffuse manner. Transmitting diffusers are used for such applications as rear projection screens and to produce even illumination.

Quartz glass is employed to manufacture components operating in the ultraviolet range and exhibiting low temperature expansion, and also for heat resistant elements operating in the infrared.

Soviet made quartz glass has the following type designations: KU1 glass of high transmission in the region 170 to 250 nm; KU2, having appreciable attenuation in the region 170 to 250 nm; KV, transparent in the visible spectrum; KV-R, resistant to gamma rays; and KI, a grade without appreciable absorption band up to 2800 nm.

All the aforelisted types of quartz glass exhibit the coefficient of linear expansion 2×10^{-7} per °C in the range of temperatures from $+20$ to -60 °C, and 5.2×10^{-7} per °C from $+20$ to $+120$ °C.

Another material of low linear expansion used for optical purposes is ceramic glass. This is a fine crystalline material with crystal size not exceeding a half-wavelength in the visible range. Some physical characteristics of this type of material are given in Table A2.

The SO 115M grade of ceramic glass goes to manufacture mirrors for astronomic instruments, laser giros, etc. The SO 156 grade exhibits high transparency and goes to manufacture astronomic mirrors and test glasses. The SO 21 grade exhibits a negative coefficient of linear expansion, high heat resistance (maximum admissible temperature) and is used for making monitoring hole glasses, streamlined housings operating at high temperatures, etc.

Crystalline optical materials are transparent in the ultraviolet and especially in the infrared. For instance, calcium fluoride or fluorite is transparent in the region 180 nm to 10 μm, and germanium transmits well in the infrared at wavelengths 2 to 15 μm and 40 to 60 μm.

Plastic optical materials (organic glasses) are also used for manufacturing noncritical components such as magnifying glasses, single lenses, viewfinders, Fresnel lenses, and such. These materials include polymethyl metacrylate, polyethylene, various fluoroplastics, polystyrene, to name just a few familiar materials. These materials are attractive in their low cost and mass manufacturing charges (predominantly moulding and pressing), low density and brittleness. However, they exhibit a high coefficient of linear expansion, at around 70 to 200 $\times$ 10^{-6} per °C, low optical homogeneity, low hardness, trend to natural ageing and accumulation of static electricity.

The refractive index for polymeric materials lies in the range 1.49 to 1.58, and the V-values range from 57.6 to 29.9.

Liquids such as water, benzene, kerosene, are also used as optical media of specific optical parameters. Monobrominenaphthalene, cedar oil and some other liquids are used as immersion media in microscopes, refractometers, and the like. Refractive indices for some liquids are summarized in Table 5.3.

Table 5.3. Refractive Indices of Liquids at 20 °C

Liquid	n_D	Liquid	n_D
Distilled water	1.33299	Oils	
Ethyl alcohol	1.361	paraffin	1.440
Carbon tetrachloride	1.460	olive	1.467
Benzene	1.500	terpentine	1.470
Carbon bisulphide	1.620	cedar	1.504-1.516
Monobrominenaphthalene	1.650	cloves	1.532-1.544
		anise	1.547-1.553
Methylene iodine	1.7275	cinnamon	1.585-1.619

5.2 Single Lenses

A lens is a piece of transparent material (commonly glass, plastic, quartz, etc.) bounded by two refracting surfaces of regular curvature, usually being axially symmetric and centred. Most popular lenses are bounded by two spheric surfaces. If one of the surfaces is a plane, it must be perpendicular to the optical axis.

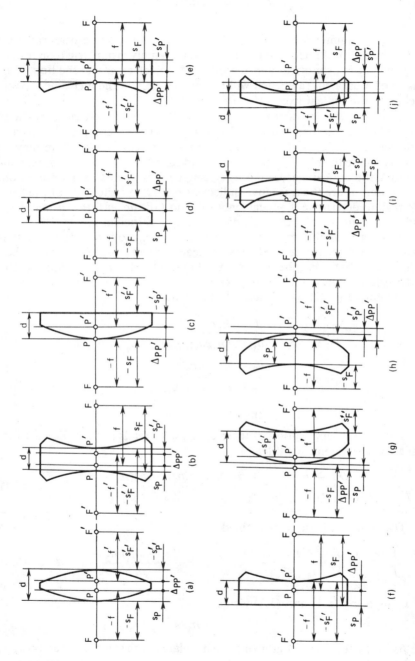

Fig. 5.2. Lens types

5 — 7391

Axially symmetric and centred lenses conserve a homocentric ray bundle in the paraxial region. When a lens is not axially symmetric (one refracting surface is cylindrical and the other plane, for example) the homocentricity of a paraxial ray pencil will be ensured only in the meridional plane including the optical axis.

Consider the refraction at a single lens with spherical surfaces (Fig. 5.2a) on the basis of zero (quasiparaxial) ray treatment. The structural parameters of such lens with spherical surfaces (one of them may be plane) include the radii of curvature r_1 and r_2, the lens thickness, d, reckoned along the optical axis, and the refractive index n_2 of lens material; note that n_1 and n_3 are the refractive indices of media on either side of the lens. We wish to determine the focal length f' and f of the lens, the distances $s_{F'}'$ and s_F, s_P' and s_P, and the spacing $\Delta_{PP'}$ of the principal points.

It is worth emphasizing that the evaluation that follows relates in equal measure to aspheric centred lenses because in the paraxial region they behave essentially as spheric lenses with the radii of curvature equal to the vertex radii of the aspheric surfaces.

In view of (4.22), (4.19) and (4.20) we have

$$f' = h_1/\sigma_3$$
$$\sigma_3 = (n_2/n_3)\sigma_2 + h_2(n_3 - n_2)/n_3 r_2$$
$$h_2 = h_1 - \sigma_2 d$$
$$\sigma_2 = h_1(n_2 - n_1)/n_2 r_1$$
$$\sigma_1 = 0$$

Consecutive application of these expressions leads to the following formula for the second focal length

$$\frac{1}{f'} = \frac{1}{n_3}\left(\frac{n_2 - n_1}{r_1} + \frac{n_3 - n_2}{r_2}\right) - \frac{(n_2 - n_1)(n_3 - n_2)}{n_2 n_3 r_1 r_2} d \qquad (5.1)$$

The first focal length is defined as

$$\frac{1}{f} = \frac{1}{n_1}\left(\frac{n_1 - n_2}{r_1} + \frac{n_2 - n_3}{r_2}\right) + \frac{(n_1 - n_2)(n_2 - n_3)}{n_1 n_2 r_1 r_2} d \qquad (5.2)$$

Comparing (5.1) and (5.2) yields

$$f'/f = -n_3/n_1$$

that is to say the focal lengths of the lens relate in the same manner as those of the perfect optical system (see Eq. (3.3)). In optical system design, the power of a lens

$$\phi = n_3/f' = -n_1/f$$

is one of its key characteristics, as the power of the system is a measure of optical action of any multilens system. The higher it is in absolute value, the closer to the lens the image appears (see Eq. (3.10)). For a lens in air $(n_1 = n_3 = 1)$, $\phi = 1/f'$.

A unit to express the power of a lens (especially spectacle lens) is the *dioptre* being the reciprocal of the focal length in metres of a lens in air, $1/f'$. Obviously, for a focal length in millimetres D (dioptres) $= 1000/f'$ (mm).

In view of (4.13), the first and second focal lengths of each of the lens refracting surfaces are as follows

$$f_1' = n_2 r_1/(n_2 - n_1) \qquad\qquad f_1 = -n_1 r_1/(n_2 - n_1)$$
$$f_2' = n_3 r_2/(n_3 - n_2) \qquad\qquad f_2 = -n_2 r_2/(n_3 - n_2)$$

Substituting the right-hand sides of these expressions into (5.1) and (5.2) yields $\phi = n_2/f_1' + n_3/f_2' - n_3 d/f_1'f_2'$, or, denoting the powers of the first and second refracting surfaces by ϕ_1 and ϕ_2,

$$\phi = \phi_1 + \phi_2 - \phi_1\phi_2 d/n_2 \tag{5.3}$$

It is worthwhile to compare (5.3) with (3.25) defining the power of two-element optical system filled with a medium of refractive index n_2. This comparison indicates that a single lens can be represented as a two-element system where ϕ_1 will be the power of the first refracting surface and ϕ_2 the power of the second refracting surface of the lens.

The back focal length of the lens follows from (4.23) as

$$s_{F'}' = \frac{h_2}{\sigma_3} = \frac{h_2}{h_1}\frac{h_1}{\sigma_3} = \frac{h_1[1 - (n_2 - n_1)d/n_2 r_1]}{h_1}f'$$

$$= f'\left(1 - \frac{n_2 - n_1}{n_2 r_1}d\right) \tag{5.4}$$

Reverting the direction of the ray we obtain the front focal distance

$$s_F = -f'\frac{n_1}{n_3}\left(1 + \frac{n_2 - n_1}{n_2 r_2}d\right) \tag{5.5}$$

The relevant notation is obvious from Fig. 5.2a.

Now we wish to find the distances $s_{P'}'$ and s_P defining the positions of the principal planes with respect to the vertices of the refracting surfaces. Fig. 5.2 a again shows that $s_{P'}' = s_{F'}' - f'$ and $s_P = s_F - f$. Then from

(5.4) and (5.5) we get

$$s_{P'}' = -f' \frac{n_2 - n_1}{n_2 r_1} d \qquad (5.6)$$

$$s_P = -f' \frac{n_1}{n_3} \frac{n_2 - n_3}{n_2 r_2} d \qquad (5.7)$$

The spacing of the principal planes is derived to be

$$\Delta_{PP'} = d + s_{P'}' - s_P$$

$$= \left[1 - \frac{f'}{n_2} \left(\frac{n_2 - n_1}{r_1} - \frac{n_1}{n_3} \frac{n_2 - n_3}{r_2} \right) \right] d \qquad (5.8)$$

Example 5.1. Determine the focal lengths for a biconvex lens of $r_1 = 20$ mm, $r_2 = -15$ mm, $d = 15$ mm, $n_2 = 1.5$, interposed between air, $n_1 = 1$, and water, $n_3 = 1.33$.

Substitution into Eqs. (5.1) and (5.2) yields $f' = 40$ mm and $f = -30$ mm. The back and front focal lengths are obtained with Eqs. (5.4) and (5.5) equal to $s_{F'}' = 30$ mm, and $s_F = -26.67$ mm. The distance between the principal planes $\Delta_{PP'} = 1.67$ mm.

For a lens in air ($n_1 = n_3 = 1$, $n_2 = n$) Eqs. (5.1)-(5.8) reduce to

$$\frac{1}{f'} = (n - 1) \left(\frac{1}{r_1} - \frac{1}{r_2} \right) + \frac{(n - 1)^2}{n r_1 r_2} d \qquad (5.9)$$

$$\frac{1}{f} = (1 - n) \left(\frac{1}{r_1} - \frac{1}{r_2} \right) - \frac{(n - 1)^2}{n r_1 r_2} d \qquad (5.10)$$

$$f'/f = -1; f = -f' \qquad (5.11)$$

$$\phi = \phi_1 + \phi_2 - \phi_1 \phi_2 \, d/n \qquad (5.12)$$

$$s_{F'}' = f'[1 - (n - 1) \, d/n r_1] \qquad (5.13)$$

$$s_F = -f'[1 + (n - 1) \, d/n r_2] \qquad (5.14)$$

$$s_{P'}' = -f'(n - 1) \, d/n r_1 \qquad (5.15)$$

$$s_P = -f'(n - 1) \, d/n r_2 \qquad (5.16)$$

$$\Delta_{PP'} = \left[1 - \frac{f'}{n}(n - 1) \left(\frac{1}{r_1} - \frac{1}{r_2} \right) \right] d \qquad (5.17)$$

All single lenses may be divided into three groups as follows.

(i) Lenses with the radii of curvature of opposite sign — these are biconvex and biconcave lenses shown in Fig. 5.2 *a*, *b*.

(ii) Lenses with one plane surface — these are planoconvex and planoconcave lenses shown in Fig. 5.2 *c-f*.

(iii) Lenses having radii of curvature of the same sign for both surfaces — these are concave-convex lenses with the axial thickness exceeding that at the edges, shown at (*g*) and (*h*); and those thinner at the axis than at the edges, shown at (*i*) and (*j*). The latter type of lens is called a *meniscus*.

As a rule lenses are axially symmetric pieces of transparent material. However, cylindrical lenses, described in Chapter 20, proved to be advantageous for some applications. Below we briefly dwell on the features of various types of lens with spherical and plane surfaces operating in air.

Biconcave lens. This lens, shown in Fig. 5.2*b*, has $r_1 < 0$ and $r_2 > 0$. The second focal length f' is negative for any lens thickness d, thus defining its divergent action. This is a negative lens.

Convex-plane lens. This lens, shown in Fig. 5.2*c*, has $r_1 > 0$ and $r_2 = \infty$. The principal lens data: f' and f, $s'_{F'}$ and s_F, $s'_{P'}$ and s_P, and $\Delta_{PP'}$ can be determined with equations derivable from Eqs. (5.9)-(5.17), namely,

$$f' = -f = r_1/(n - 1)$$
$$s'_{F'} = r_1/(n - 1) - d/n, \qquad\qquad s_F = f$$
$$s'_{P'} = -d/n, \qquad\qquad\qquad\quad s_P = 0 \qquad\qquad (5.18)$$
$$\Delta_{PP'} = d\,(n - 1)/n$$

From these expressions it will be seen that the focal lengths f and f' are independent of the lens thickness d, and the first principal plane is tangent to the convex refracting surface.

Planoconvex lens. This lens, shown in Fig. 5.2*d*, has $r_1 = \infty$ and $r_2 < 0$. The design expressions are similar to (5.18), namely,

$$f' = -f = -r_2/(n - 1)$$
$$s'_{F'} = f', \qquad\qquad\qquad s_F = r_2/(n - 1) + d/n$$
$$s'_{P'} = 0, \qquad\qquad\qquad\quad s_P = d/n \qquad\qquad (5.19)$$
$$\Delta_{PP'} = d\,(n - 1)/n$$

This type of lens having one plane refracting surface and the other convex is termed collecting (converging) or positive.

Concave-plane lens. This lens, shown in Fig. 5.2*e*, has $r_1 < 0$ and $r_2 = \infty$. The expressions of numerical raytracing for this lens are as follows

$$f' = -f = r_1/(n - 1)$$

$$s'_{F'} = r_1/(n - 1) - d/n, \qquad\qquad s_F = f$$
$$s'_{P'} = -d/n, \qquad\qquad\qquad\qquad s_P = 0$$

$$\Delta_{PP'} = d\,(n - 1)/n$$

Comparison of these expressions with those in (5.18) for the convex-plane lens shows their complete numerical identity.

Planoconcave lens. This type of lens, shown in Fig. 5.2f, is described by the following expressions

$$f' = -f = -r_2/(n - 1)$$

$$s'_{F'} = f', \qquad\qquad\qquad s_F = r_2/(n - 1) + d/n$$
$$s'_{P'} = 0, \qquad\qquad\qquad\qquad s_P = d/n$$

$$\Delta'_{PP'} = d\,(n - 1)/n$$

These expressions are seen to completely coincide with those in (5.19) for the plano-convex lens.

Convex-concave meniscus. This lens, shown in Fig. 5.2g, has $r_1 > 0$ and $r_2 > 0$ with $r_1 < r_2$. It is a converging (positive) lens as $f' > 0$. Both $s'_{P'}$ and s_P are seen to be negative, consequently, the first principal plane is situated in front of the lens.

Concave-convex meniscus. This lens, shown in Fig. 5.2h, has $r_1 < 0$, and $r_2 < 0$ with $|r_1| > |r_2|$. It is also referred to as a positive lens. The second principal plane is always situated behind the lens for this type of optical components.

Concave-convex meniscus. This lens, shown in Fig. 5.2i, has $r_1 < 0$ and $r_2 < 0$ with $|r_1| < |r_2|$. It refers to the negative lenses as $f' < 0$. The first principal plane of this meniscus is always in front of the lens.

Convex-concave meniscus. This lens, shown in Fig. 5.2j, also belongs to the negative lenses as $f' < 0$. The second principal plane of this meniscus is behind the lens.

In general, when the structural lens parameters appearing in Eq. (5.9) are related as $|r_2 - r_1| < d(n - 1)/n$, the concave-convex and convex-concave lenses will be positive, i. e. have $f' > 0$. In other words, a negative meniscus can be converted into a positive one by increasing the thickness d.

Telescopic lens. This type of lens converts a parallel pencil of rays incident on the lens into another parallel pencil emerging from the lens. The structural parameters of this lens are determined with the expression (5.9) letting $f' = \infty$, viz.,

$$r_1 - r_2 = d(n - 1)/n \qquad\qquad (5.20)$$

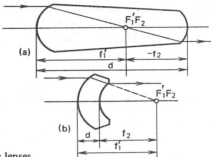

Fig. 5.3. Telescopic lenses

Equation (4.13) yields for the second focal length of the front spherical surface

$$f_1' = r_1 n/(n - 1) \tag{5.21}$$

and for the first focal length of the rear surface

$$f_2 = r_2 n/(n - 1) \tag{5.22}$$

After a little algebra Eqs. (5.20)-(5.22) lead us to

$$f_1' - f_2 = d \tag{5.23}$$

Figure 5.3 shows two telescopic lenses corresponding to (5.23) — a biconvex lens, shown at (a), and a convex-concave lens (b) which may be classified as a meniscus.

Concentric spherical lens (Fig. 5.4). In this lens the spacing of the principal points $\Delta_{PP'}$ is zero as follows from (5.17), consequently,

$$f' \frac{n - 1}{n} \left(\frac{1}{r_1} - \frac{1}{r_2} \right) = 1 \tag{5.24}$$

Substituting in this expression f' from (5.9) yields after some rearrange-

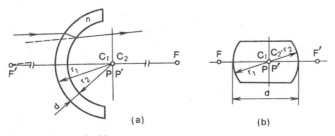

Fig. 5.4. Concentric spherical lenses

ment $r_1 - r_2 = d$, which may be viewed as a constraint defining the structural parameters of this type of lens.

The focal length of the concentric lens are determined as

$$\frac{1}{f'} = -\frac{1}{f} = \frac{n-1}{n}\left(\frac{1}{r_1} - \frac{1}{r_2}\right)$$

Lenses with spherical surfaces of equal radii. Substituting in (5.9) $r_1 = r_2 = r$ gives

$$\frac{1}{f'} = \frac{(n-1)^2}{nr^2}d$$

If $r_1 = -r_2 = r$ and the lens thickness $d = 2r$, the lens is a sphere with

$$\frac{1}{f'} = 2\frac{n-1}{nr}$$

Lenses with inverted principal planes (Fig. 5.5). In these lenses the spacing between the principal planes is negative, i. e. a ray traversing the lens from left to right meets first the second principal plane, and then the first principal plane. Equation (5.17) suggests that this case occurs when

$$f'\frac{n-1}{n}\left(\frac{1}{r_1} - \frac{1}{r_2}\right) > 1$$

Eliminating f' between this inequality and (5.9), after some algebra, yields another condition for this type of lens

$$0 < n + \frac{n-1}{r_2 - r_1}d < 1$$

Subject to this condition the spacing $\Delta_{PP'}$ computed with (5.17) will be negative.

Lenses with aspheric refracting surfaces. Recent advances in manufacturing technology and testing of aspheric surfaces makes aspheric lenses economically attractive propositions in optical system design. Physically,

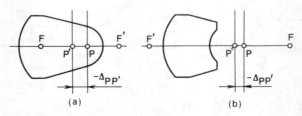

(a)   (b)

Fig. 5.5. Lenses with inverted principal planes

aspherical surfaces are advantageous in that they improve the quality of imagery, enlarge the field coverage and relative aperture of the system, and simplify the system proper owing to fewer components necessary, thus reducing system's size and weight.

Refracting surfaces manufactured as second or higher order surfaces are used in lenses of illumination systems, objectives and eyepieces. The illumination system of the microscope, for example, involves a biconvex lens in which one of the surfaces has a profile of a paraboloid of revolution. The hydroobjectives designed by M. M. Rusinov and P. D. Ivanov use a lens with a paraboloid or ellipsoid surface.

A spheroelliptic lens, shown in Fig. 5.6 is advantageous in that it ensures a homocentric bundle of rays in image space. Normally this lens faces the object with its ellipsoid surface, the spherical surface being centred on the second focal point of the lens. In a coordinate system having its origin at the surface pole, the equation of ellipse, being the meridional section of an ellipsoid, has the form

$$y^2 = 2(s'_{F'} + d) z \frac{n-1}{n} - z^2 \frac{n^2-1}{n^2}$$

Here z and y are the coordinates in the meridional section, $s'_{F'}$ is the back focal length of the lens equal to r_2, i. e., to the radius of the spherical surface.

The structural parameters of lenses governing their performance can be determined in optical system design. These parameters include the optical constants of the lens material (commonly optical glass), the radii of spherical surfaces or equations for aspheric surfaces, thickness along the optical axis, and clear aperture (rim) diameters.

Not infrequently lens specifications contain certain requirements concerning surface roughness, material quality (say, a class of the glass), blooming of the lens (antireflection coating), and tolerancing on structural parameters.

To facilitate lens manufacture and mounting, the Soviet optical in-

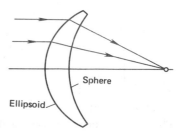

Ellipsoid

Sphere

Fig. 5.6. A spheroelliptic lens

dustry recommends that the lens diameter D, axial thickness d, and edge thickness t should be related:

(1) for positive lenses as

$$4d + 10t \geqslant D$$

with $t \geqslant 0.05D$, and

(2) for negative lenses as

$$12d + 3t \geqslant D$$

with $d \geqslant 0.05D$.

For a computed clear aperture, the lens diameter D depends on the method of lens mounting. Allowances for all structural parameters are derivable in the respective computational procedures (to be discussed in Section 21.16), and may be rounded off to meet the relevant standard [21].

5.3 Plane-Parallel Plates

Protective glasses, reticles on glass supports, filters, cover glasses and other optical elements bounded by parallel planes are essentially plane-parallel plates. Any normal to a surface of such plate may be regarded as an optical axis, therefore, if in a system, the plate's axis is taken to coincide with the common optical axis of the system.

Figure 5.7 illustrates the course of a ray in a plane-parallel plate. This ray makes with the optical axis an angle σ_1 in object space. The point A where this ray would intersect the axis in the absence of the plane plate is regarded as a virtual object point.

Since $\varepsilon_1 = \sigma_1$, then $\sin \varepsilon_1' = (n_1/n_2) \sin \sigma_1$. From Fig. 5.7a, $\varepsilon_2 = \varepsilon_1'$,

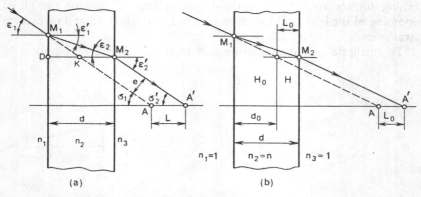

Fig. 5.7. Refraction at a plane parallel plate

therefore

$$\sin \varepsilon_2' = \sin \sigma_2' = (n_2/n_3) \sin \varepsilon_1' = (n_1/n_3) \sin \sigma_1$$

If the plane-parallel plate is in air or has the same medium on both sides so that $n_3 = n_1$, the angles σ_1 and σ_2' are equal.

The axial (longitudinal) shift L of the refracted ray at a plate inserted in a uniform medium can be readily computed from Fig. 5.7a as

$$L = d - DK = d - M_1 D \cot \varepsilon_1 = (1 - \tan \varepsilon_1'/\tan \varepsilon_1)\, d$$

For a case of small angles ε_1 and ε_1', $\tan \varepsilon_1'/\tan \varepsilon_1 \approx \varepsilon_1'/\varepsilon_1 \approx n_1/n_2$. Consequently, for a plane plate of refractive index n in air (Fig. 5.7b)

$$L_0 = d\,(n - 1)/n \qquad (5.25)$$

The lateral displacement e of a ray by a plane-parallel plate having the same medium on both sides can be also deduced from Fig. 5.7a as

$$e = d \sin (\varepsilon_1 - \varepsilon_1')/\cos \varepsilon_1'$$

Substituting for ε_1' in agreement with the refraction law at $n_1 = 1$ (air) and $n_2 = n$ we get

$$e = \sin \varepsilon_1 \left(1 - \cos \varepsilon_1/\sqrt{n^2 - \sin^2 \varepsilon_1}\right) d \qquad (5.26)$$

This expression relates the deviation angle of the plate $\sigma_1 = \varepsilon_1$ and the lateral displacement e of the ray.

The above consideration of a ray course in a plane-parallel plate indicates that having been interposed in the way of a parallel ray pencil this plate will introduce identical axial and transverse shifts for all the rays of the pencil.

Assume now that the back face of the plate is shifted to the left parallel to itself so that the ray $M_2 A'$ coincides with the direction of the ray $M_1 A$ (Fig. 5.7b). Then, obviously, A' will coincide with A and the plate thickness d reduces by a value of L. Let $L = L_0$, then in the plate so obtained the ray suffers no refraction, hence in air its refractive index has to be unity. Thereby we have reduced the optical medium of the plate to the refractive index of air. The thickness of the reduced plate (Fig. 5.7b) is

$$d_0 = d - L_0 \qquad (5.27)$$

hence $d - d_0 = L_0$ and $h_0 = h$. Incorporating L_0 into (5.27) from (5.25) yields $d_0 = d/n$, n being the actual refractive index of the plate.

Replacing a plane-parallel plate by its counterpart reduced to air simplifies size design. On recovering from the reduced plates to actual thicknesses the designer has to reconsider the shift L_0 introduced in the course of reduction.

Among the factors influencing the choice of plate thickness are allowable deformation (bend or sag of the plate), available accuracy of machining optical planes, a need for correcting for the optical path length of a ray, to name just a few commonly met causes.

High accuracy plates inserted in front of long-focus objectives must have a thickness of 0.1-0.125 diameter or diagonal. Plates of moderate accuracy — equalizing (corrector) glasses, reticle glasses, and filters inserted in the image plane — are normally as thick as 1/15 to 1/12 diameter or diagonal.

Protecting glasses, microscope slides, and cover plates are made of K8 glass. High accuracy plates are manufactured of LK5 glass, ceramic glass, or quartz if heat resistance is a requirement.

In determining the clear aperture of a plane plate, the respective refraction has to be allowed for, but on reduction this is no longer necessary.

5.4 Plane, Spherical and Aspheric Mirrors

A *plane mirror* is an optical element having a plane reflecting surface used to redirect the optical axis. A combination of mirrors set at an angle may be used to invert the image.

To make a plane mirror, a reflecting coating is applied either on the front or rear surface of a plane-parallel plate, thus producing first surface mirrors and second surface mirrors. In high accuracy mirrors the reflecting surface is at the front plane, as shown in Fig. 5.8a. This method, above all, excludes the appearance of ghost reflections. It also circumvents the influence of manufacturing inaccuracies of the rear surface with respect to the front, for example a wedge profile. Second surface mirrors may also introduce asymmetry in the beam structure.

Precise mirrors are manufactured of optical glass, say, of the K8 grade. For noncritical applications, technical glass or plastics may be used. Light

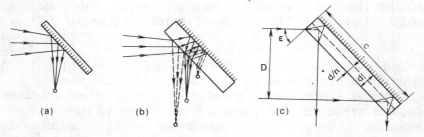

Fig. 5.8. Plane mirrors: (*a*) first surface mirror, (*b*) second surface mirror, (*c*) notation to determine the mirror size *c*

plastics help to diminish the moment of inertia for mirrors rotated at high angular speeds as in rapid cine cameras. The thickness of a mirror is usually a function of its size, accuracy of the surface required, and the method of mounting.

The size c of a second surface mirror (Fig. 5.8c) can be determined as

$$c = \frac{D}{\cos \varepsilon} + 2 \frac{d}{n} \tan \varepsilon$$

where D is the diameter of the light beam defining the mirror width, ε is the angle of ray incidence, defining the mirror position, and d/n is the reduced thickness of the plane-parallel plate.

Reflecting coatings applied to a glass surface can be of silver, aluminium, chromium, and rhodium. A widely used technology consists in applying a layer of silver onto the back surface of a plane-parallel plate and covering it by another layer of copper or laquer for protection purposes.

Semitransparent mirrors have a light dividing coating of silver (calls for additional protection against oxidation), aluminium, chromium, niobium, and gold. Most coatings of this type allow division of reflected and transmitted light in any desired proportion.

Ghosting in second surface mirrors can be suppressed by making the plane-parallel plate slightly wedge-shaped. An example of correction of this type introduced by a wedge θ is presented in Fig. 5.9.

In *spherical mirrors*, as the name implies, the reflecting surface has a spherical profile. As with plane mirrors, the reflecting surface is silvered or some other metal is deposited for reflection. Mirrors with curvilinear aspheric surfaces also find their way to optics offering the same advantages as the aspheric refracting surfaces outlined in Section 2.5.

In an optical system, spherical mirrors produce effects equivalent to that of lenses. The advantages they offer include (1) higher transmission, (2) absence of distortions incurred by refracting surfaces due to dispersion (chromatic aberrations), (3) lower size and weight, (4) possibilities of compact size arrangements, and (5) better utilization of the source power in illumination systems.

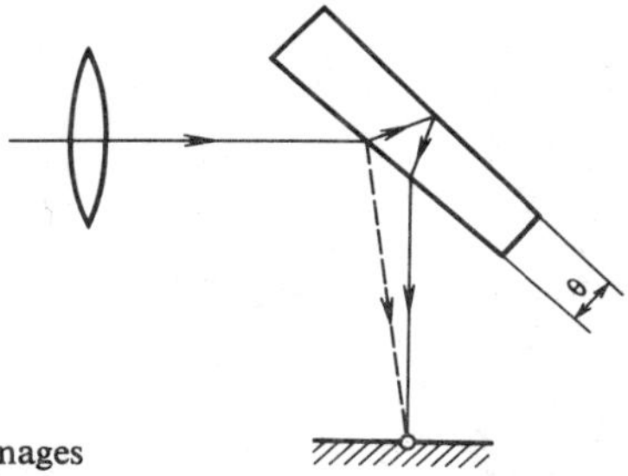

Fig. 5.9. Wedge-shaped mirror to cope with ghost images

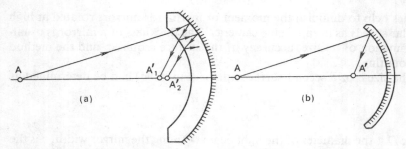

Fig. 5.10. Second surface (*a*) and first surface (*b*) spherical mirrors

Disadvantages of the mirrors, including plane mirrors, are more rigorous requirements of the accuracy of reflecting surfaces as on reflection surface defects quadruple the distortion of the incident wavefront over that produced by the defects of the refracting surface, and also the screening of a portion of the beam by the preceding mirror as this is the case, for example, with the two-mirror system.

Both spherical and aspheric mirrors are used to advantage in photographic cameras, projection systems, telescopes, microscopes, and illumination systems.

Figure 5.10 shows spherical mirrors with front and back reflecting surfaces. Second surface spherical mirrors are disadvantageous, in very much the same manner as plane mirrors, in that they are apt to ghosting illustrated in the diagram at (*a*).

Precise spherical and aspherical mirrors are manufactured of optical glass, for instance of the K8 grade. Applications calling for minimal temperature variations receive mirrors of quartz. Ceramic glass is a material of large size mirrors.

In noncritical illumination applications, the surface receiving a reflecting coating may be a metal, for example, a brass or aluminium alloy. Metal mirrors are used to advantage as reflectors in illumination systems of projection instruments. Such a reflector, shown in Fig. 5.11, is set concen-

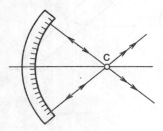

Fig. 5.11. Retro-reflector of an illumination system

tric with the light source thereby increasing its effective brightness by 20 to 50 per cent.

In two-mirror systems (see Fig. 1.6), the mirror light meets first in its propagation has a central hole of diameter deduced by raytracing analysis. The diameter, D_{eq}, of the continuous mirror equivalent to the mirror with the aperture is determined from the condition that their areas must be equivalent, i. e.,

$$D_{eq}^2 = D_{ex}^2 - D_{in}^2$$

where D_{ex} is the external diameter, and D_{in} is the aperture diameter of the mirror.

With regard to the outside diameter, D, the thickness d of a glass concentric mirror may be selected anywhere from $D/25$ to $D/5$. Rough mirrors may have a smaller thickness, while precision mirrors, for example those in reflecting objectives and catadioptric objectives, must have greater thicknesses.

To cut down weight of precision mirrors, I. I. Kryzhanovsky has suggested that a thin glass plate be fused into a titanium base providing the mirror the necessary rigidity. The glass layer is then ground to a surface which receives a reflecting coating.

5.5 Reflecting Prisms

Optical elements made as pieces of transparent material with reflecting and refracting surfaces making dihedral angles with one another are called *prisms*. Owing to reflecting surfaces (faces) a prism can be referred to as reflecting if for a beam subjected to its action the dependence of the angle of deflection on wavelength may be neglected, and a disturbance introduced into the homocentricity of a monochromatic beam may be disregarded.

In reflecting prisms the angle of refraction at the last face is equal to the angle of incidence at the front face. These prisms are installed to change the direction of the optical axis and reverse the image in the desired sense. Of course these tasks can be fulfilled with the help of plane mirrors but the resultant system will be more involved and large.

Prisms are superior to mirror systems in that (1) the angles between the prism faces are invariable whereas the included angles between the mirrors call for adjustment, and (2) no light losses are incurred when the phenomenon of total internal reflection is utilized.

Reflecting faces without mirror coating must ensure total reflection of incident light. If the angle of incidence on the reflecting face is smaller than the critical angle, ε_{cr}, of total internal reflection, then this face must receive a reflecting coating. As a rule, prisms are made of K8 and BK10 glass for which ε_{cr} amounts to 41°16′ and 39°36′ (for the refractive index corresponding to the D line).

If the angle of incidence at the front refracting face of the prism is other than 90°, then refraction must be precluded when the ray encounters the next face. For this purpose the angle of incidence, ε_1, at the front face must be limited so that, as will be seen from Fig. 5.12, $\varepsilon'_1 = \varepsilon_{cr} - \theta$. Therefore, $\sin \varepsilon_1 = n \sin(\varepsilon_{cr} - \theta)$.

For a 90° isosceles prism, shown by its principal section in Fig. 5.12, the *refracting angle* $\theta = 45°$, therefore, for K8 glass $\varepsilon_1 = 5°40′$ and for BK10 glass $\varepsilon_1 = 8°28′$. Taken twice these angles give the largest values of the fields of view for those parts of the instrument where the prism operates, provided of course that there is no reflecting coating on the reflecting face. Whether or not these ultimate angles ε_1 can be used in full depends on the admissible disturbance of ray bundle homocentricity inflicted by the action of the prism.

For converging (diverging) ray bundles, a given prism setting does not violate the admissible homocentricity of rays so long as the prism may be replaced by an equivalent plane-parallel plate. The possibility of replacement can be verified by developing (unfolding) the prism into a plane-parallel plate to find its images relative to a reflecting face. If such faces are a few, the images are to be evaluated consecutively for each face. The procedure is illustrated in Fig. 5.13. It is based on the fact that the disturbance of homocentricity for a ray bundle in the prism is the same as that inflicted by the action of the plane-parallel plate in which the prism can be unfolded. The major purpose of the unfolding and the subsequent reduction consists in evaluating the clear aperture of the prism's front face handling convergent ray bundles. The ray displacement L_0 (see Eq. (5.25)) incurred in the reduction should not be overlooked.

Some authors characterize a prism by the prism coefficient c being the ratio of the optical path length in the prism, d, to the front face clear aper-

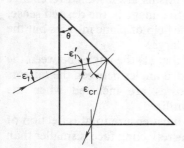

Fig. 5.12. Confining the angle of incidence for a reflecting prism

ture D, i. e. $c = d/D$. For the prisms unfolded in Fig. 5.13, $c = 1, 2$, and 2 for (a), (b), and (c) respectively.

Prisms may be made or set so that they have one, two, or three reflecting faces, one of their faces may be made into a roof, the reflecting system can be a single prism or a composite prism system. A prism with odd number of reflecting faces (can be replaced with the respective number of mirrors) produces a mirror image of the object, whereas that with an even number forms an erect image. This rule breaks down on reflection of rays of one ray bundle in different planes.

One reflecting prism can be converted into another with a 'roof' by replacing one of the reflecting faces with two faces with an included right angle between them, for example, the hypotenuse surface in a right-angle prism (Fig. 5.13a) can be replaced with a 'roof' to become the roof prism (Amici prism) shown in Fig. 5.14. Whereas a right angle prism with one reflecting surface (hypotenuse) gives a mirror image, on conversion into the roof prism, it will produce images inverted upside down and left to right.

Figure 5.14 illustrates step by step how a roof prism images a horizontal arrow. We assume that the roof prism is of a perfect make. The ray through points *1-2-3* has only one reflection. The ray traced through points *4-5-6-7* suffers its first reflection at the surface labelled *I*, and the second reflection at point *6* on the surface labelled *II*. The ray through points *8-9-10-11* suffers its first reflection at point *9* on the surface labelled *II*, and the second reflection at point *10* on surface *I*. The course of marginal rays indicates that the roof produces a left-to-right inversion which, combined with the mirror inversion, produces the complete reversion of the image with respect to the object.

The Soviet nomenclature for prisms uses Russian (here transliterated) uppercase letters and a figure for type indication. The letter index consists of two upper case letters, the first letter coding the number of reflecting

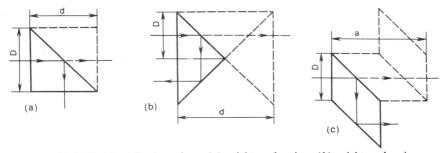

Fig. 5.13. Unfolding the reflecting prisms: (a) a right-angle prism, (b) a right angle prism used with hypotenuse as entrance and exit face, (c) a rhomboid prism

surfaces (A, for one; B, for two; and V, for three); and the second letter
coding the prism geometry: R, isosceles; S, rhombus; P, penta prism; U,
semipenta prism; M, range finder; L, Leman prism. The numerical index
indicates the angle of deflection for the axial ray. Roof prisms are indicated
by inserting a letter 'k' in the letter prefix. To illustrate, the prisms
schematised in Fig. 5.13 are coded as (a) AR = 90°, (b) BR = 180°, (c)
BC = 0°; the roof prism in Fig. 5.14 will be designated as AkR = 90°.

Simple prisms discussed above may constitute a prism system or be a
part of a prism with a rather involved geometry. Modern optical engineering abandons in such prisms. By way of example, Fig. 5.15 shows Porro
prisms of (a) first type and (b) second type. These prisms involve two and
three right angle prisms, respectively, and completely invert the image with
respect to the object. To do justice, these designs should be called
Malafeyev prisms in deference to a Russian engineer Malafeyev who
originally invented them as far back as 1827. These prisms are widely used
in binocular systems.

Prism systems can be constituted by a prism proper and a compensating
wedge incorporated in order to unfold the entire system into a plane-
parallel plate.

To give more examples of prism design, Fig. 5.16 depicts a penta prism
and its unfolding into a plane plate. This prism has two reflecting faces
which are silvered as the angles of incidence for these faces are under the
critical angle of total internal reflection. The angle of deflection for the axial ray is 90° and it is independent of the angle of incidence on the front
face, therefore revolving the prism about the axis perpendicular to the page
at C, where the planes of two faces intersect, will leave the image unmoved.
This follows in particular from Eq. (2.7) describing the effect of a double-

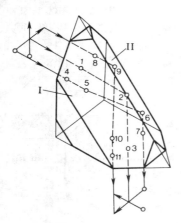

Fig. 5.14. Roof (Amici) prism

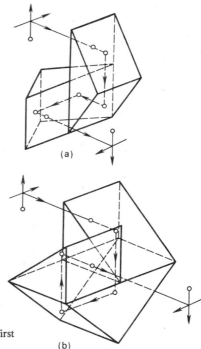

Fig. 5.15. Porro prism systems: (a) first type, (b) second type

(a)

(b)

mirror system. The effect of a penta prism is 'erecting' — indeed the number of reflecting faces is even. Replacing one of the faces with a roof produces a mirror effect.

With reference to Fig. 5.16, the ray path length in the penta prism is

$$d = D + D\sqrt{2} + D = 3.414D \qquad (5.28)$$

therefore the prism coefficient $c = 3.414$.

Figure 5.17 shows a Dove prism (sometimes called half-speed prism) whose front and rear faces are inclined at 45°. This design retains the direction of any incident ray which emerges from the prism without displacement. The prism produces a mirror effect as it has only one reflecting face. If we rotate the prism about the longitudinal axis, the image of an unmoved object will swing with the double speed. This feature of the prism is illustrated in Fig. 5.17. A 90° turn from the position shown at (a) to that in (b) swings the image through 180°. Another 90° turn to the position shown at (c) gives the image another 180° swing. Therefore, the image swings through twice the angle of prism rotation.

6*

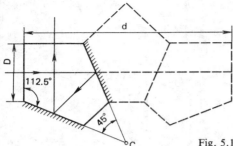

Fig. 5.16. Unfolding a penta prism

Referring to Fig. 5.17a, by the law of sines

$$\frac{a}{2\sin(90° - \varepsilon')} = \frac{D}{2\sin 45° \sin(45° + \varepsilon')} = \frac{d}{2\sin 45°}$$

where a is the base of the prism, and d is the path length of a ray in the prism. After small rearrangement and account of $\sin \varepsilon' = 1/n\sqrt{2}$ in this case we get

$$a = D \frac{2\sqrt{2n^2 - 1}}{\sqrt{2n^2 - 1} - 1} \tag{5.29}$$

$$d = D \frac{2n}{\sqrt{2n^2 - 1} - 1} \tag{5.30}$$

For K8 glass, $a = 4.23D$, $d = 3.337D$; for BK10 glass, $a = 4.04D$ and $d = 3.20D$.

The Dove prism is set to operate with parallel light beams only, as otherwise the angle of incidence for symmetric rays in a pencil will not be identical and cause asymmetry in emergent rays.

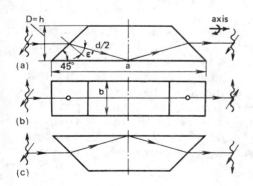

Fig. 5.17. Orientation of an image by a Dove prism (a) original position, (b) prism rotated through 90°, image swings through 180°, (c) prism rotated through 180°, image swings through 360°

Expressions (5.28)-(5.30) indicate that the diameter of the ray pencil incident on the front face of the prism is the key initial parameter in prism design.

Appendix 1 lists the basic types of reflecting prism designs, their relative dimensions, and key properties.

5.6. Refracting Prisms and Wedges

A refracting prism as the name implies is a piece of refracting medium bounded by plane surfaces which produce in intersection a *refracting edge*. These surfaces make with each other a dihedral angle θ, called the *refracting angle* of the prism. This angle is measured in the *principal section* of the prism perpendicular to its refracting edge. This region is also called the *apex* of the prism. One of the purposes of such a prism is to deviate the oncoming beam from its original direction. The angle between the emergent and incident rays is known as the *angle of deviation* (or simply *deviation* usually expressed in degrees), or the *power* of the prism when expressed in prism dioptres which are units of deviating power of a prism based on a tangent measure in centimetres effected on a scale placed one metre away from the prism. With reference to Fig. 5.18, prism dioptres $= 100 \tan \omega$.

Let us consider the course of a ray in the principal section of a refracting prism. The angle of refracting at the first face can be obtained as

$$\sin \varepsilon_1' = \sin \varepsilon_1 / n \qquad (5.31)$$

on the assumption that the prism is in air. From Fig. 5.18

$$\varepsilon_2 = \theta + \varepsilon_1' \qquad (5.32)$$

then

$$\sin \varepsilon_2' = n \sin \varepsilon_2 = n \sin (\theta + \varepsilon_1') \qquad (5.33)$$

Equations (5.31)-(5.33) lead us to the deviation

$$\omega = - \varepsilon_1 + \varepsilon_1' + \varepsilon_2' - \varepsilon_2 = \varepsilon_2' - \varepsilon_1 - \theta \qquad (5.34)$$

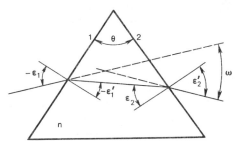

Fig. 5.18. Deviation of a light ray by a refracting prism

It is of interest to determine the angle of incidence resulting in the least deviation. For this purpose in view of (5.34) we write

$$-\frac{d\omega}{d\varepsilon_1} = \frac{d\varepsilon_2'}{d\varepsilon_1} - 1 = 0$$

or

$$d\varepsilon_2' = d\varepsilon_1 \tag{5.35}$$

From (5.32) we also have

$$d\varepsilon_2 = d\varepsilon_1' \tag{5.36}$$

Differentiating (5.31) and (5.33) yields

$$n\cos\varepsilon_1' d\varepsilon_1' = \cos\varepsilon_1 d\varepsilon_1$$

$$n\cos\varepsilon_2 d\varepsilon_2 = \cos\varepsilon_2' d\varepsilon_2'$$

whence, observing (5.35) and (5.36), we obtain

$$\frac{\cos\varepsilon_1'}{\cos\varepsilon_2} = \frac{\cos\varepsilon_1}{\cos\varepsilon_2'} \tag{5.37}$$

Multiplication of the right- and left-hand sides of Eqs. (5.31) and (5.33) results in the equality

$$\frac{\sin\varepsilon_1'}{\sin\varepsilon_2} = \frac{\sin\varepsilon_1}{\sin\varepsilon_2'} \tag{5.38}$$

Equations (3.37) and (3.38) may exist simultaneously only on the condition that

$$\varepsilon_1' = -\varepsilon_2 \quad \text{and} \quad \varepsilon_1 = -\varepsilon_2' \tag{5.39}$$

Because the second derivative $d^2\omega/d\varepsilon_1^2 > 0$, equations (5.39) define the condition for obtaining the minimum deviation for a given refracting angle of the prism θ. From these equalities it follows that ω_{min} is achievable with a symmetric course of refracted ray in the prism, i. e., perpendicular to the bisector of the prism angle θ.

From (5.33) with account of (5.39) and (5.32) we get the expression to determine ω_{min}

$$\sin\frac{\omega_{min} + \theta}{2} = n\sin\frac{\theta}{2} \tag{5.40}$$

This expression is useful in determining the refractive index of the prism material with the angles θ and ω_{min} being measured, for example, with a goniometer.

Another function of the refracting prism is to disperse the incident polychromatic light into its component colours. We consider the effect of the refractive index of prism material, being a function of monochromatic wavelength, on the angle of deviation of the refracted ray. If a beam incident on the prism is not monochromatic, its monochromatic components emerge on refraction deviated through various angles. This phenomenon of the decomposition of a beam of white light into coloured beams which spread out to produce spectra is called *dispersion*.

The *angular dispersion* or dispersive power for a prism is defined as the derivative of angle of deviation with respect to wavelength, i. e. as $d\omega/d\lambda$. We determine this derivative for the case described by the expression (5.40). Upon differentiation we have

$$\frac{d\omega_{min}}{d\lambda} = \frac{2\sin(\theta/2)}{\sqrt{1 - n_{av}^2\sin^2(\theta/2)}}\frac{dn}{d\lambda}$$

where $dn/d\lambda$ is the dispersion of the prism material, and n_{av} is the average refractive index over the wavelength interval $d\lambda$.

For a given difference of refractive indices at the ends of a wavelength interval $dn = n_{\lambda_2} - n_{\lambda_1}$ the dispersion can be determined as

$$d\omega_{min} = \frac{2\sin(\theta/2)}{\sqrt{1 - n_{av}^2\sin^2(\theta/2)}}dn \tag{5.41}$$

For $\theta = 60°$, $dn = n_{F'} - n_{C'} = 0.00812$, and $n_{av} \approx n_e = 1.5183$ (K8 glass), this expression gives $d\omega_{min} = 0.0123$ rad $\approx 40'$.

Because we have calculated angular dispersion for ω varying around the value of ω_{min}, for other $\omega \neq \omega_{min}$ greater values of dispersion will result. Formula (5.41) indicates that dispersion increases for greater refracting angles of prisms. This process, however, is not without limit, the respective constraint on θ is imposed by the same formula as $\sin(\theta/2) < 1/n_{av}$, the n_{av} again being the average value of the refractive index for a given material in the specified interval of wavelengths. If this condition is violated, there occurs total internal reflection at the other refracting face of the prism.

Refracting prisms are predominantly used in spectral devices. In the normal adjustment they are set for minimal deviation ω_{min}, because with growing dispersion for $\omega > \omega_{min}$ the resolution of the instrument decreases. Most popular designs are prisms with the principal refracting angle 60°.

Figure 5.19 shows schematically three compound prisms of spectral instruments: (a) a Rutherford prism, (b) an Amici prism, (c) an Abbe prism, and a three-prism system at (d).

The salient feature of the Rutherford prism is a large dispersion value achievable by way of increasing the refracting angle θ of the main prism, labelled 2, made of flint. The flanking prisms 1 and 2 are of crown, consequently, $n_1 = n_3 < n_2$. For such a prism the angular spread $d\omega_{\min}$ of the spectrum (see Eq. (5.41)) can be determined by differentiating the expression

$$n_2\sin(\theta/2) = n_1\sin\left[(\omega_{\min} + \theta)/2\right].$$

with respect to λ. This leads to

$$d\omega_{\min} = 2\left(\frac{dn_2}{n_{2,av}} - \frac{dn_1}{n_{1,av}}\right)\frac{n_{2,av}\sin(\theta/2)}{\sqrt{n_{1,av}^2 - n_{2,av}^2\sin^2(\theta/2)}} \qquad (5.42)$$

This formula also helps in arriving at a conclusion that for a prism immersed in a medium of refractive index over unity ($n_1 = n_3 > 1$) the limiting value of the principal refracting angle θ is dictated by the inequality

$$\sin(\theta/2) < n_{1,av}/n_{2,av}$$

where $n_{1,av}$ is the average value of the refractive index for prisms *1* and *3* in the wavelength range under examination, and $n_{2,av}$ the average index for prism *2*. For existent materials, the angle of the prism θ can be as wide as 120-125°.

Another name of the Amici prism (Fig. 5.19b) is the direct vision prism, for the direction of the incident and refracted rays coincide when the prism handles monocromatic beams. This feature is used to advantage in optical instrument design. In a specific case the setting may be such that the same direction is kept for the incident ray and the bisector of the disper-

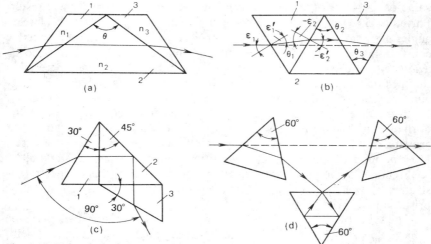

Fig. 5.19. Refracting prism systems

sion angle. In this design prisms *1* and *3* are made of crown, and prism *2* of flint so that $\theta_1 = \theta_3$ and $n_1 = n_3$.

Subject to the direct vision condition (see Fig. 5.19b), assuming an isosceles prism *2*, we arrive at the following relation between the refracting angles and refractive indices

$$n_2 \sin \frac{\theta_2}{2} = \pm\; \theta_1 \sqrt{n_1^2 - \sin^2(\theta_1 - \theta_2/2)} - \cos \theta_1 \sin \left(\theta_1 - \frac{\theta_2}{2} \right)$$

To conclude the topic, we note that the angular dispersion has no maximum for direct vision prisms.

The Abbe prism shown in Fig. 5.19c relates to the group of constant deviation prisms having $\omega = 90°$. To cope with losses, the total internal reflection prism is made of crown.

The prismatic system shown in Fig. 5.19d is constituted by three refracting prisms and also relates to direct vision designs.

Narrow angle prisms of refracting angle $\theta \leqslant 6°$ are often referred to as *wedges*. For these prisms Eq. (5.33) becomes

$$n\varepsilon_2 = n\theta \cos \varepsilon_1' + n \sin \varepsilon_1'$$

Using (5.34) we have

$$\sin \varepsilon_2' = \sin(\omega + \theta + \varepsilon_1) = (\omega + \theta)\cos \varepsilon_1 + \sin \varepsilon_1$$

These two expressions together with (5.31) yield the narrow angle prism deviation as

$$\omega = \theta \left(n \frac{\cos \varepsilon_1'}{\cos \varepsilon_1} - 1 \right)$$

Assuming that the angle of incidence, ε_1, and hence the angle of refraction, ε_1', is small, we obtain for the angle of deviation

$$\omega = \theta(n - 1) \tag{5.43}$$

Accordingly, the dispersion of a wedge can be expressed as $d\omega = \theta dn$. To exemplify, for a wavelength interval, corresponding to the range from the blue, F', to the red, C', colours, for which the mean dispersion of wedge material is $n_{F'} - n_{C'}$ often quoted in the respective standards or catalogues, the dispersion of a thin prism is

$$d\omega_{F'C'} = \theta(n_{F'} - n_{C'})$$

Optical wedges are often employed in optical systems as compensating elements in adjustments and measurements. More often than not a wedge shaped profile appears as an error of grinding the plane-parallel plates. Tolerable limits for this error are estimated and accounted for in the optical system design.

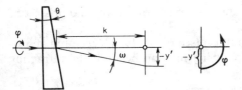

Fig. 5.20. Rotation of a wedge

Let us examine the use of a narrow angle prism as a compensator. When the wedge shown in Fig. 5.20 rotates, the image of an axial point traces out a circle of radius y' proportional to the angle of deviation ω and the distance from the wedge to the image plane, viz.,

$$y' = k \tan\omega \approx k\omega = k(n-1)\theta$$

Unfortunately, rotational motion cannot be adapted for either measurement or error compensation. With a sufficient accuracy for ordinary applications this motion may be made rectilinear by keeping another aligned wedge in opposite rotation through the same angle, as schematized in Fig. 5.21. The largest total deviation will be achieved when the principal sections of these narrow angle prisms are coplanar and the refracting angles turned to one side, namely,

$$\omega_{\Sigma_0} = 2\omega = 2(n-1)\theta$$

When the wedges are set in rotation, the deviation in the meridional plane will be as follows

$$\omega_{\Sigma} = \omega_{\Sigma_0}\cos\varphi = 2(n-1)\cos\varphi$$

Relations of spherical trigonometry may be used to prove for this case that the lateral component of ray deviation from the meridional plane may be neglected.

Image motion can be also made rectilinear by translating the wedge parallel to itself in the direction of incident ray, as shown in Fig. 5.22.

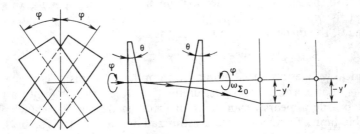

Fig. 5.21. The effect of two oppositely rotated wedges

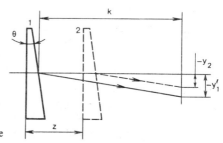

Fig. 5.22. Translation of a wedge

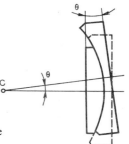

Fig. 5.23. A thin prism with variable refracting angle

Shifting the wedge a distance z deflects the image by

$$\Delta y' = y_1' - y_2' = z(n - 1)\theta \tag{5.44}$$

The shift of image is seen to be proportional to wedge translation.

Unlike rotating wedges, a parallel shifted wedge can be set to operate in converging rays. To compensate for or measure small angular or linear quantities, the rotating wedge pair is a more attractive proposition. To a rather wide angle of rotation there corresponds a small change in position of the refracted ray. The shifted wedge is a less accurate means of compensation and measurement of small linear quantities.

A wedge of refracting angle θ can be made up of two lenses, a planoconvex and a concave-plane lens, forming a plane-parallel plate in normal adjustment, but free to move with respect to each other into a wedge arrangement, as shown in Fig. 5.23.

5.7 Optical Lightguides

A glass fibre of round or other cross section with polished surface and ends can be used as a lightguide to transmit light energy in otherwise inaccessible areas without any transport of heat from the light source. When

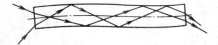

Fig. 5.24. Travel of rays in a lightguide

hot, light fibres can be bent to curvature radii of 20 to 50 their section diameters and, after cooling, to 200-300 diameters.

Figure 5.24 illustrates schematically the way of light propagation through a lightguide. A light ray incident on an end face travels in the lightguide by suffering multiple total internal reflections from the side surface. To ensure better conditions for internal reflection lightguide cores are made of dense flints cladded by crown or metal reflecting layer.

To determine the aperture angle σ'_A confining the solid angle of rays which still can be launched into the lightguide we refer to Fig. 5.25. We assume that the end face of the lightguide is perpendicular to its axis and the refractive index of the core, n_c, exceeds that of the cladding, n_{cl}.

For a ray totally reflected at the interface between the core and the cladding we have

$$n_c \sin \varepsilon_c = n_{cl} \qquad (5.45)$$

The refraction at the input face is described by

$$\sin \sigma_A = n_c \sin \varepsilon'_1 \qquad (5.46)$$

Since $\varepsilon'_1 = 90° - \varepsilon_c$, equations (5.45) and (5.46) yield the following expressions defining the input aperture angle for rays launched from air ($n_1 = 1$)

$$\sin \sigma_A = \pm \sqrt{n_c^2 - n_{cl}^2}$$

If the refractive index of the medium in which the output end of the lightguide is immersed is also unity (the situation in Fig. 5.25), the output angle σ'_A is exactly the input angle σ_A, whereas for $n_2 > 1$

$$n_2 \sin \sigma'_A = \sin \sigma_A$$

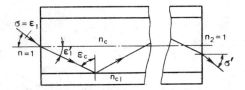

Fig. 5.25. Notation to define the aperture angle of a lightguide

Multifibre bundles are used to transmit light patterns such as facsimile. The smallest fibre diameter is 5 to 6 micrometres. Smaller diameters impair the image quality because of diffraction. In multiple-fibre lightguides, enveloped fibres are laid parallel and fused together to transmit light patterns from one polished end to the other. The resolving power of such a bundle depends on fibre diameter and interfibre distance. The most reliable figure is as high as 100 mm^{-1}. Attempts of further increase in resolution reduce the contrast and brightness of the image.

The range of light fibre elements for optical system is expanding. Nowadays it includes rigid lightguides, flexible lightguides, discs, focon lenses, anamorphots, ring-line transducers, to name but a few popular components [3, 25].

Rigid multifibre lightguides are used to transmit light patterns in instruments designed to monitor and photograph the inner walls of pipings, to bring out the readings of measuring devices onto display boards, and to serve as light conduits in microscope needles.

Flexible lightguides are mounted in flexible periscopes and gastroscopes. These lightguides are constructed of parallel fibres fused together at the input and output ends. Flexible-fibre bundles in which one end is formed into a band of one optical fibre diameter in height finds use for rapid motion picture recording as fast as 10^7 frames a second.

For contact photoprinting from convex TV screens or convex electron-optical transducers, it has been suggested to fit to them fibre optical discs one surface of which matches the profile of the screen to take up the light pattern, while the other is plane. Such discs can consist of as many as half a billion fibres and provide a resolving power of up to 100 mm^{-1} at $\sin\sigma_A = 0.54$, where σ_A is the aperture input angle.

Focons and focon lenses are multifibre elements constituted by light fibres of variable cross section, of cone profile to be more precise, which enable en route transverse magnification of the image to be transmitted.

Pressing one end of a rigid multifibre lightguide, flexible fibre bundle, or focon transforms them into anamorphots which compress or expand the image in one direction. In general, forming the input and output ends of a light-guiding device into a desired form serves the purpose of coupling, for example, in ring-line branching.

In the Soviet Union optical fibres are manufactured of the following glass types: (a) for cores TK16 ($n_e = 1.6152$), F8($n_e = 1.6291$), VS586 ($n_e = 1.5893$), VS682 ($n_e = 1.6855$); (b) for cladding VO488 ($n_e = 1.4898$), and VO513 ($n_e = 1.5150$).

5.8 Fresnel Lenses. Axikons

Fresnel lenses are optical elements with step-profiled surface, as depicted in Fig. 5.26. The closer the steps are spaced, the better the condition is satisfied for diminishing the effect of residual aberrations (imperfections of imagery in non-perfect lenses, to be discussed later) at small lens thicknesses. The smallest spacing achievable is 0.05 mm. The steps can be separated by concentric, spiral or parallel grooves. In the first two cases the steps are just segments of conic or spheric surfaces, and in the last case these are segments of plane or cylinder areas. Technologically such surfaces are readily effected by plastic moulding.

A popular material for such lenses is polymethylmetacrylate having the following characteristics: $n_D = 1.4903$, $V_D = 57.8$, the temperature coefficient of the refractive index $dn_D/dt = \beta_D = -16 \times 10^{-5}$, the coefficient of linear expansion $\alpha_t = 70\text{-}190 \times 10^{-6}$ per °C, and the softening temperature is at 72 °C. This material exhibits good transmittance in the ultraviolet.

Plastic Fresnel lenses have proved useful as magnifying glasses, condensors, prisms, mirrors, and other optical components in optical systems where compact size is a requirement.

Figure 5.27 presents a construction to illustrate the idea of step-profiled axially symmetrical surface separating two media of refractive indices $n_1 = 1$ and $n_2 = n$. We assume that each step of the profile is an infinitely narrow element. In imaging an object point A the ray AM meets a narrow effective step profile at a point M elevated a distance h above the optical axis and on refraction intersects the axis at a point A'. The normal to the profile at M intersects the optical axis at a point C making with the axis an angle φ. This angle measured from the vertical axis defines the position of the profile element under consideration.

To describe the profile, we determine the angles φ for different heights

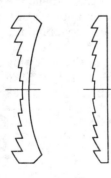

Fig. 5.26. Fresnel lenses

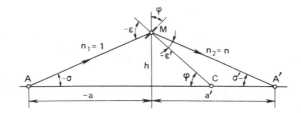

Fig. 5.27. Construction to illustrate refraction at a step-element of a Fresnel lens

of incidence of rays with a given position of the object point A and its image A' separated from the profile by a and a', respectively. With reference to Fig. 5.27, we have

$$-\varepsilon = -\sigma + \varphi \quad \text{and} \quad \varphi = -\varepsilon' + \sigma'$$

From the refraction law, $\sin \varepsilon = n \sin \varepsilon'$ we obtain

$$\sin (\varphi - \sigma) = n \sin (\varphi - \sigma') \tag{5.47}$$

After rearranging this equation we arrive at the following dependence suitable for calculating angle φ defining the slope of cone ring segments constituting the stepped refracting profile

$$\tan \varphi = (n \sin \sigma' - \sin \sigma)/(n \cos \sigma' - \cos \sigma) \tag{5.48}$$

where the angles σ and σ' are to be calculated in advance by the given a and a' for different h. This expression may be used to design this Fresnel lens with plane second surface whose aberrations may be neglected.

The second focal length of the Fresnel lens will be determined with the value of σ' at $\sigma = 0$. Under this condition, Eq. (5.48) yields

$$\tan \varphi_0 = n\sigma'/(n - 1)$$

so that

$$\sigma' = \tan \varphi_0 (n - 1)/n$$

Thus, for a small height h

$$f' = h/\sigma' = hn/(n - 1) \tan \varphi_0$$

where $\tan \varphi_0$ is obtained from (5.48) at the small h.

The clear aperture of the lens, D_c, results for the angle of incidence $\varepsilon_m = -90°$. With reference to Fig. 5.28

$$\tan \sigma'_{A'} = \tan (\varphi + \varepsilon'_m) = D_c/2a' \tag{5.49}$$

where ε'_m is the ultimate value of the refraction angle. Hence,

$$D_c = 2a'(\tan \varphi + \tan \varepsilon'_m)/(1 - \tan \varphi \tan \varepsilon'_m)$$

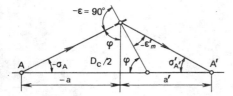

Fig. 5.28. Construction to determine the clear aperture of a Fresnel lens

Observing that $\tan \varphi = -2a/D_c$, we put down

$$D_c^2 + 2D_c(a - a') \tan \varepsilon_m' + 4aa' = 0 \qquad (5.50)$$

where $\tan \varepsilon_m'$ is defined with the refraction law ($\sin \varepsilon_m' = 1/n$) as $\tan \varepsilon_m' = -(n^2 - 1)^{-1/2}$. Solving the quadratic equation (5.50) yields the aperture of the Fresnel lens

$$D_c = [a - a' + \sqrt{(a - a')^2 - 4aa'(n^2 - 1)}]/\sqrt{n^2 - 1}.$$

Axicon is a collective name for optical elements or systems causing a considerable distortion of the homocentric property of a ray bundle emanating from an object point. The resultant image of an axial point appears as a segment of the optical axis in the image space, whereas in an image plane it appears as a circle of a substantial diameter. Axicons find use in optical systems where no focusing is required when the object changes its position with respect to the system, and also to provide a given illumination in the image plane, and to compensate for disturbance of homocentricity incurred in other components of the system.

Figure 5.29 schematizes an optical, axially symmetric, lens-type element. It has a plane front surface and a cone in the second surface. In the meridional section, this conic lens may be represented as an isosceles refracting prism of refracting angle θ and optical axis width d. We find the image A' of an axial point A distanced s_1 away from the front face as a function of slope angle σ_1.

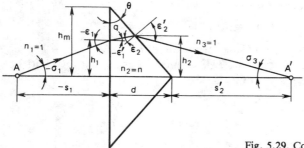

Fig. 5.29. Cone-shaped axicon

For an incidence height h_1 we have

$$\tan \sigma_1 = h_1/s_1 \tag{5.51}$$

By the law of refraction

$$\sin \varepsilon_1' = \sin \varepsilon_1/n = \sin \sigma_1/n \tag{5.52}$$

From Fig. 5.29 we observe that

$$\varepsilon_2 = \varepsilon_1' + \theta \tag{5.53}$$

Repeated use of the refraction law yields for the second face

$$\sin \varepsilon_2' = n \sin \varepsilon_2 = n \sin (\varepsilon_1' + \theta) \tag{5.54}$$

And, finally, the angle between the refracted ray and the optical axis is

$$\sigma_3 = \varepsilon_2' - \theta \tag{5.55}$$

The line section s_2', defining the position of A', is measured from the vertex of the refracting surface — from the apex of the cone in this case,

$$s_2' = h_2/\tan \sigma_3 - h_2 \tan \theta = h_2(\cot \sigma_3 - \tan \theta) \tag{5.56}$$

where h_2 is the height of incidence of the conic surface.

Denote the ray path length within the lens by q, then

$$h_2 - h_1 = -q \sin \varepsilon_1' \tag{5.57}$$

where

$$q = (h_M - h_1) \sin \theta/\cos \varepsilon_2 \tag{5.58}$$

and h_M is a given distance from the optical axis to the rim of the lens.

Observing in Eq. (5.56) Eqs. (5.57), (5.58) and (5.52) we obtain

$$s_2' = [h_1 - (h_M - h_1) \sin \theta \sin \sigma_1/n \cos \varepsilon_2](\cot \sigma_3 - \tan \theta) \tag{5.59}$$

Let us look at this expression for the case of $\sigma_1 = 0$, i.e. when the axial object point is at infinity. Now $\varepsilon_1 = \varepsilon_1' = 0$ and $\sin \varepsilon_2' = n \sin \theta$ and

$$s_2' = h_1(\cot \sigma_3 - \tan \theta)$$

For a paraxial ray (h_1 very near zero), $s_0' = 0$, and for a marginal ray, incident at the rim of the lens where $h_1 = h_M$,

$$s_M' = h_M(\cot \sigma_3 - \tan \theta)$$

Consequently, the greatest longitudinal spread of the point image (this imperfection of imagery is known as spherical aberration) for the case of $\sigma_1 = 0$ is

$$\delta s_M' = s_M' - s_0' = h_M(\cot \sigma_3 - \tan \theta) \tag{5.60}$$

To illustrate we estimate this quantity for specific structural parameters of a conical lens, namely, $h_M = 20$ mm, $\theta = 20°$, $n_e = 1.5183$. With the help of Eqs. (5.53), (5.55) and (5.60) we obtain $\delta s_M' = 93.37$ mm. This spread obviously renders the lens an axicon.

Popular conic axicons have the angle θ not exceeding $6°$ ($\tan \theta \approx \sin \theta \approx \theta$). For small σ_1, $\sin \sigma_1 \approx \sigma_1$, $\sin \varepsilon_1' \approx \varepsilon_1'$, $\sin \varepsilon_2' \approx \varepsilon_2' \approx n\varepsilon_2$. Observing that in this case $\cos \varepsilon_2 \approx 1$ and $\cot \sigma_3 \approx 1/\sigma_3$, we get from Eq. (5.59)

$$s_2' = [h_1 - (h_M - h_1)\theta\sigma_1/n](1/\sigma_3 - \theta)$$

and at $\sigma_1 = 0$

$$s_2' = h_1(1/\sigma_3 - \theta)$$

Recalling that $\sigma_3 = \omega = (n - 1)\theta$ (see Eq. (5.43) for a narrow angle prism) we get

$$s_2' = \delta s' = h_1[1/(n - 1)\theta - \theta] \approx h_1/(n - 1)\theta$$

Determine also the closest distance $s_{\min}$ from the axicon to the object point for which incident rays still do not suffer total internal reflection. For $\sin \varepsilon_2' = 1$, Eq. (5.54) yields

$$\sin \varepsilon_{2m} = 1/n = \sin(\varepsilon_1' + \theta)$$

where ε_{2m} is very near the critical angle of total internal reflection.

Because each of the angles ε_{2m} and $\varepsilon_1' + \theta$ does not exceed $90°$, with account of Eq. (5.53) we may write

$$\varepsilon_1' = \varepsilon_{2m} - \theta$$

From Eqs. (5.52) and (5.53) it follows that

$$\sin \sigma_1 = n \sin (\varepsilon_{2m} - \theta)$$

and

$$s_{\min} = h_M \cot \sigma_1$$

Other designs related to the class of axicons are shown in Fig. 5.30.

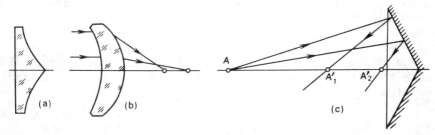

Fig. 5.30. Axicons (a) conoid lens, (b) positive meniscus, (c) cone-shaped mirror

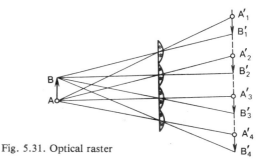

Fig. 5.31. Optical raster

The name 'optical raster' will be used in this book to describe an array of lens or mirror elements having optical effects and producing brightness patterns. The spacing between the axes of any two adjacent elements, measured normal to their symmetry axes, will be called the raster period. Each element of an optical raster forms its own image of the object. Therefore, the number of images equals the number of elements in the raster.

If all the elements in a raster have the same powers, then, with reference to Fig. 5.31, the images $A_k' B_k'$ of an object AB appear in the same plane (we assume the perfect imagery of raster elements).

Figure 5.32 illustrates operation of a raster system providing a practically uniform illumination. An oblique bundle of rays emanating from the light source C is used to illuminate the entrance pupil, 3, of a subsequent optical system. Each lobe-shaped element of the raster, 1, forms an image of the source C at C_1'. In turn, each element of the raster labelled 2 directs the bundle of light into the entrance pupil 3.

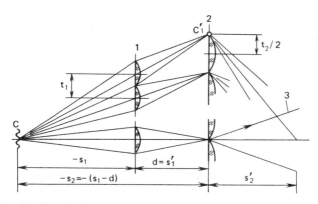

Fig. 5.32. A raster illumination system

Assume the period of raster 1 is t_1, then with reference to Fig. 5.31, the period of raster 2 is

$$t_2 = t_1(1 - d/s_1)$$

where d is the spacing between the rasters, and s_1 is the distance of the light source C from raster 1.

The number of elements in both rasters must be the same.

The focal lengths of the axial elements in rasters 1 and 2 can be determined by the formula (3.7) as follows

$$f_1' = s_1 d/(s_1 - d) \quad \text{and} \quad f_2' = (s_1 - d)s_2'/(s_1 - d - s_2')$$

where s_2' is the distance of the entrance pupil 3 from the second raster.

Another example illustrating the application of optical rasters is given by directional screens. The mirror elements of such a screen can be spherical and cylindrical shapes.

Figure 5.33 shows the meridional section through an element of a directional screen. In this section the reflected light is scattered within a given angle $2\sigma'$. With the notation of Fig. 5.33,

$$\sin \sigma' = \frac{D}{2r^2} \sqrt{4r^2 - D^2} \tag{5.61}$$

where r is the radius of element's spherical or cylindrical surface, and D is the transverse size of an element.

The expression just derived indicates that a raster screen is equivalent to a diffusing screen ($\sigma' = 90°$ at $D = r/2$). An apparent brightness of the image in cinematographic projection and microscope work can be increased by decreasing the angle $2\sigma'$ which can be achieved with a larger radius of the convex cylindrical surface, i.e. with a greater focal length of a raster element.

Numerous examples of optical raster design and applications can be found in the handbook (in Russian) edited by Rusinov [3].

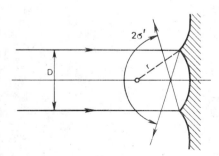

Fig. 5.33. Element of a raster reflecting screen

CONFINING RAY BUNDLES IN OPTICAL SYSTEMS

6.1 Apertures

Lenses, mirrors, plane-parallel plates and prisms constituting optical systems have finite dimensions and usually are set in mounts and rims limiting their apertures. In addition, many optical systems have circular aperture stops, or simply stops, centred on the optical axis which are to confine ray bundles traversing the system in the same way as mounts and rims, i.e. lens apertures do. A direct sequel of this confinement is, first, that the optical system receives only a portion of radiation flux emanating from every point of the object, and, second, only a portion of object space can be imaged.

The action of optical element apertures and aperture stops themselves, which may be of variable size as in iris diaphragms, can influence a number of optical characteristics. These in particular include (1) integral illumination of the image, (2) distribution of illumination over the image field, (3) angular field or transverse coverage within the boundaries of satisfactory image quality, (5) image contrast and other characteristics related to image quality.

An aperture in an opaque screen (diaphragm) limiting the bundle of rays emanating from an axial object point and thereby controlling the illumination of the image is called an *aperture stop*, whereas that placed in the object plane or in one of the planes conjugated with this and confining the transverse size of the field in image space is called a *field aperture* or *field stop*.

Figure 6.1 demonstrates the action of aperture stops. The axial ray pencil emanating from the point A at the first focal point F_1 of the front group (lens), labelled 1, traverses the optical system consisting of two groups (components). The values of *aperture angles*, or *angular apertures*, σ_A, for the entrance angle, and σ_A', for the exit angle, are seen to depend on the aperture stop diameter, all other conditions being equal at (a) and (c).

An oblique ray pencil emanating from the off-axis point B is limited by the apertures, say rims, of components 1 and 2 (as in the diagram at (a)), or by an aperture stop, as in the diagram at (c). The central or representative ray of the oblique pencil traversing the central point of the aperture stop is referred to as the *principal or chief ray*.

Inspection of Fig. 6.1 indicates that the diameter of the stop dictates both the integral illumination of the image and the distribution of the intensity over the image field. The illumination in the vicinity of A' is determined, other conditions being the same, by the values of entrance aperture angles σ_A (or exit angles σ'_A). The illumination intensity near point B' will also be different as the section of the oblique ray pencil in the aperture plane for situation at (a), shown at (b), exceeds that for the situation in the diagram at (c), shown at (d), accordingly the angular dimensions of ray bundles entering the optical system will be also larger.

Placement of apertures influences the course of oblique ray pencils. As a rule, the aperture stop is placed between components of the system, as, for example, in Fig. 6.1. In some situations it is desirable to place an aperture stop before the system or behind it. Fig. 6.2a shows a stop placed in the first focal plane, consequently the principal ray will be parallel to the optical axis in image space. When the stop is at one of the principal foci, it is said to be a *telecentric stop*. In the diagram at (b) the aperture stop is in the second focal point, and the principal ray in object space is parallel to the optical axis.

Ray pencils are referred to as *telecentric* if their principal rays travel parallel to the optical axis either in object or image space. Telecentric ray bundles in image space are used in photographic objectives (often referred

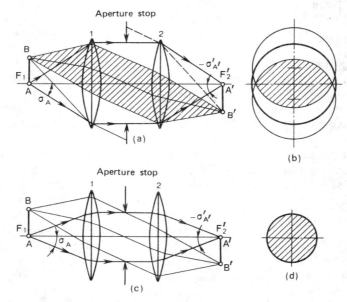

Fig. 6.1. Illustrating the effect of an aperture stop

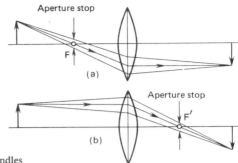

Fig. 6.2. Telecentric ray bundles

to as photographic lenses) operating with colour multiple-layer film, and in TV objectives with the sensitive layer of the photocathode having a considerable thickness. In object space, telecentric ray pencils are encountered in measuring microscopes.

6.2 Entrance and Exit Pupils

When the positions of the aperture stop are known, its image in object space formed by all optical system elements between the object plane and the stop in the reverse travel of rays, i.e. from right to left, is known as the *entrance pupil* of the optical system. The entrance pupil subtends a cone of light rays emanating from the axial object point and passing through the system. The image of the stop formed by all optical elements between the stop and the image plane in the direct travel of rays is known as the *exit pupil* of the system.

Because the entrance and exit pupils are images of the stop, they can be real and virtual. If an aperture stop is in front of the lens, as in Fig. 6.2a, the entrance pupil coincides with it and is real. Virtual pupils can be seen in Fig. 6.3 where the stop is set between components *1* and *2* of the optical system. This situation occurs in photographic and projection objectives. Being images of the stop, the entrance and exit pupils are conjugate images, i.e., the exit pupil is the image of the entrance pupil. The principal ray, accordingly, will pass through the centre C of the entrance pupil, the centre of the stop C_s, and the centre C' of the exit pupil.

The placement of a stop and, consequently, the entrance and exit pupils, depends therefore on the specific arrangement of an optical instrument at hand and will be established in examining particular designs.

Let us solve a practical problem. Given a system having several aperture stops and mounts of optical elements. It is required to determine which of

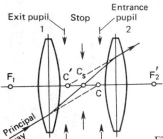

Fig. 6.3. Entrance and exit pupils

the apertures will decide the light collecting properties of the instrument, i.e., which gives the least entrance pupil.

The general approach in solving this problem will be as follows. We shall seek the position and size of the images of all apertures and stops in the system in reversed rays as formed by the preceding system components. The image subtending in the meridional plane the least angle at the axial object point will be the entrance pupil. The stop or aperture whose image gives the entrance pupil decides the optical power of the instrument. The image of this stop or mount formed by the subsequent optical components of the system will be the exit pupil.

Figure 6.4 illustrates the solution of such problem for a system constituted by thin lenses 1 and 2 and a stop 3. The images of mounts of components 1 and 2 and of the stop 3 in reversed rays, i.e. by the lens 1, are denoted, respectively, as $\overline{1}{}'$, $\overline{2}{}'$, and $\overline{3}{}'$. Note that the image $1{}'$ of the rim of lens 1 coincides with the rim proper. The entrance pupil sought after will

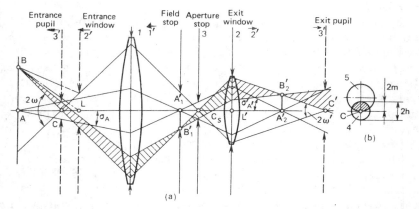

Fig. 6.4. Schematic sketch of an optical system to illustrate the relationships between pupils, stops, fields, and windows

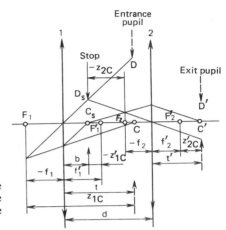

Fig. 6.5. Notation to derive analytical expressions yielding the position and diameter of the entrance pupil

be the image which subtends the least angle at the point A. In this example, the entrance pupil will be the image $\bar{3}'$ of the stop 3. Consequently, the amount of light passing through this system will be limited by the aperture in the stop 3. The image of this aperture stop formed in the direct rays, i.e. by the component 2, will be the exit pupil of the system.

The entrance aperture angle σ_A is connected with the exit aperture angle σ_A' by the dependence derived from (3.15) and (3.16)

$$\tan \sigma_A' = \frac{n}{n'} \frac{1}{\beta} \tan \sigma_A \qquad (6.1)$$

where n and n' are the refractive indices of object and image spaces, respectively, and β is the lateral magnification of the system for conjugate axial points A and A_2'.

Now we derive analytical expressions to determine the position and entrance pupil diameter in a two-element system specified in Fig. 6.5.

(i) The entrance pupil is given by its distance t' from the component 2. The graphical construction to locate the centres of the stop, C_s, and entrance pupil, C, is indicated in the figure. The line section b defines the position of the stop, and the segment t locates the entrance pupil with respect to the component 1.

The positions of the entrance pupil and the stop with respect to the foci of the component 1 are defined as follows

$$z_{1C} = f_1' + t \quad \text{and} \quad -z_{1C}' = f_1' - b$$

Substituting these into Newton's formula (3.5) yields

$$b = t f_1' / (f_1' + t) \qquad (6.2)$$

Similarly, for z_{2C} and z'_{2C} defining the positions of the stop and the entrance pupil relative to the foci of the component 2 as

$$-z_{2C} = d - f'_2 - b \quad \text{and} \quad z'_{2C} = t' - f'_2$$

d being the spacing between the components, we obtain the Newton formula in the form

$$(f'_2 + b - d)(t' - f'_2) = -f'^2_2$$

whence

$$b = d - t'f'_2/(t' - f'_2) \tag{6.3}$$

From (6.2) and (6.3) we obtain for the sought position of the entrance pupil

$$t = f'_1 \frac{f'_2 d - t'(d - f'_2)}{t'(d - f'_1 - f'_2) - f'_2(d - f'_1)} \tag{6.4}$$

(ii) Given the diameter of the exit pupil D' we wish to find the diameters of the stop, D_s, and entrance pupil, D, other conditions being the same as stated for the previous problem.

For the system of Fig. 6.5 in air,

$$D'/D_s = t'/(d - b)$$

whence the diameter of the stop aperture

$$D_s = D'(d - b)/t' \tag{6.5}$$

where b is to be defined by (6.2).

Similarly, we have $D_s/D = b/t$, whence, in view of (6.5), we obtain the entrance pupil diameter as

$$D = D'(d - b)t/t'b \tag{6.6}$$

with t computed by (6.4).

Finally, the transverse magnification between the pupils of this optical system

$$\beta_p = D'/D = t'b/t(d - b) \tag{6.7}$$

where the parameters are substituted from (6.2) and (6.4).

6.3 Field of View. Vignetting. Entrance and Exit Windows

The largest dimension in object plane a finite distance away from the optical system will be called in this text the linear field of view or the linear (transverse) coverage of the system. Similarly, the largest dimension in the image plane a finite distance away from the system will be referred to as the linear field of the system in image space.

A more popular measure of the system coverage is the *field of view*, or simply the field, being the angle subtended by the image of the field stop in object space at the centre of the aperture stop image in this space. This angle may be traced in object space as twice that between the optical axis and the ray passing through the centre of the aperture stop and just grazing the edge of the field stop. Not infrequently, the double value of the angle this ray makes with the optical axis in the image space is referred to as the field of view of the system *in image space*.

These definitions can be better understood with reference to Fig. 6.4. In agreement with the definition of the field stop it will be placed at the plane of the intermediate image $A_1'B_1'$. Then the linear coverage in the meridional plane of object space will be twice the length AB, and the field of view will be the angle 2ω subtended by this length at the point C, the centre of the entrance pupil. Hence another definition of the field of view is the angle subtended by the image of the field stop in object space (called the entrance port elsewhere) at the centre of the entrance pupil. Similarly, the linear coverage of the system in the meridional plane of image space is twice the length $A_2'B_2'$, and the back field of view is the angle subtended by this line segment, $2\omega'$, at the centre of the exit pupil C'.

In the general case the semiangles ω and ω' are related by the formula, derived similarly to (6.1), namely,

$$\tan \omega' = \frac{n}{n'} \frac{1}{\beta_p} \tan \omega$$

where β_p is the transverse magnification between the conjugate pupils, i.e. D'/D, the ratio of the entrance and exit pupil diameters, and n and n' are the refractive indices of the media on either side of the system.

Referring to Fig. 6.4, we note that in constructing the entrance pupil we obtained the images denoted $\bar{1}'$, $\bar{2}'$, and $\bar{3}'$. The last of them as will be recalled, is the entrance pupil. Let us look at the action of the other two. For this purpose we examine the point B, the farthest of the linear coverage in object space. The bundle of rays emanating from this point and passing through the end point of the entrance pupil diameter will be narrowed by the image $\bar{2}'$ of the mount of lens *2*, i.e. the rim which has been recognized as the aperture stop for this system. In the meridional section the system passes from B only the cross-hatched portion of the bundle. In this system, the mount of lens *1* (its image is labelled $\bar{1}'$) has no bearing on the bundle narrowing.

It would be quite natural to conclude that the illumination of the image plane depends on how much the bundles of rays travelling from conjugate object points have been cut off and narrowed by the system. In the case of Fig. 6.4a, the illumination at point B_2' will be less intense than at point A_2'. The point B is in this case the marginal point of the linear transverse field of coverage. If we enlarged the field stop, the marginal point of the enlarged image field would have a lesser illumination. Consequently, the field stop should be as large as to ensure the desired illumination of marginal points of the image (the relevant radiometric analysis will be given in the next chapter). In addition, enlargement of image field entails reduction of image quality due to imagery imperfections to be discussed in ensuing chapters.

Of the four apertures evaluated with reference to Fig. 6.4 — the aperture stop, field stop, and two lens mounts — the mount of lens 2 limits the ray bundles emanating from extra-axial object points. This progressive reduction in the cross-sectional area of a beam traversing an optical system as the obliquity of the beam is increased is known as *vignetting*, and the aperture causing this reduction is referred to as a vignetting aperture. More often than not this is a field stop.

The image of the vignetting aperture (field stop) in object space is called the *entrance window* or *entrance port*, and the image in image space is called the *exit window* or *exit port*.

It is worth noting that some systems exhibit two-side vignetting, as is the case, for example, with two-element photographic objectives (doublets) having an aperture stop in between the elements, as in Fig. 6.1a. This is a system with two entrance windows (two vignetting apertures).

In considering the phenomenon of vignetting we assumed that with a field stop of certain size the image was formed by principal rays (see Fig. 6.4). We note, however, that principal rays may not take part in imaging marginal object points. This case is illustrated in Fig. 6.6 for a system whose aperture stop, and hence the entrance and exit pupils, are of finite size. Therefore, the boundaries of the object space to be imaged are traced out by the marginal, rays passing through the system rather than by principal rays. The actual field of view $2\bar{\omega}$ in this case exceeds the field of view defined by the principal rays.

If the entrance window of diameter D_w is situated at a distance $|c|$ from the entrance pupil of diameter D, then

$$\tan \omega = D_w/2|c| \quad \text{and} \quad \tan \bar{\omega} = (D_w + D)/2|c|$$

whence

$$\tan \bar{\omega} = \tan \omega + D/2|c| \tag{6.8}$$

The absolute value symbol for the line segment c in (6.8) implies that

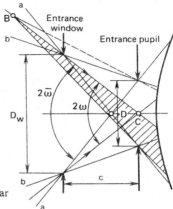

Fig. 6.6. Construction to determine the angular field of view in object space

the entrance window can be situated (or can be given) both in front of the entrance pupil and behind it. It will be noted that the vertex of the angle 2ω does not coincide with the centre of the entrance pupil.

Rays b touching the edges of the entrance window and entrance pupil in the meridional plane confine the domain in object space where every point can be the vertex of an angle subtended by the entire diameter of the entrance pupil.

The points in the region between the cones of rays a and b cannot be vertices of angles subtended by the entrance pupil diameter. In Fig. 6.6, the rays emanating from B on the principal ray confine an angle subtended by only half the entrance pupil diameter. Hence, for principal rays confining the field of view 2ω, vignetting in the meridional plane amounts to 50 per cent.

For most practical applications a 50% vignetting is regarded as admissible, therefore the angle 2ω is adopted as the field of view in object space. Some applications, however, admit greater vignetting with the aim to enlarge the field of view. It will be noted that for object points on rays a in Fig. 6.6, the amount of vignetting is 100 per cent.

6.4 Effective Aperture of Entrance Pupil

If oblique light beams are subjected to vignetting in a system, then, as has been demonstrated in the previous section, the entrance pupil is not used in full as a portion of each beam is obstructed by the entrance window (vignetting stop). The area of the entrance pupil illuminated by an oblique beam passing through the system is referred to as an *effective aperture* of the entrance pupil. The ratio of the area of the effective aperture for the given field to the entire area of the entrance pupil (relative effective aperture) will be referred to as the vignetting coefficient, $k_A = A_{eff}/A_0$.

The vignetting coefficient may be derived as the ratio of sectional areas for the oblique beam over the axial beam taken in any plane perpendicular to the optical axis of the system. In Fig. 6.1, for example, this is the plane of the aperture stop.

If vignetting is within 20 to 65 per cent, which is the case for most systems, then the areas in the ratio of vignetting coefficient may be replaced with their linear measures. With reference to Fig. 6.4b, this will be the ratio of the line segment $2m$ perpendicular to the optical axis to the height of the axial beam section $2h$, both taken in the meridional plane. This ratio accordingly may be termed the 'linear vignetting coefficient', $k_\omega = 2m/2h$. Referring to Fig. 6.6, for point B, $2m = D/2$ and $2h = D$, i.e. $k_\omega = 0.5$.

For the figures of vignetting indicated (20% to 65%), $k_A \approx k_\omega - 0.1$.

By way of example consider the evaluation of vignetting coefficients for two systems.

(i) In Fig. 6.4b the area labelled 4 refers to the section of the axial beam by the entrance pupil plane, i.e. the pupil proper of area A_0, and the section labelled 5 belongs to the oblique beam from point B through the entrance window, also in the plane of the entrance pupil (see Fig. 6.4a). The axis of the oblique beam passes through the centre of the entrance window L. In this case the entrance window, and consequently the vignetting stop 2, obstruct the ray bundle from point B. The amount of vignetting in the meridional plane is 50%. The area of the effective aperture of entrance pupil is cross-hatched in Fig. 6.4b, $k_A \approx 0.4$ and $k_\omega = 0.5$.

(ii) Figure 6.7 represents an example of two-side vignetting. The optical system at hand has the positive lens component 1 of $f_1' > 0$, and the negative lens component 2 of $f_2' < 0$, and an aperture stop placed in between. The diagram at (a) shows the second focal point F' of the equivalent system, the centres C, C' and C_2 of the entrance and exit pupils, and the entrance window, respectively, with the respective diameters constructed for the given focal lengths of lens components, spacings, and location and clear apertures of the stop, and lens components.

The entrance and exit pupils are seen to be between the system components. For an infinitely distant axial point, the course of marginal rays, i.e. the rays grazing the edges of the entrance pupil, aperture stop and exit pupil, respectively, is constructed. The construction for the principal ray locates the image B' of the infinitely distant extra-axial object point. The cross-hatched meridional section belongs to a ray pencil emanating from this point and traversing the entire system. The upper marginal ray of this pencil in object space may be regarded as a generating element of the cylin-

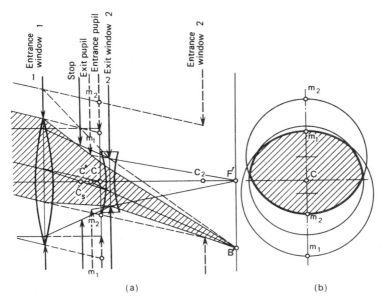

Fig. 6.7. Two-side vignetting

drical ray pencil grazing the edges of the mount of component 1 of diameter $m_1 m_1$. The bottom ray of this pencil in object space is a generating element of a cylindrical ray pencil grazing the diameter $m_2 m_2$ of the entrance window 2 centered on C_2. Having passed through component 1, in the absence of the aperture stop this pencil would be a converging cone filling the free aperture of component 2.

The point B' is formed by the ray pencil common for both aforementioned cylindrical bundles and passing through the entrance pupil. The bottom portion of the pencil grazing the rim of component 1 (entrance window 1) is partially cut off by the aperture stop at the section $m_1 m_2$ and totally cut off by the rim of component 2, being the vignetting diaphragm and exit window 2. The top portion of the pencil grazing the entrance window 2 is cut off by the rim of component 1 at the section $m_2 m_1$.

Figure 6.7b shows the cross sections of all the aforementioned pencils in the plane of the entrance pupil. The shadowed portion represents the effective aperture of the entrance pupil. The vignetting coefficient k_A is determined as the ratio of the areas of the effective aperture and the entire entrance pupil.

7

ENERGY FLOW IN OPTICAL INSTRUMENTS

7.1 Light Flux

Radiation commonly referred to as 'optical' covers electromagnetic wavelengths in the range 1 nm to 1 mm. On the short-wave side this range is flanked by x-rays, and on the long-wave side by microwaves. Fig. 7.1 illustrates the placement of the optical range in the electromagnetic radiation spectrum. The visible range is seen to extend from a wavelength of 0.38 μm to that of 0.77 μm. It should be noted that the boundaries between the ranges shown are fairly conditional. Ultraviolet radiation, for instance, overlaps x-rays, while the infrared overlaps the microwave band.

The electromagnetic radiation spectrum of a self-luminous body is always a distribution of radiant-power over wavelengths. If a source emits at one wavelength, as is the case with most lasers, the light is called *monochromatic*. A spectrum constituted by a few monochromatic emissions is known as *line spectrum*. Its power distribution versus wavelength looks like that shown in Fig. 7.2a. Most sources, however, exhibit spectra in which radiant power varies with wavelength continuously, as shown in the radiant intensity distribution at (b). In such a *continuous spectrum* the range of wavelengths is from zero to infinity.

Continuous electromagnetic spectra are found in heated solids and liquid, line EM spectra in hot gases and vapours. Some lasers are also exhibit line spectra. In fact, there exists no natural source of monochromatic radiation. In practice, therefore, the term monochromatic radiation is applied to a narrow interval of wavelengths such that may be characterized by a single wavelength.

The German physicist Fraunhofer (1787-1826) measured optical wavelengths which appeared dark in the spectrum of the Sun because of the absorption of these wavelengths in the Earth's atmosphere. Called Fraunhofer lines, these lines are scattered all over the visible spectrum and, if reproduced as bright lines, i.e. emitted by a luminous source, can provide a convenient reference system in measuring, say, transmittance of optical glasses in various regions of the spectrum. Emission in these lines can be obtained from glued gases of some chemical elements put in bulbes with electric arc discharge.

Refractive indices of optical media are also measured at certain Fraunhofer lines indicated, as will be recalled, by the symbol of the respec-

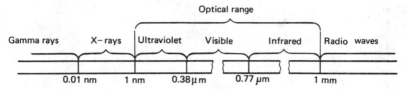

Fig. 7.1. The electromagnetic spectrum

tive line in subscript. Table 7.1 shows the letter symbols of some Fraunhofer lines with the respective wavelength in nanometres, the colour of the spectrum band where they appear, and the chemical element whose spectrum contains this line.

The total rate of energy flow on to a surface of area A receiving light is called the flux on A (or through A if A is the area of an aperture, such as the pupil of an optical system). Light flux is denoted by the symbol Φ_e, where the subscript 'e' is used to indicate that this is an *energy* quantity, and it has the dimensions of power; it can thus be measured in watts.

If we denote the flux within a narrow interval of wavelengths $d\lambda$ by $d\Phi_e$ then the ratio

$$\frac{d\Phi_e}{d\lambda} = \varphi_e(\lambda) \tag{7.1}$$

Table 7.1. Fraunhofer Lines

Symbol	Wavelength, nm	Colour	Element	Symbol	Wavelength, nm	Colour	Element
i	365.0	ultraviolet	Hg	e	546.07	green	Hg
h	404.66	violet	Hg	d	587.56	yellow	He
g	435.83		Hg	D	589.29		Na
F'	479.99	blue	Cd	C'	643.85		Cd
F	486.13		H	C	656.27	red	H
				r	706.52		He

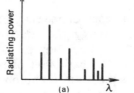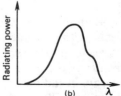

Fig. 7.2. Line spectrum (*a*) and continuous spectrum (*b*)

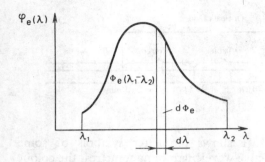

Fig. 7.3. The spectral distribution of a radiant flux

is the spectral density of the flux. Fig. 7.3 shows the spectral density plotted versus the wavelength for a continuous spectrum. The elementary flux $d\Phi_e$ is seen to be the area of the strip $d\Phi_e = \varphi_e(\lambda)d\lambda$.

For a continuous spectrum between the wavelengths λ_1 and λ_2 the flux is given as

$$\Phi_e = \int\limits_{\lambda_1}^{\lambda_2} \varphi_e(\lambda)d\lambda \qquad (7.2)$$

The total (integral) flux over the entire spectrum is as follows

$$\Phi_e = \int\limits_{\lambda=0}^{\lambda=\infty} \varphi_e(\lambda)d\lambda \qquad (7.3)$$

For a line spectrum, the total flux is the sum of fluxes for all emission wavelengths λ_i

$$\Phi_e = \sum\limits_{\lambda=0}^{\lambda=n} \Phi_e(\lambda_i)$$

7.2 Radiometry and Photometry. Symbols and Units

Radiometry is the science of measurement of radiation. *Radiometry* deals with the radiant energy (i.e. electromagnetic radiation) of any wavelength. *Photometry* is the study of light energy flow in the *visible region* of the spectrum. The basic unit of power (i.e. rate of transfer of energy) in radiometry is the watt; in photometry the corresponding unit is the lumen. Watts and lumens have the same dimensions, namely energy per time. In general, to estimate the radiant energy and its effect on radiation detectors such as photoelectric devices, thermal and photochemical detectors, and the eye, both radiometry and photometry use a number of derivable quantities.

Most frequently the light is not monochromatic and the fact that most light detectors, e.g. photoelectric cells, photographic plates, the eye, are not equally sensitive to all wavelengths complicates the problem of photometry. Accordingly, the quantity which corresponds to power or flux in photometry is the *luminous flux*, denoted by Φ_v, the subscript 'v' indicating that this quantity relates to the set of 'visible' quantities. We omit this subscript in this context and denote this quantity simply by Φ.

For simplicity we consider first monochromatic light and introduce polychromatic light at the end of the section. The definitions that follow will be given for photometric units but are applicable throughout the spectrum and the comparison table of radiometric units will also be given.

To measure the radiant power of two sources of visible radiation, one may compare their luminant fluxes intercepted by the same surface. If one of them is adopted as a standard source of unit luminance, then by comparison the luminous intensity of the other will be derived. The *Système International* (SI) unit of luminous intensity is the *candela*, symbol cd, defined as the luminous intensity, in the perpendicular direction, of a surface of 1/600 000 square metre of a black body at the temperature of freezing platinum under a pressure of 101 325 newtons per metre squared.

Mathematically, *luminous intensity* or flux solid-angle density is given as

$$I = \frac{d\Phi}{d\Omega} \tag{7.4}$$

where $d\Phi$ is the luminous flux in a certain direction confined within a solid angle $d\Omega$.

As will be recalled, unit solid angle (one steradian) is subtended at the centre of a sphere of radius r by an area r^2 of any shape on the surface of the sphere. If the symbol Ω is used for solid angle, then with reference to Fig. 7.4 any area A on the surface of the sphere of radius r is

$$A = \Omega r^2 \tag{7.5}$$

If a source of radiation is at the vertex of a right circular cone, as is usually the case in photometric considerations, then the solid angle confined by this cone can be determined with the semiangle of the cone σ.

Consider the case of axially symmetric distribution of luminous intensity most frequent in applications. On the surface of a hemisphere of radius r (Fig. 7.5) the incremental annular area of radius ρ confined between two

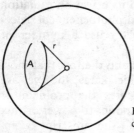

Fig. 7.4. Area A on the surface subtends a solid angle at the centre of the sphere

cones separated by the semiangle dσ is

$$dA = 2\pi\rho \, d\rho$$

From Fig. 7.5, $\rho = r \sin \sigma$, $d\rho = r \, d\sigma$, therefore

$$dA = 2\pi r^2 \sin \sigma \, d\sigma$$

whence the solid angle subtended by this ring from the centre of the hemisphere is

$$d\Omega = 2\pi \sin \sigma \, d\sigma \qquad (7.6)$$

The solid angle confined by a cone of semiangle σ_n is as follows

$$\Omega = \int_0^{\sigma_n} 2\pi \sin \sigma \, d\sigma = 2\pi(1 - \cos \sigma_n) \qquad (7.7)$$

It is quite obvious (see Eq. (7.5)) that for a hemisphere the solid angle is 2π, and for a sphere 4π.

From Eq. (7.4) it follows that luminous flux into a solid angle Ω is

$$\Phi = \int_0^{\Omega} I \, d\Omega \qquad (7.8)$$

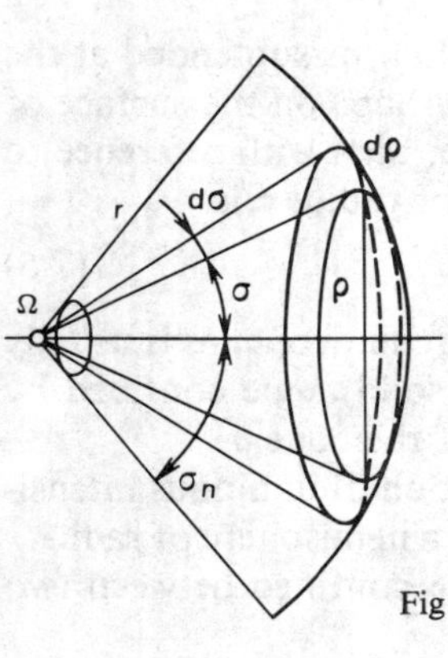

Fig. 7.5. Flux within a solid angle

If the luminous intensity of a source is invariable on changing from one direction to another, then the total flux into the solid angle Ω is

$$\Phi = I\Omega \qquad (7.9)$$

where the solid angle may be given by an area A a distance r from the source as $\Omega = A/r^2$.

If a source of light of luminous intensity I radiates into a solid angle, then the same luminous flux Φ will fall on areas subtending this angle at various distances from the source. These areas grow as the squared distance from the source, r. The flux that they intercept is obviously proportional to their size, A, that is, the flux per unit area will vary inversely as the squared distance from the source. Then we may write $\Phi/A = I/r^2$, i.e. we arrive at Eq. (7.9) again.

In the above argument we tacitly assumed that the source we deal with is a point source, that is, the distance from the source to the illuminated area is sufficiently large compared with the size of the source itself, and also the medium between the source and illuminated area neither absorb nor scatter the light energy.

In photometry the basic unit of luminous flux is the *lumen* (symbol lm) which is the flux per solid angle of one steradian (sr) from a point source of one candela. Note that watts and lumens have the same dimensions, namely energy per time. It is useful to observe that the lumen is simply radiant power as modified by the relative spectral sensitivity of the eye.

The *illumination* or *illuminance* at a surface A normal to the incident light is the flux falling on A per unit area, symbol E. Thus

$$E = \Phi/A \qquad (7.10)$$

This formula as well as Eq. (7.9) is valid on the condition that the luminous intensity I under examination does not vary on changing from one direction to another within the given solid angle. If it is not true, this formula may be used for small areas dA only so that

$$E = \frac{d\Phi}{dA} \qquad (7.11)$$

When a light beam of cross section A makes an angle ε with the normal to the illuminated surfaces, formulas (7.10) and (7.11) must be changed as the lit area will be larger. To account for this fact

$$E = \frac{\Phi}{A} \cos \varepsilon = \frac{1}{r^2} \cos \varepsilon \qquad (7.12)$$

$$E = \frac{d\Phi}{dA} \cos \varepsilon \qquad (7.13)$$

Illumination exhibits the property of additivity, namely, for n sources sending light to an area

$$E = \sum_{i=1}^{n} E_i \qquad (7.14)$$

i.e., the total illumination is the sum of individual illuminations received by the area from each source.

The SI unit of illumination is the *lux*, symbol lx, which is the illumination of one square metre normal to the incident light intercepting a luminous flux of one lumen. An alternative unit is the *phot* equal to one lumen per square centimetre.

If the size of the source cannot be neglected, then a number of problems require that the distribution of the luminous flux over its surface be given. The ratio of a luminous flux emitted from a source to the area of the source is called *emitted luminous exitance* or *luminous emittance* measured in lumens per square metre (lm m^{-2}). This concept may be used also to characterize the distribution of a reflected luminous flux. Mathematically luminous exitance is defined as

$$M = \frac{d\Phi}{dA} \qquad (7.15)$$

The luminous intensity in a given direction per unit area of the source projected on a plane normal to this direction is the *luminance* (photometric brightness), symbol L. Thus

$$L = \frac{dI}{dA \cos \varepsilon} \qquad (7.16)$$

where ε is the angle between the normal to the area element dA and the direction of the incident light dI. Incorporating $I = d\Phi/d\Omega$ from (7.4) we get the expression for luminance in terms of luminous flux

$$L = \frac{d^2\Phi}{d\Omega \, dA \cos \varepsilon} \qquad (7.17)$$

The unit of luminance is the candela per square metre (cd m^{-2}).

A measure of the total amount of light energy incident on a surface per unit area over a time interval is called the *light exposure*. An important parameter in photography, it can be measured as the integral of the surface light concentration over the period of illumination when E is variable

$$H = \int_0^t E(t)dt \qquad (7.18)$$

This concept therefore defines the amount of luminous energy in lumen-seconds incident on one square metre of the illuminated area, the appropriate unit being the lux second (lx s).

It should be noted that radiometric calculations may be carried out exactly as photometric calculations with the expressions (7.4) through (7.18). If watts are substituted for lumens in all the expressions, the computations are straightforward. Table 7.2 summarizes the basic photometric and radiometric quantities along with the respective relationships and SI units. It should not be overlooked that these expressions describe monochromatic situations or those where the spectral content of the radiant energy is of no importance — not infrequently the case in visual optical instruments.

The distribution of light power over wavelengths is of significance in computing the radiant quantities for the case of selective receivers of radiant power. To account for the spectral distribution, the aforementioned quantities may be restricted to a narrow wavelength band by adding the word *spectral* in their definitions and indicating the wavelength. The corresponding symbols are changed for lowercase letters (or by adding a subscript λ) for a spectral concentration and a λ in parentheses for a function of wavelength. We recall that radiometric functions have a subscript 'e' (for energy). Examples are the spectral density of irradiance

$$e_e(\lambda) = \frac{dE_e}{d\lambda} \qquad (7.19)$$

Table 7.2. Photometric and Radiometric Quantities

Quantity	Equation	Unit	Quantity	Equation	Unit
Luminous intensity	$I = \dfrac{d\Phi}{d\Omega}$	cd	Radiant intensity	$I_e = \dfrac{d\Phi_e}{d\Omega}$	W/sr
Luminous flux	$\Phi = \dfrac{dQ}{dt}$	lm	Radiant flux	$\Phi_e = \dfrac{dQ_e}{dt}$	W
Illumination	$E = \dfrac{d\Phi}{dA}$	lx	Irradiance	$E_e = \dfrac{d\Phi_e}{dA}$	W/m^2
Luminous exitance	$M = \dfrac{d\Phi}{dA}$	lm/m^2	Radiant exitance	$M_e = \dfrac{d\Phi_e}{dA}$	W/m^2
Luminance	$L = \dfrac{dI}{dA \cos \varepsilon}$	$\dfrac{cd}{m^2}$	Radiance	$L_e = \dfrac{dI_e}{dA \cos \varepsilon}$	$\dfrac{W}{sr\ m^2}$
Luminous exposure	$H = \displaystyle\int_0^t E(t)dt$	lx · s	Radiant exposure	$H_e = \displaystyle\int_0^t E_e(t)dt$	$\dfrac{J}{m^2}$

the spectral density of radiant emittance (exitance)

$$m_e(\lambda) = \frac{dM_e}{d\lambda} \qquad (7.20)$$

the spectral density of radiance

$$l_e(\lambda) = \frac{dL_e}{d\lambda} \qquad (7.21)$$

The total (integral) irradiance in the general form

$$E_e = \int_0^\infty e_e(\lambda)d\lambda \qquad (7.22)$$

the integral radiant emittance

$$M_e = \int_0^\infty m_e(\lambda)d\lambda \qquad (7.23)$$

and the integral radiance

$$L_e = \int_0^\infty l_e(\lambda)d\lambda \qquad (7.24)$$

In engineering applications more often than not these quantities are computed by integrating over finite wavelength intervals so that

$$E_{e,\lambda_1,\lambda_2} = \int_{\lambda_1}^{\lambda_2} e_e(\lambda)d\lambda \qquad (7.25)$$

$$M_{e,\lambda_1,\lambda_2} = \int_{\lambda_1}^{\lambda_2} m_e(\lambda)d\lambda \qquad (7.26)$$

$$L_{e,\lambda_1,\lambda_2} = \int_{\lambda_1}^{\lambda_2} l_e(\lambda)d\lambda \qquad (7.27)$$

Equations (7.19) through (7.27) are applicable in the visible range of the spectrum, the subscript 'e' is of course replaced with 'v'.

7.3 Polychromatic Measurements in Photometry

In many applications it is absolutely necessary to take into account the spectral characteristics of sources, detectors, optical systems, filters, and the like. In this section we demonstrate how this is done in photometry, that is, for the eye. If a source of radiation has a spectral power function $P(\lambda)$ in watts per micrometre of wavelength (the spectral density of power at every wavelength), the visual effect of this radiation is commonly obtained by multiplying it by the visual response function of the eye, $K(\lambda)$. The eye sensitivity function looks like the function in Fig. 7.3 with a maximum sensitivity at a certain wavelength. The effective visual power of the source is then obtained as the integral of $P(\lambda)K(\lambda)d\lambda$ over the appropriate wavelength interval. The eye sensitivity or *responsivity function* can be introduced as the ratio of a flux $d\Phi$ as sensed by the eye in a narrow wavelength interval centred on a wavelength λ, related to the same small band, to the incident radiant flux $d\Phi_e$. The ratio is called the *spectral luminous efficacy*

$$K(\lambda) = \frac{d\Phi}{d\Phi_e} \qquad (7.28)$$

The visual effect of the power of a source emerges thus weighted by the variation in responsivity of the measuring instrument, the eye. The responsivity is usually reported as a relative, i.e. normalized, function $V(\lambda)$, given in Table 7.3, the most common normalization technique being normalization to the peak of instrument sensitivity, K_m. For the eye, the maximum

Table 7.3. Normalized Visual Response for Photopic Vision *

λ, nm	$V(\lambda)$	λ, nm	$V(\lambda)$	λ, nm	$V(\lambda)$	λ, nm	$V(\lambda)$
380	4×10^{-5}	490	0.208	554	1.000	680	17×10^{-3}
390	1×10^{-4}	510	0.503	580	0.870	700	41×10^{-4}
400	4×10^{-4}	520	0.710	600	0.631	720	1×10^{-3}
420	4×10^{-3}	530	0.862	620	0.381	740	25×10^{-5}
440	23×10^{-3}	540	0.954	640	0.175	760	6×10^{-5}
450	38×10^{-3}	550	0.995	660	61×10^{-3}	770	3×10^{-5}

* Note that $V(\lambda)$ is customarily the photopic (normal level of illumination and luminance) visual response curve. Under the conditions of complete dark adaptation, the visual response for scotoptic vision should be used.

visual sensitivity is at the 0.554- μm wavelength. The normalized visual response, called the *luminous efficiency* is thus

$$V(\lambda) = K(\lambda)/K_m \qquad (7.29)$$

From (7.28) and (7.29) the weighted luminous flux is thus

$$d\Phi = K_m V(\lambda)d\Phi_e$$

or for luminance

$$dL = K_m V(\lambda)dL_e \qquad (7.30)$$

From the definition of the lumen, it can be determined that one watt of radiant energy at the wavelength of maximum visual sensitivity (0.554 μm) is equal to 680 lumens, i.e. $K_m = 680$ lm/W.

Now Eq. (7.30) rewrites as $dL = 680\ V(\lambda)dL_e$ and in view of (7.21) the integral luminance in the wavelength interval 0.38 to 0.77 μm is as follows

$$L = 680 \int_{0.38}^{0.77} V(\lambda)l_e(\lambda)d\lambda$$

where the units of the spectral radiance $l_e(\lambda)$ are W m^{-2} sr^{-1} μm^{-1}.

This procedure can be used to obtain the luminous flux emitted by a source with a spectral flux density $\varphi_e(\lambda)$ defined in (7.1), namely

$$\Phi = 680 \int_{0.38}^{0.77} V(\lambda)\varphi_e(\lambda)d\lambda$$

The luminous exitance and illuminance produced by this source can be deduced in the same manner.

The luminous efficiency of a given source can be defined as the ratio of the luminous flux Φ emitted by the source to the total radiant flux Φ_e of the source. Thus

$$n_v = \int_{0.38}^{0.77} V(\lambda)\varphi_e(\lambda)d\lambda \Big/ \int_{0}^{\infty} \varphi_e(\lambda)d\lambda$$

The luminous efficacy of a given radiant power source is defined as the luminous flux over the radiant flux in lumens per watt, viz.,

$$K = \Phi/\Phi_e = 680 \int_{0.38}^{0.77} V(\lambda)\varphi_e(\lambda)d\lambda \Big/ \int_{0}^{\infty} \varphi_e(\lambda)d\lambda$$

7.4 Radiative Transfer in Optical Systems

The luminance of a self-luminous area generally depends on the orientation of the area and on the direction of radiation (see Eq. (7.16)). It is an experimental fact, however, that for many light sources, L in Eq. (7.16) is nearly independent of ε, the angle at which it is viewed. In other words, the surface appears to be equally bright from whatever angle it is viewed. From Eq. (7.16) it can be seen that for such sources the luminous intensity I must be proportional to the cosine of the angle of viewing: $dI = dI_0 \cos \varepsilon$, where dI_0 is the luminous intensity of the elementary area dA in the normal to the area. This is *Lambert's cosine law*. It is not, of course, strictly true, but many sources, including secondary light sources such as diffusing surfaces, comply with it approximately; these are called Lambertian emitters or diffusers.

Consider radiation of such a source dA, shown in Fig. 7.6, into a hemisphere. The differential solid angle confined between two cones of semiangles σ and $\sigma + d\sigma$ as shown in the figure (see Eq. (7.6)) is $d\Omega = 2\pi \sin \sigma \, d\sigma$. By virtue of (7.17) the radiant flux from the differential area dA within the solid angle $d\Omega$ is as follows

$$d^2\Phi_e = L_e \, dA \cos \sigma \, d\Omega = 2\pi L_e \, dA \sin \sigma \cos \sigma \, d\sigma \qquad (7.31)$$

The total flux from the area dA into the hemisphere

$$d\Phi_e = \int_0^{\pi/2} d^2\Phi_e = \pi L_e \, dA \qquad (7.32)$$

Dividing both sides of this expression by dA we obtain the radiant exitance for a plane Lambertian emitter in the form

$$M_e = \pi L_e \qquad (7.33)$$

The same is true of a Lambertian diffuser; accordingly if an irradiance E_e is obtained on its diffusing surface, and as it diffuses the light

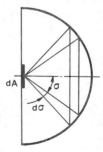

Fig. 7.6. Geometry of a Lambertian source of area DA radiating into a hemisphere

backwards, this irradiance equals the radiant exitance so that the radiance of this surface in view of (7.33) is

$$L_e = E_e/\pi \tag{7.34}$$

Let us determine the radiant flux $d\Phi_e$ within a solid angle $d\Omega$ confined by a circular cone of angle 2σ fanning with the normal to the small area dA as the axis, as shown in Fig. 7.7. We assume that this area is a part of a Lambertian diffuser of constant L_e.

The desired radiant flux from dA is obtained by integrating (7.31) between 0 and σ, namely,

$$d\Phi_e = \pi L_e dA \int_0^\sigma 2\sin\sigma\cos\sigma\,d\sigma = \pi L_e dA \sin^2\sigma \tag{7.35}$$

Suppose further that this flux falls on an area dA' and determine the flux from the circular area dA' on to the parallel area dA, propagating within a solid angle $d\Omega'$ subtended by dA at dA'. Denote the angle of the solid angle cone by $2\sigma'$. Assume that the radiance of dA' is the same in all directions.

In lossless transmission of radiation the flux incident on dA' and dA must be the same by energy conservation. Therefore, this flux can be determined from (7.35) upon changing dA for dA' and σ for σ'. Thus the flux falling from dA' on to dA is

$$d\Phi_e = \pi L_e dA' \sin^2\sigma' \tag{7.36}$$

Notice that the radiance of both radiated and radiating areas is the same.

The irradiance of dA' is obtained as

$$E_e = d\Phi_e/dA' = \pi L_e \sin^2\sigma'$$

Let us now determine the flux $d\Phi_e$ oncoming from an elementary Lambert emitter dA_1 on to a small area dA_2 oblique to the first surface. We assume the geometry of Fig. 7.8 with the centre to centre line being l long and the normals to the areas making angles ε_1 and ε_2 with this line.

The flux incident on to dA_2 can be determined by the formula deduced from (7.31)

$$d^2\Phi_e = L_e dA_1 \cos\varepsilon_1 d\Omega_1 \tag{7.37}$$

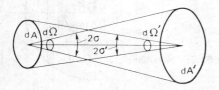

Fig. 7.7. Geometry of radiative transfer between two parallel coaxial circular Lambertian diffusers

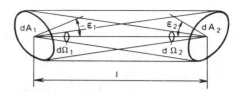

Fig. 7.8. Geometry of radiative transfer between two oblique elementary areas

From Fig. 7.8 it follows that $d\Omega_1 = (dA_2/l^2)\cos\varepsilon_2$ so that

$$d^2\Phi_e = L_e\, dA_1\, dA_2 \cos\varepsilon_1 \cos\varepsilon_2/l^2 \qquad (7.38)$$

This formula is valid so long as Lambert's cosine law holds.

By conservation of energy, i.e. when the flux is conserved,

$$d^2\Phi_e = L_e\, dA_2 \cos\varepsilon_2\, d\Omega_2 \qquad (7.39)$$

where $d\Omega_2 = (dA_1/l^2)\cos\varepsilon_1$.

Consider a specific case of the elementary areas dA_1 and dA_2 being parallel to each other but their normals do not coincide, as shown in Fig. 7.9. From this figure it follows that $\varepsilon_2 = \varepsilon_1 = \varepsilon$ and $l = l_0/\cos\varepsilon$. Substituting this data in (7.39) we get

$$d^2\Phi_e = L_e \frac{dA_1\, dA_2}{l_0^2} \cos^4\varepsilon$$

For $\varepsilon = 0$

$$d^2\Phi_e = L_e\, dA_1\, dA_2/l_0^2$$

The respective irradiances are

$$E_e = L_e \frac{dA_1}{l_0^2} \cos^4\varepsilon$$

and

$$E_{e0} = L_e\, dA_1/l_0^2$$

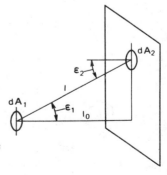

Fig. 7.9. Construction to deduce the irradiance in the area dA_2 parallel to an emitter of area dA_1

Hence,

$$E_e = E_{e0} \cos^4 \varepsilon \tag{7.40}$$

Thus the irradiance diminishes outwards as the fourth power of the cosine of the angle of view.

From (7.37) and (7.39) it follows that

$$dA_1 d\Omega_1 \cos \varepsilon_1 = dA_2 d\Omega_2 \cos \varepsilon_2 \tag{7.41}$$

that is, the product of the projected area ($dA \cos \varepsilon$) of the light tube and the solid angle subtended from this cross section is an invariant throughout the light tube.

Consider a more general case of the light tube constituted by two non-colinear legs of refractive indices n_1 and n_2 separated by an interface as shown in Fig. 7.10. Denote the area of the interface by dA, the solid angles subtended by the end areas dA_1 and dA_2 of the light tube by $d\Omega_1$ and $d\Omega_2$, and the angles between the normals to dA_1 and dA_2 and the broken axis of the light tube by ε_1 and ε_2, then

$$d\Omega_1 = \frac{dA_1}{l_1^2} \cos \varepsilon_1 \quad \text{and} \quad d\Omega_2 = \frac{dA_2}{l_2^2} \cos \varepsilon_2$$

where l_1 and l_2 are the centre to centre distances between dA and dA_1 and dA and dA_2, respectively.

The solid angles subtended by dA from the centres of dA_1 and dA_2 are by definition

$$d\Omega = (dA/l_1^2) \cos \varepsilon \quad \text{and} \quad d\Omega' = (dA/l_2^2) \cos \varepsilon'$$

where ε and ε' are the angles of incidence and refraction at the centre of dA.

The invariant derived in (7.41) leads us to the following equations

$$dA_1 d\Omega_1 \cos \varepsilon_1 = dA d\Omega \cos \varepsilon$$

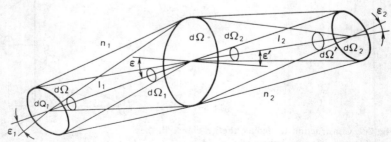

Fig. 7.10. A broken-axis light tube

and

$$dA \, d\Omega' \cos \varepsilon' = dA_2 \, d\Omega_2 \cos \varepsilon_2$$

whence

$$dA_1 \, d\Omega_1 \cos \varepsilon_1 / d\Omega \cos \varepsilon = dA_2 \, d\Omega_2 \cos \varepsilon_2 / d\Omega' \cos \varepsilon' \qquad (7.42)$$

We continue our evaluation of invariants involved in light propagating through light tubes with reference to Fig. 7.11. It shows two elementary areas dA_1 and dA spaced at l_1. The solid angle $d\Omega$ subtended by dA_1 from dA cuts from the sphere of radius l_1 an area $dA_1 \cos \varepsilon_1$, where the angle ε_1 is between the normals to the respective areas. The nomenclature is evident from the figure. The angle $d\varphi$ remains the same for the refracted portion of the light tube beyond dA_1, not shown in the figure, as on refraction the rays remain in the plane of incidence.

By definition of solid angle

$$d\Omega = dA_1 \cos \varepsilon_1 / l_1^2$$

The area cut by this solid angle on the sphere may be approximated as the rectangle

$$dA_1 \cos \varepsilon_1 = l_1 \sin \varepsilon \, d\varphi \, l_1 \, d\varepsilon$$

consequently,

$$d\Omega = \sin \varepsilon \, d\varepsilon \, d\varphi \qquad (7.43)$$

Similarly, for the portion of the light tube beyond the interface

$$d\Omega' = \sin \varepsilon' \, d\varepsilon' \, d\varphi \qquad (7.44)$$

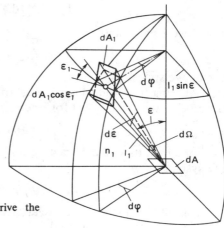

Fig. 7.11. Geometry to derive the Straubel invariant

Differentiating the refraction law $n_1 \sin \varepsilon = n_2 \sin \varepsilon'$ we get $n_1 \cos \varepsilon \, d\varepsilon = n_2 \cos \varepsilon' \, d\varepsilon'$. Multiplying these two last equations sidewise and by $d\varphi$ yields

$$n_1^2 \sin \varepsilon \cos \varepsilon \, d\varepsilon \, d\varphi = n_2^2 \sin \varepsilon' \cos \varepsilon' \, d\varepsilon' \, d\varphi \qquad (7.45)$$

Substitution of equations (7.42)-(7.44) into (7.45) leads to the *Straubel invariant*

$$n_1^2 \, dA_1 \, d\Omega_1 \cos \varepsilon_1 = n_2^2 \, dA_2 \, d\Omega_2 \cos \varepsilon_2 \qquad (7.46)$$

This invariant takes place for a light tube with any number p of refracting interfaces, namely,

$$n_1^2 \, dA_1 \, d\Omega_1 \cos \varepsilon_1 = n_{p+1}^2 \, dA_{p+1} \, d\Omega_{p+1} \cos \varepsilon_{p+1} \qquad (7.47)$$

For a reflecting surface and $|n_1| = |n_2|$ and $|\varepsilon| = |\varepsilon'|$, Eqs. (7.43) and (7.44) tell us that $d\Omega' = d\Omega$, that is, on reflection of an elementary light tube the solid angles subtended by the reflecting area in both image and object space are conserved, and Straubel's invariant written for one reflecting surface gives the equality

$$dA_1 \, d\Omega_1 = dA_2 \, d\Omega_2 \qquad (7.48)$$

The Straubel invariant represented by the relationships (7.46)-(7.48) holds for constant radiant fluxes both on refraction and on reflection.

If the light tube is filled with an optically homogeneous medium, the radiance, L_e in Eqs. (7.35)-(7.38), remains the same.

Given the radiant flux $d^2\Phi_e$ is constant within the light tube and the refractive indices in front of and behind the refracting interface are n_1 in object space and n_2 in image space, then

$$d^2\Phi_e = L_{e1} \, dA_1 \cos \varepsilon_1 \, d\Omega_1$$

for object space, and

$$d^2\Phi_e = L_{e2} \, dA_2 \cos \varepsilon_2 \, d\Omega_2$$

for image space.

Equating the right-hand sides of these expressions and observing Straubel's invariant (7.46) for one refracting surface yields

$$L_{e2} = (n_2/n_1)^2 L_{e1}$$

For a light tube of p refracting surfaces, the brightness at the exit end of the system

$$L_{e,\,p+1} = (n_{p+1}/n_1)^2 L_{e1} \qquad (7.49)$$

It should be noted that all relationships deduced in this section apply to photometric quantities as well.

7.5 Energy Flow in Optical Instruments

In the previous section we assumed the radiant flux to be constant at any section of the light tube. Actually light losses due to reflection at refracting interfaces, absorption at reflecting interfaces, and absorption and scattering in the bulk of the optical medium take place in optical systems. In computing the flux transmitted by an optical system these losses are accounted for as the respective fractions of the incident (radiant or luminous) flux Φ_i, known as *reflectance* or *reflection factor* $\rho = \Phi_r/\Phi_i$, *absorptance* or *absorption factor* $\alpha = \Phi_a/\Phi_i$, and *transmittance* or *transmission factor* $\tau = \Phi_t/\Phi_i$. Here Φ_r is the flux reflected at the refracting surface or the secondary flux if the surface is reflecting, Φ_a is the flux absorbed and scattered in the medium, or absorbed at the surface if it is reflecting, and Φ_t is the flux transmitted through the optical system under investigation. By energy conservation,

$$\Phi_r + \Phi_a + \Phi_t = \Phi_i$$

and $\rho + \alpha + \tau = 1$. All these factors are optical characteristics of the medium and are wavelength dependent, i.e. spectral characteristics.

In integral form these factors can be defined as

$$\rho = \int_{\lambda_1}^{\lambda_2} \varphi_e(\lambda)\rho(\lambda)d\lambda \bigg/ \int_{\lambda_1}^{\lambda_2} \varphi_e(\lambda)d\lambda \qquad (7.50)$$

for a radiant flux of spectral density $\varphi_e(\lambda)$, and as

$$\rho = \int_{0.38}^{0.77} V(\lambda)\varphi_e(\lambda)\rho(\lambda)d\lambda \bigg/ \int_{0.38}^{0.77} V(\lambda)\varphi_e(\lambda)d\lambda \qquad (7.51)$$

for a luminous flux of the same spectral density.

To determine the transmission of an optical system we examine the light losses due to reflection and absorption in the system. To this end, we recall that the portion of the light reflected from the surface of an ordinary dielectric material (such as glass) is given by the expression for Fresnel surface reflection

$$\rho = \frac{1}{2}\left[\frac{\sin^2(\varepsilon - \varepsilon')}{\sin^2(\varepsilon + \varepsilon')} + \frac{\tan^2(\varepsilon - \varepsilon')}{\tan^2(\varepsilon + \varepsilon')}\right] \qquad (7.52)$$

where ε and ε' are the angles of incidence and refraction respectively. At

9 — 7391

normal incidence (for small angles of incidence) this expression reduces to

$$\rho = \left(\frac{n' - n}{n' + n}\right)^2 \qquad (7.53)$$

where n and n' are the refractive indices of the media.

The variation of reflection from an air-glass interface as a function of the angle of incidence, ε, is shown in Fig. 7.12a. It will be seen that for angles as large as 30° to 40° the reflectance increases only insignificantly, so that for many optical systems engineers assume ρ being independent of the angle and estimate this factor by Eq. (7.53).

Figure 7.12b shows the variation of reflection from an air-glass interface as a function of the refractive index n' of glass, computed with Eq. (7.53).

When elements in a optical system are in contact or cemented with Canada balsam of $n = 1.52$, then owing to a small difference in refractive indices (up to 0.2) the light losses due to reflection are insignificant and may be neglected. To illustrate, for $n = 1.52$ and $n' = 1.72$, $\rho = 0.0038$, i.e. the loss of flux due to reflection amounts to 0.4 per cent.

For optical glasses in air, $\rho = 0.05$ on the average, i.e. reflectance losses are as high as 5 per cent. In sophisticated optical systems these losses can be as high as 30 to 40 per cent because the overall reflectance is in this case

$$\rho = \prod_{k=1}^{k=N} \rho_k$$

where N is the number of air-glass interfaces contributing to reflectance losses ρ_k.

It is worth noting also that secondary reflections from refracting surfaces are apt to diminish the contrast of the image, as illustrated in Fig. 7.13 where images of a bright point A appear both at the conjugate point A' and at a point A_1' coinciding with the image B' of a dark point B.

To cope with reflectance at refracting surfaces these are bloomed by applying anti-reflecting (quarter-wavelength) coating which is made as one or a few thin films diminishing, owing to destructive interference, the reflect-

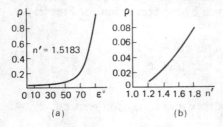

(a)   (b)

Fig. 7.12. Reflectance as a function of (a) angle of incidence, and (b) refractive index

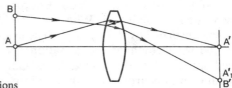

Fig. 7.13. Effect of secondary reflections

ed portion of the incident flux. The appropriate film thickness is determined as

$$d \approx (2k + 1)\lambda/4n_f \cos \varepsilon'$$

where λ is the wavelength at which reflectance is to be suppressed, n_f is the refractive index of the film, ε' is the angle of refraction, and k is a positive integer. For polychromatic radiation the least reflection factor will be achieved at $k = 0$. With $k = 0$ and $\varepsilon' = 0$ (normal incidence), $d = \lambda/4n_f$. For air-glass interfaces where either $n = 1$ or $n' = 1$,

$$n_f = \sqrt{n_{glass}} \tag{7.54}$$

Observing that for optical glasses the refractive index varies between 1.47 and 1.80, the index for anti-reflecting films is selected in the range from 1.21 to 1.34. Materials for the films are magnesium fluoride (MgF_2) and cryolite deposited by sedimentation from vapour phase in vacuum. However, the mechanical strength of these films is insignificant and limits their application. For many applications therefore the film is deposited by sedimenting silicon dioxide or titanium from the solution in ethanol. This technology results in a hard durable film which, however, exhibits a somewhat greater refractive index (about 1.45) than needed, thus reducing the anti-reflecting effect.

It should be emphasized that reflection from quater-wave low-reflection coated surfaces is wavelength selective and so is the transmission of the optical system with this element.

Two- or three-layer quater-wave films on refracting surfaces cut reflection losses down to 0.5 per cent providing at the same time high strength of the coating and a constant spectral distribution.

Reflecting surfaces are coated with aluminium, gold, silver, rhodium, and other metals. The spectral reflectance of these coatings can be determined with the formula

$$\rho(\lambda) = 1 - 365\sqrt{1/\sigma\lambda}$$

where σ is the conductivity in siemens per metre, and λ is the wavelength in micrometres. For aluminium coatings deposited in vacuum, $\rho(\lambda) = 0.93$ at $\lambda = 0.5\ \mu m$. The reflectance is seen to improve towards greater wavelengths.

9*

We turn now to discuss absorption within the optical element and the associated transmission. Within the optical element some of the radiation can be absorbed by the material. The extent to which it is absorbed depends on the wavelength of the radiation and the thickness and nature of the medium. If $d\Phi$ is the change in radiant or luminous flux of a collimated beam of monochromatic radiation in passing through a small thickness dl of an absorbing medium, then the initial flux Φ_0 after a thickness of l units of length becomes Φ_l such that

$$\Phi_l/\Phi_0 = e^{-\alpha l}$$

where $\alpha = d\Phi/\Phi dl$ is the linear absorption coefficient in inverse metres. This equation is known as *Bouguer's law* or *Lambert's law* of absorption enunciated by Bouguer and Lambert.

Abstracting for a moment from absorption and reflection of intensity at the surfaces of optical elements we can state that if the portion of intensity absorbed per unit length is α_1 then the portion transmitted through is $1 - \alpha_1$, that is, the transmittance after a unit of length is $\tau_1 = 1 - \alpha_1$. Assuming this portion is 40 per cent, the next unit of length will transmit 40 per cent of this 40 per cent, therefore the transmission through a thickness of l units will be

$$\tau_\alpha = \tau_1^l = (1 - \alpha_1)^l \qquad (7.55)$$

The units of measurements for the absorption coefficient and the length of the optical medium must be the same.

Now the total transmission through an optical element is a product of its surface transmissions and its internal transmission derived above. The transmission of a completely non-absorbing plate in air, including all internal reflections, can be shown to be $T = 2n/(n^2 + 1)$. The total transmission of an optical element of refractive index n in air may be given as

$$\tau = T\tau_\alpha$$

The concept of *optical*, or *transmission*, *density*, symbol D, is convenient to use in estimating the spectral internal transmittance of optical elements of various length. The density of an optical element is the log of its *opacity* (the reciprocal of transmittance τ), thus

$$D(\lambda) = \log \frac{1}{\tau_\alpha(\lambda)} = -\log \tau_\alpha(\lambda) \qquad (7.56)$$

If the density of unit length of material is D_1, then from (7.55) the total density of a uniform material of thickness l is

$$D(\lambda) = lD_1(\lambda) \qquad (7.57)$$

If the system consists of p media of optical density D_i, being either spectral densities at certain wavelengths or integral over a range of wavelengths, to a fair approximation the density of such a 'stack' is the sum of the individual densities,

$$D(\lambda) = \sum_{i=1}^{i=p} D_i(\lambda)$$

The total transmittance of an optical system constituted by p media can be given as

$$\tau = (1 - \rho)(1 - \alpha) = \prod_{k=1}^{k=p+1} (1 - \rho_k) \prod_{m=1}^{m=p} (1 - \alpha_{1m})^{l_m}$$

For an optical system constituted by refracting and reflecting surfaces, the transmittance is found as the product

$$\tau = \prod_{k=1}^{k=p+1} (1 - \rho_k) \prod_{i=1}^{i=p} (1 - \alpha_{1i})^{l_i} \prod_{n=1}^{n=N_m} \rho_{mn} \prod_{q=1}^{q=N_b} \tau_{bq}$$

where ρ_m is the reflectance of mirrors and beam-splitting coatings of semitransparent mirrors, N_m is the number of mirrors, τ_b is the transmittance of beam-splitting coatings, and N_b is the number of beam-splitting coatings.

It should be noted that in approximate estimations of optical system transmission, the designer must take into account only those refracting surfaces which contact with air. For all surfaces of glass elements of refractive indices 1.4 to 1.6 (crown glass) the reflectance may be assumed to be $\rho_c = 0.05$, while for elements of flint with n over 1.6 $\rho_f = 0.06$. The absorption coefficient for glass of any type may be taken to be $\alpha_1 = 0.01$, that is, the transmittance per unit length is $\tau_1 = 0.99$. The losses at surfaces with total internal reflection may be neglected.

With the above allowances, for an optical system without anti-reflecting and beam-splitting coatings the transmission can be estimated as

$$\tau = 0.95^{N_c} 0.94^{N_f} 0.99^d \rho_m^{N_m}$$

where N_c is the number of non-cemented crown surfaces, N_f is the number of non-cemented flint surfaces, N_m is the number of mirror reflecting surfaces, and d is the total thickness of all glass elements, all in consistent units — more frequently in centimetres.

For silvered second surfaces, $\rho_m = 0.85$; for aluminized first surfaces, $\rho_m = 0.87$; and for oxidized aluminium surfaces, $\rho_m = 0.8$ to 0.84 [3].

7.6 Absorption Filters

Absorption filters are made of materials which transmit light selectively, that is, they transmit certain wavelengths better than others. More often than not this element is a plane parallel plate of coloured glass, or plastic, or a thin gelatin film or other optical materials such as gases and liquids coloured by dyeing.

The material preferred for absorption filter fabrication is coloured optical glass. The filters are named according to the spectral band where the filter transmission is largest. Accordingly the glasses for filters are referred to as ultraviolet, violet, blue, blue-green, green, yellow-green, yellow, orange, red, infrared, purple, neutral, dark, and colourless. Neutral glass filters attenuate the incident light almost uniformly. Colourless glass filters transmit throughout the optical portion of the spectrum — from infrared through visible to ultraviolet.

Spectral characteristics of absorption filters can be stated in a number of ways. More specifically, the spectral absorption factor can be tabulated for various wavelengths, optical density $D(\lambda)$ curves can be plotted, but a more popular method of characterization is by plotting the spectral transmission versus wavelength in micrometres. We recall that the transmission density and transmittance are related by the formula combined from (7.56) and (7.57):

$$D(\lambda) = -\log \tau_\alpha(\lambda) = D_1 d \qquad (7.58)$$

With account of losses for reflection at two surfaces of the filter, the total transmission at a given wavelength will be given by

$$\tau(\lambda) = (1 - \rho)^2 \tau_\alpha(\lambda) \qquad (7.59)$$

where ρ is the reflection factor of the filter glass. Incorporating Eq. (7.58) we obtain that the optical density of an absorption filter accounting for the reflection losses is

$$D'(\lambda) = -\log \tau_\alpha(\lambda) - 2\log (1 - \rho)$$
$$= D(\lambda) + D(\rho)$$

Soviet catalogues of glass filters quote spectral absorption coefficients and optical densities $D(\lambda)$ and depict transmission curves versus wavelength for each type of filter glass taken at a certain thickness. By way of example, Fig. 7.14 shows such a plot for a blue-green filter 2 mm thick. It absorbs infrared radiation and has the largest transmission at $\lambda = 500$ nm. The wavelengths at which the transmittance reduces by half the peak value are referred to as cut-off wavelengths. The range between these wavelengths is quoted as the transmission band of the filter.

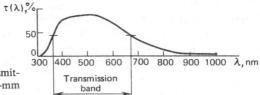

Fig. 7.14. Spectral transmittance of a blue-green 2-mm glass filter

In photography, filters receive one more characteristic — the filter factor indicating how many times the exposure must be increased (or the image illuminance must be increased by an iris diaphragm) when the given absorption filter is used as against the exposure without the filter.

To close this section we note that in addition to absorption filters, which selectively absorb radiation as their name implies, there exists a variety of almost 100 per cent effective *interference filters*. They harness the phenomenon of interference for their effect and are manufactured as thin films (coatings) deposited on a transparent substrate. Among the types of interference coatings which are readily available are long or short wavelength transmission filters (note that these filters are classified as in radio engineering according to the portion of the spectrum they transmit through), band-pass filters, and narrow bandpass (spike) filters. The characteristics of an interference filter depend on the thickness of the films in its coating, therefore once a combination of films has been designed to produce a desired characteristic, the operable wavelength region can be shifted at will by simply increasing or decreasing all the film thicknesses in due proportion.

7.7 Image Illuminance in Optical Systems

Suppose an optical system forms an image of a self-luminous object of area dA on the optical axis and the entrance pupil of the system subtends an angle σ_A at the object of radiance L_e, then the system collects the radiant flux (see Eq. (7.35))

$$d\Phi_e = \pi L_e \, dA \, \sin^2 \sigma_A$$

We learned in the previous sections that the flux will be attenuated in the optical system, so that the image $dA\,'$ will have a lesser radiance in the proportion dictated by Eq. (7.49). Accordingly the radiant flux illuminating the image, after account of the transmission of the system τ, will be written as follows

$$d\Phi_e' = \tau (n\,'/n)^2 \pi L_e \, dA\,' \, \sin^2 \sigma_A'. \qquad (7.60)$$

where n' and n are the refractive indices of object and image media, respectively, and σ_A' is the aperture angle subtended by the exit pupil of the instrument at the image.

On the other hand, Eq. (7.35) indicates that the flux leaving the optical system of transmittance τ must be

$$d\Phi_e' = \tau\pi L_e\, dA\, \sin^2\sigma_A \tag{7.61}$$

Equating the fluxes in the penultimate and this equation we get

$$dA\,\sin^2\sigma_A = (n'/n)^2 dA'\,\sin^2\sigma_A'.$$

or

$$n^2\sin^2\sigma_A/n'^2\sin^2\sigma_A'. = dA'/dA$$

Observing that the ratio of the conjugate areas dA'/dA equals the squared ratio of the image and object heights y'/y, which is the transverse magnification β, we arrive at the *Abbe sine condition*

$$\frac{n\,\sin\sigma_A}{n'\,\sin\sigma_A'.} = \beta$$

This condition must be satisfied by any optical system imaging an axial elementary area perpendicular to the optical axis at any finite aperture angles σ_A and $\sigma_A'.$.

The image irradiance $E_e' = d\Phi_e'/dA'$ can be computed by one of the following formulas

$$\begin{aligned}
E_e' &= \tau(n'/n)^2\pi L_e\,\sin^2\sigma_A'. \\
&= \tau(dA/dA')\pi L_e\,\sin^2\sigma_A \\
&= \tau\pi L_e\,\sin^2\sigma_A/\beta^2
\end{aligned} \tag{7.62}$$

Since most optical systems are immersed in air and $n = n' = 1$, the image irradiance becomes

$$E_e' = \tau\pi L_e\,\sin^2\sigma_A'. \tag{7.63}$$

Let us represent $\sin\sigma_A'.$ in a form convenient for computation. For the optical system represented in Fig. 7.15 the approximation holds

$$\tan\sigma_A'. \approx \sin\sigma_A'. \approx D'/2(z' - z_p')$$

where D' is the diameter of the exit pupil, and z_p' and z' define the locations of the exit pupil and image plane with respect to the second focal point of the system.

Given the transverse magnification of the system β, and the magnification in the pupils (pupil diameter ratio) $\beta_p = D'/D$, we have

$$z' = -f'\beta, \quad z_p' = -f'\beta_p$$

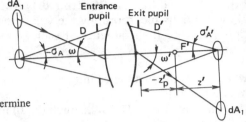

Fig. 7.15. Construction to determine
the exit aperture angle

Substituting these expressions in (7.64) yields

$$\sin \sigma'_A \cdot \approx D\beta_p / 2f'(\beta_p - \beta) \qquad (7.65)$$

Now we incorporate this expression for $\sin \sigma'_A \cdot$ into (7.62) to get

$$E'_e = \frac{1}{4} \tau \left(\frac{n'}{n}\right)^2 \pi L_e \left(\frac{D}{f'}\right)^2 \frac{\beta_p^2}{(\beta_p - \beta)^2} \qquad (7.66)$$

For often met symmetric systems of $\beta_p = 1$ this reduces to

$$E'_e = \frac{1}{4} \tau \left(\frac{n'}{n}\right)^2 \pi L_e \left(\frac{D}{f'}\right)^2 \frac{1}{(1 - \beta)^2} \qquad (7.67)$$

and when this system is in air

$$E'_e = \tau \pi L_e (D/f')/4(1 - \beta)^2 \qquad (7.68)$$

For the object at infinity when $\beta = 0$,

$$E'_e = \frac{1}{4} \tau \left(\frac{n'}{n}\right)^2 \pi L_e \left(\frac{D}{f'}\right)^2 \qquad (7.69)$$

or with $n' = n$

$$E'_e = \frac{1}{4} \tau \pi L_e (D/f')^2 \qquad (7.70)$$

Now we recall from Eq. (7.34) that the radiance of a diffuse surface in which an irradiance E_e is produced is $L_e = E_e/\pi$. With account of the reflection factor of such a Lambertian diffuser

$$\pi L_e = \rho E_e \qquad (7.71)$$

and we may substitute this expression for πL_e in Eqs. (7.66)-(7.70) to describe the situations where the object is illuminated rather than self-luminous.

If the luminant area dA_1 is extra-axial, as depicted in Fig. 7.15, then to determine the irradiance of the image area dA'_1 one has to take into account that the irradiance falls away from the axis as $\cos^4 \omega'$ (see

Eq. (7.40)) and that the flux from dA_1 is vignetted. The last factor may be allowed for by incorporating the vignetting coefficient k_l, defined in Section 6.4. Thus the irradiance of such an extra-axial area is

$$E_e' = k_\omega E_e' \cos^4 \omega' \tag{7.72}$$

where E_e' is the irradiance in the axial area computed with one of the formulae (7.66)-(7.70).

Analysis of Eqs. (7.66) through (7.70) indicates that in all situations examined the irradiance is proportional to the squared ratio of the entrance pupil diameter to the effective focal length. This ratio, by other definition equal to $2n' \sin \sigma_A'$, is called the *aperture ratio*. The inverse quantity, i.e. the ratio of the focal length to the clear aperture of a lens system is called the *relative aperture*, *f-number*, *stop number* or *focal ratio* (symbol F); for a ratio of say 2.5 it is customarily written $f/2.5$, $f2.5$, or $f{:}2.5$. Accordingly, as the f-number is diminished, the irradiance of the image improves.

We mention in passing one more important characteristic, called *numerical aperture*, abbreviated to NA, which is the index of refraction (of the medium in which the image lies) times the sine of the half angle of the cone of illumination, i.e. NA $= n' \sin \sigma_A$. We shall return to this characteristic below in discussing the resolving power of optical instruments. Thus far we confine ourselves to noting only that for systems with infinite object distances f-number $= 1/2$NA.

Numerical aperture and f-number are obviously two methods of describing the same characteristic of a system. Numerical aperture is more conveniently used for systems that work at finite conjugates (such as in the case of microscope objectives) and f-number is applied to systems for use with distant objects (such as camera lenses and telescope objectives).

The effect of the magnification between the pupils β_p on the illumination of the image can be gleaned from Fig. 7.16 which plots $\beta_p^2/(\beta_p - \beta)^2$ proportional to the image irradiance as a function of the lateral magnification β for two values of pupil diameter ratio $\beta_p = 0.7$ and $\beta_p = 1.5$, other conditions being equal. As can be seen at $\beta = -1$ the irradiance increases 2.2 times as the pupil diameter ratio increases from 0.7 to 1.5.

To close the section we determine the irradiance of an axial image for an entrance pupil in the shape of a ring. This geometry is typical of reflecting instruments and catadioptric systems with the obscured central portion of the pupil (see Fig. 1.6).

In view of the ring geometry Eq. (7.61) gives for the image irradiance

$$E_e' = \tau \pi L_e (\sin^2 \sigma_{A'\text{ex}}' - \sin^2 \sigma_{\text{in}}') \tag{7.73}$$

where $\sigma_{A'\text{ex}}'$ is the aperture angle in image space formed by the ray grazing the external edge of the exit pupil, and σ_{in}' is the angle made by the ray graz-

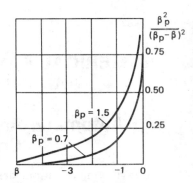

Fig. 7.16. The effect of the lateral magnification between the conjugate pupils, β_p, and the transverse magnification, β, on the image irradiance proportional to the quantity in the ordinate

ing the internal edge of the exit pupil with the optical axis.

In similarity with Eq. (7.64) we may write

$$\sin \sigma'_{A'ex} \approx D'_{ex}/2(z' - z'_p)$$

and

$$\sin \sigma'_{in} \approx D'_{in}/2(z' - z'_p)$$

where D'_{ex} and D'_{in} are the external and internal diameters of the exit pupil, and z'_p and z' define the position of the exit pupil and image plane with respect to the second focal point of the system.

For a power equivalent system, the diameter of the circular unobscured exit pupil results from the equality of pupil areas, namely,

$$D'_{eq} = \sqrt{D'^2_{ex} - D'^2_{in}}$$

and accordingly the image space aperture angle can be obtained from

$$\sin\sigma'_{A'eq} = \sqrt{\sin^2 \sigma'_{A'ex} - \sin^2 \sigma_{in}}$$

Note in conclusion that although the argument of this section dealt with radiometric quantities, the resultant equations hold for photometric quantities with the appropriate substitutions.

NUMERICAL RAYTRACING THROUGH AN OPTICAL SYSTEM

8.1 Automatic Raytracing by Electronic Computer

The raytracing equations derived in Chapters 2 and 3 hold either for a perfect optical system or in the paraxial domain. The real image of an object as formed by an actual optical system is produced by a number of real rays traversing the system. A real ray trace is needed therefore to derive a correct indication of system performance.

Before the advent of computer the image quality was judged from raytracing computations for meridional rays, more seldom for sagittal rays, and very seldom for skew rays. This could be attributed to rather modest performance of the systems being designed and image quality desired and also to the designers' wish to cut down the time of computations which amounts to about 50 to 70 per cent of the total time of the optical system design.

Modern optical establishments extensively employ computers for automatic numerical raytracing. Feder [27] has devised raytracing formulae which proved to be more convenient for automatic design. Of the Soviet publications discussing these raytracing formulae in detail we would point out to books by Volosov [2], Rodionov [17], and Slyusarev [26]. In what follows we outline briefly this approach to solving the numerical raytracing problem.

The position of a ray entering an optical system is defined by the direction cosines $\nu_1 = \cos \gamma$, $\mu_1 = \cos \beta$, $\lambda_1 = \cos \alpha$. To compute these cosines it is sufficient to know the position of the object plane, defined by s_1, the plane of the entrance pupil, t, and the coordinates of the ray in these planes, y_1, x_1, and m_1, M_1, as illustrated in Fig. 8.1. From geometric considerations we have

$$\nu_1 = -(s_1 - t)/R_3$$
$$\mu_1 = (m_1 - y_1)/R_3 \qquad (8.1)$$
$$\lambda_1 = (M_1 - x_1)/R_3$$

where $R_3 = \sqrt{(s_1 - t)^2 + (m_1 - y_1)^2 + (M_1 - x_1)^2}$, the sign of the radical coinciding with the sign of the difference $s_1 - t$.

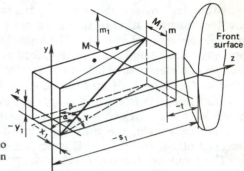

Fig. 8.1. Construction to determine the direction cosines

For a two-dimensional ray fan, $M_1 = 0$, $x_1 = 0$, and $\lambda_1 = 0$, and the direction cosines reduce to

$$\nu_1 = -(s_1 - t)/R_2$$
$$\mu_1 = (m_1 - y_1)/R_2$$

where $R_2 = \sqrt{(s_1 - t)^2 + (m_1 - y_1)^2}$.

For an infinitely distant object, the position of ray is defined if the values of t, m_1, M_1 and the field of view 2ω are specified. Then

$$\nu_1 = \cos \omega$$
$$\mu_1 = \cos (90° + \omega) = -\sin \omega$$
$$\lambda_1 = 0$$

The optical system design is specified by its radii of curvature r_1, r_2, ... , r_p; element thicknesses and spacings $d_1, d_2, ... , d_{p-1}$; and refractive indices $n_1, n_2, ... , n_{p+1}$. The respective geometry is illustrated in Fig. 8.2 by the trace of a ray through surfaces k and $k + 1$ of curvature

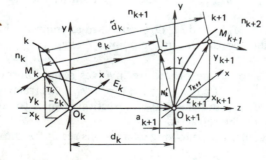

Fig. 8.2. Notation to deduce raytracing equations

radii r_k and r_{k+1} spaced a distance d_k along the optical axis.

Numerical raytracing for any ray is a two-step procedure. The first step involves the evaluation of the coordinates of ray intercept with an optical surface (say number $k + 1$) by the known coordinates of intercept with the previous surface and the direction cosines of this ray after this surface k. The second step is devoted to computing the direction cosines of the ray after surface $k + 1$. From Fig. 8.2 it can be seen that in each space of sequential images there exists a special system of coordinate axes whose origin is at the vertex of the optical system, the z axis is along the optical axis of the system, the y axis is vertical and lies in the meridional (or tangent) plane, and the x axis is perpendicular to the meridional plane.

The formulae for this raytracing procedure are derived by analytical geometry in vector form. Point M_k where the ray meets the kth surface has the known coordinates z_k, y_k, x_k and the ray is specified by the direction cosines $\nu_{k+1}, \mu_{k+1}, \lambda_{k+1}$. The steps involved in the derivation of these formulae are graphically illustrated in Fig. 8.3 and are as follows.

The computation is started by initiating the vector T_k by the known coordinates of point M_k, and initiating the vector $O_k O_{k+1}$. These two vectors yield the third vector $M_k O_{k+1} = O_k O_{k+1} - T_k = E_k$.

Geometrically (see Fig. 8.2) the point L is located to define the vector $M_k L$ by tracing the normal N_k' to the ray through O_{k+1}. In Fig. 8.3 this vector is denoted by e_k. The same procedure can be repeated mathematically. The vectors $M_k O_{k+1}$ and $M_k L$ are invoked to determine the vector N_k', its squared magnitude $|N_k'|^2 = A_{k+1}^2$, and the projection a_{k+1} of this vector on the z axis.

The angle γ is computed as a function of $\alpha_k, \nu_{k+1}, |N_k'|^2, r_{k+1}$ and the vector $L M_{k+1}$ is computed as a matter of N_k'. This vector along with $M_k L$ yields the vector $M_k M_{k+1}$ whose absolute value is denoted $\tilde{d}_k$ and will be referred to as a skew thickness.

The cosine of the angle of incidence, $\cos \varepsilon_{k+1}$, denoted by q_{k+1}, is computed by $A_{k+1}^2, a_{k+1}, r_{k+1}$, and ν_{k+1}, then by the law of refraction $q_{k+1}' = \cos \varepsilon_{k+1}'$ is computed.

The vectors $M_k M_{k+1}$ and $M_k O_{k+1}$ are used to compute the vector T_{k+1} and determine the coordinates of the point $z_{k+1}, y_{k+1}, x_{k+1}$ where the ray meets surface $k + 1$.

The second step of the procedure involves some intermediate values $g_{k+1} = q_{k+1}' + (n_{k+1}/n_{k+2})q_{k+1}; c_{k+1} = 1/r_{k+1}$, refractive indices, the coordinates of M_{k+1}, and the known direction cosines to compute ν_{k+2}, μ_{k+2}, and λ_{k+2} for the refracted ray.

Collected together the following expressions result for the evaluation of a ray trace through an optical system of a few spherical surfaces

$$e_k = -[(z_k - d_k)\nu_{k+1} + y_k\mu_{k+1} + x_k\lambda_{k+1}]$$

$$a_{k+1} = e_k\nu_{k+1} + (z_k - d_k)$$

$$A_{k+1}^2 = (z_k - d_k)^2 + y_k^2 + x_k^2 - e_k^2$$

$$P_{k+1} = c_{k+1}A_{k+1}^2 - 2a_{k+1}$$

$$q_{k+1} = \sqrt{\nu_{k+1}^2 - c_{k+1}P_{k+1}}$$

$$\tilde{d}_k = e_k + P_{k+1}/(\nu_{k+1} + q_{k+1})$$

$$q'_{k+1} = \sqrt{1 - (n_{k+1}/n_{k+2})^2(1 - q_{k+1}^2)} \qquad (8.2)$$

$$g_{k+1} = q'_{k+1} - (n_{k+1}/n_{k+2})q_{k+1}$$

$$z_{k+1} = (z_k - d_k) + \tilde{d}_k\nu_{k+1}$$

$$y_{k+1} = y_k + \tilde{d}_k\mu_{k+1}$$

$$x_{k+1} = x_k + \tilde{d}_k\lambda_{k+1}$$

$$\nu_{k+2} = (n_{k+1}/n_{k+2})\nu_{k+1} - g_{k+1}(z_{k+1}c_{k+1} - 1)$$

$$\mu_{k+2} = (n_{k+1}/n_{k+2})\mu_{k+1} - g_{k+1}y_{k+1}c_{k+1}$$

$$\lambda_{k+2} = (n_{k+1}/n_{k+2})\lambda_{k+1} - g_{k+1}x_{k+1}c_{k+1}$$

The numerical raytracing with these formulae terminates by deriving the coordinates z_p, y_p, x_p of the ray intercept with the last surface number p, and the direction cosines $(\nu_{p+1}, \mu_{p+1}, \lambda_{p+1})$ of the ray leaving the system.

If s' is the distance from the last surface to the plane where the image quality is estimated, the coordinates where the ray meets this plane can be obtained as

$$y' = y_p + (\mu_{p+1}/\nu_{p+1})(s' - z_p)$$
$$x' = x_p + (\lambda_{p+1}/\nu_{p+1})(s' - z_p) \qquad (8.3)$$

We note the advantages of the aforementioned formulae over the trigonometric raytracing set. They contain no trigonometric functions, no variables which may become infinite, no dependences corrupting the accuracy, and there is no need for check-out computations as the set has the self-test relations:

$$(z_{k+1}c_{k+1} - 1)^2 + (y_{k+1}c_{k+1})^2 + (x_{k+1}c_{k+1})^2 = 1$$
$$\nu_{k+1}^2 + \mu_{k+1}^2 + \lambda_{k+1}^2 = 1$$

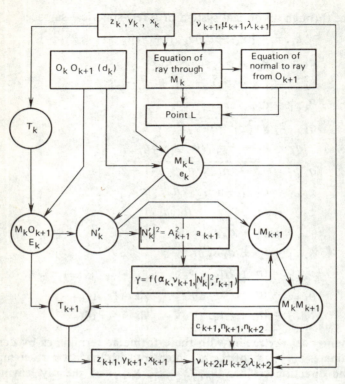

Fig. 8.3. Flow chart to derive raytracing formulae

8.2 Optical Computations with Very Narrow Astigmatic Ray Pencils

In infinitely narrow ray pencils the rays diverge at a very small angle to one another. These ray pencils are also called elementary pencils as they are able to fill only elementary areas in the pupils. For an axial object point A in Fig. 8.4 such a pencil consists of paraxial rays. This pencil retains its homocentric property when it emerges from the optical system, that is its rays form the point (stigmatic) image A_0'.

The principal ray of any axial infinitely narrow ray pencil traverses the centre of curvature of the optical surface, therefore the surface element involved has identical radii of curvature, $r_m = r_s$ in the meridional plane, mm, and in the sagittal plane, ss.

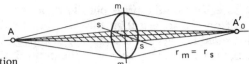

Fig. 8.4. Stigmatic image formation

For narrow ray pencils emanating from an off-axis point B, the conditions of travel will be different in the meridional and sagittal planes. The principal ray, about which the other rays in the pencil are symmetrically displaced, does not pass through the centre of curvature of the optical surface in the general case, therefore the lit surface element will have different radii of curvature, $r_m \neq r_s$, in the directions mm and ss, shown in Fig. 8.5. The emanating wavefront corresponding to this oblique elementary pencil is no longer spherical. The rays travelling in the meridional and sagittal planes meet the principal ray in distinct points B'_m and B'_s of image space at a distance from the perfect image at B'_0.

In the image plane containing the point B'_m of convergence of the meridional ray fan from the extra-axial object point B, the rays of the sagittal fan form a horizontal line segment rather than a point. Accordingly in the image plane containing the point B'_s of convergence of the sagittal ray fan, the rays of the meridional fan form a vertical line segment.

The phenomenon, in fact a defect of a lens, as a result of which the image of an off-axis point is formed as two mutually orthogonal lines lying in different planes is called *astigmatism*. Accordingly, a ray pencil producing such an image may be referred to as an elementary astigmatic ray bundle. It is quite obvious that this phenomenon is undesirable in optical systems, as it degrades the quality of images of off-axis points produced even by very narrow ray bundles. An estimator of the quality of imagery for extra-axial points can be the astigmatic difference $\Delta z'_a = z'_s - z'_m$. When this difference is zero, $z'_s = z'_m$, the narrow meridional and sagittal ray fans form a point image.

The position of the image points B'_m and B'_s is calculated by tracing narrow astigmatic ray pencils through the optical system as follows.

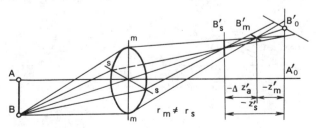

Fig. 8.5. Astigmatic image formation

Let BM be the principal ray of an elementary oblique ray pencil incident on the spherical surface of radius r from an off-axis point B, as illustrated in Fig. 8.6. Denote the distance from the point M where the principal ray meets the surface to the point B along the ray by t_m. To produce an elementary ray pencil in the meridional plane, we trace an infinitely close ray BM_1 making an angle $d\omega$ with the principal ray. After the surface these rays unite at a point B'_m on the principal ray a distance t'_m from the surface along the ray. We assume that the surface separates optical media of refractive indices n and n'.

We suppose now the quantities r, n, n', t_m, φ, ω, ε, ε', all are readily derivable or known, and evaluate the relation between t_m and t'_m with the help of the law of refraction.

Adding an infinitesimal increment $d\omega$ to the angle ω of axis crossing by the principal ray alters both the angle of incidence ε and the angle of refraction ε'. Differentiating the law of refraction yields

$$n \cos \varepsilon \, d\varepsilon = n' \cos \varepsilon' \, d\varepsilon' \qquad (8.4)$$

From Fig. 8.6 it follows that $\varepsilon = \omega - \varphi$ and $\varepsilon' = \omega' - \varphi$, consequently, $d\varepsilon = d\omega - d\varphi$, $d\varepsilon' = d\omega' - d\varphi$. By letting $d\varepsilon$, $d\varepsilon'$, $d\omega$, $d\omega'$, $d\varphi$ and other increments being infinitesimal we obtain $MM_1 = r \, d\varphi$, the angle $BM_1M = 90° - \varepsilon$, and $B'_mM_1M = 90° + \varepsilon'$. From the triangle BM_1M we have $MM_1/-d\omega = -t_m/\sin(90° - \varepsilon)$, whence $d\omega = (r \cos \varepsilon/t_m) d\varphi$ and consequently

$$d\varepsilon = (r \cos \varepsilon/t_m - 1) d\varphi \qquad (8.5)$$

From the triangle B'_mM_1M we have $MM_1/d\omega' = t'_m/\sin(90° + \varepsilon')$, whence $d\omega' = (r \cos \varepsilon'/t'_m) d\varphi$ and consequently

$$d\varepsilon' = (r \cos \varepsilon'/t'_m - 1) d\varphi \qquad (8.6)$$

Substituting (8.5) and (8.6) into (8.4) we arrive at the Abbe — Young

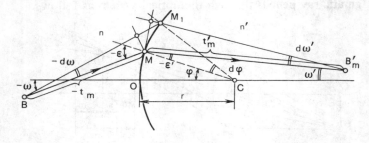

Fig. 8.6. Derivation of the Abbe — Young formula for a meridional ray fan

expression for the meridional ray fan

$$\frac{n' \cos^2 \varepsilon'}{t'_m} - \frac{n \cos^2 \varepsilon}{t_m} = \frac{n' \cos \varepsilon' - n \cos \varepsilon}{r} \tag{8.7}$$

Let BS be the principal ray of an elementary oblique ray pencil incident on a spherical surface of radius of curvature r from an off-axis point B, as shown in Fig. 8.7. Denote by t_s the distance from the point S where the principal ray meets the surface to the point B along the ray. To form an elementary ray pencil now in the sagittal plane we turn the ray BS about the line BO_1C by an infinitesimal angle $d\psi$. This yields in the sagittal plane a very close ray BS_1. Having survived the refracting surface the rays BS and BS_1 unite at a point B'_s which must lie on the line BC.

In order to evaluate the relation between t_s and t'_s we drop perpendiculars from B and B'_s on SC and obtain points N and N'. From Fig. 8.7 it can be seen that

$$BN = t_s \sin \varepsilon, \quad B'_s N' = - t'_s \sin \varepsilon' \tag{8.8}$$

From similarity of the triangles BNC and $CN'B'_s$ we have

$$BN/B'_s N' = NC/CN' \tag{8.9}$$

where

$$NC = NS + SC = -t_s \cos \varepsilon + r \atop CN' = SN' - SC = t'_s \cos \varepsilon' - r \Big\} \tag{8.10}$$

From equations (8.8)-(8.10) we obtain in view of $\sin \varepsilon/\sin \varepsilon' = n'/n$

$$- \frac{t_s n'}{t'_s n} = \frac{-t_s \cos \varepsilon + r}{t'_s \cos \varepsilon' - r}$$

Eliminating the denominator in this expression by cross multiplication and dividing both sides of the resultant equation by $t_s t'_s r$ we arrive at the

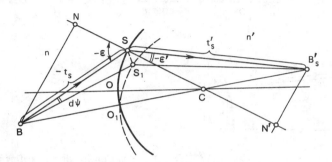

Fig. 8.7. Derivation of the Abbe — Young formula for a sagittal ray fan

10*

Abbe — Young expression for the sagittal ray fan

$$\frac{n'}{t'_s} - \frac{n}{t_s} = \frac{n' \cos \varepsilon' - n \cos \varepsilon}{r} \qquad (8.11)$$

The formulae (8.7) and (8.11) form a basis for the numerical raytracing of a narrow astigmatic ray pencil through a spherical refracting surface. In tracing such a ray pencil through an optical system of p refracting surfaces the designer has each time to allow for the skew thickness d which is the distance between sequential surfaces measured along the optical path of the principal ray. This thickness can be deduced in tracing the principal ray. To illustrate Fig. 8.8 shows the course of the principal ray between surfaces k and $k + 1$ of an optical system along with the respective nomenclature. The angle which the principal ray makes with the optical axis in the spacing between these two surfaces is seen to be denoted by ω'_k, and the heights of ray intersections with surfaces k and $k + 1$ are, respectively, h_k and h_{k+1}. Accordingly, the skew thickness between these surfaces is

$$d_k = (h_k - h_{k+1}) \operatorname{cosec} \omega'_k$$

Let $B'_{m,k}$ be the convergence point of the meridional narrow fan after surface k at a distance $t'_{m,k}$ from this surface. To transfer this ray fan through surface $k + 1$ we need to determine $t_{m,k+1}$ as

$$t_{m,k+1} = t'_{m,k} - d_k$$

Similarly for the sagittal ray fan we have

$$t_{s,k+1} = t'_{s,k} - d_k$$

In automatic tracing of a narrow astigmatic ray pencil on a computer, it would be convenient to convert the expressions (8.7) and (8.11) to another form. For this purpose we first rewrite these expressions as

$$\frac{1}{t'_{m,k}} = \frac{n_k \cos^2 \varepsilon_k}{t_{m,k} n_{k+1} \cos^2 \varepsilon'_k} + \frac{1}{r_k \cos^2 \varepsilon'_k} \left(\cos \varepsilon'_k - \frac{n_k}{n_{k+1}} \cos \varepsilon_k \right)$$

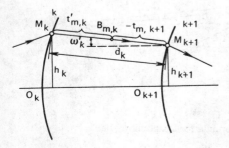

Fig. 8.8. Construction to define the skew thickness

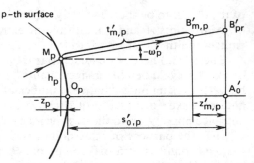

Fig. 8.9. Determining the ray
coordinates of an astigmatic
pencil in the image plane

$$\frac{1}{t'_{s,\,k}} = \frac{n_k}{n_{k+1}t_{s,\,k}} + \frac{1}{r_k}\left(\cos\varepsilon'_k - \frac{n_k}{n_{k+1}}\cos\varepsilon_k\right)$$

Let us return to the notation of the Feder scheme (Eqs. (8.2)), namely, $\cos\varepsilon_k = q_k$, $\cos\varepsilon'_k = q'_k$, $g_k = q'_k - (n_k/n_{k+1})q_k$, and augment it by the following nomenclature $\tau_{m,\,k} = 1/t_{m,\,k}$, $\tau_{s,\,k} = 1/t_{s,\,k}$, $\tau'_{m,\,k} = 1/t'_{m,\,k}$, $\tau'_{s,\,k} = 1/t'_{s,\,k}$, and $c_k = 1/r_k$. Now the Abbe — Young formulae rewrite

$$\tau'_{m,\,k} = \frac{n_k q_k^2}{n_{k+1}q_k'^2}\tau_{m,\,k} + \frac{c_k g_k}{q_k'^2}$$

$$\tau'_{s,\,k} = \frac{n_k}{n_{k+1}}\tau_{s,\,k} + c_k g_k$$

(8.12)

Sequential application of equations (8.12) in a system of p surfaces will yield the coordinates $z'_{m,\,p}$ and $z'_{s,\,p}$ indicated in Fig. 8.9 as

$$z'_{m,\,p} = \nu_{p+1}/\tau'_{m,\,p} - (s'_{0,\,p} - z_p)$$

and

$$z'_{s,\,p} = \nu_{p+1}/\tau'_{s,\,p} - (s'_{0,\,p} - z_p)$$

where $\nu_{p+1} = \nu'_p = \cos(-\omega'_p)$ is the direction cosine of the principal ray.

8.3 Initial Data for Numerical Raytracing

In actual optical systems, raytracing serves the purpose of evaluating the position and size of the image to compare the result with the perfect image. In final analysis this comparison is performed in order to assess the quality of imagery and to decide whether or not the given system suits the purpose for which it is to be designed.

Optical computations in raytracing necessitate the structural parameters of the system examined (r, d, n) and the position, s_1, and size y

of the object to be specified. Any object may be represented as a collection of axial points A and a multitude of extra-axial points B_i each of which emits a multitude of rays into the optical system.

There is no need, of course, to trace all the multitude of rays from the object to evaluate the image quality. Normally the designer confines himself to examining a limited number of rays in the meridional, sagittal and some skew planes. Fig. 8.10 shows the object plane Q, the plane of the entrance pupil Q_p, and the first surface 1 of the optical system centered on C_1. For the purpose of our examination we single out in the object plane the axial point A and off-axis points B_i, normally assigned in the meridional plane. We assume that the system under examination has a circular entrance pupil of diameter D centred on the point C on the axis. In object space the rays passing through the entrance pupil are within a cone of rays from an object point to the edges of the pupil.

For image quality assessment, more important rays are the marginal rays grazing the upper edge of the entrance pupil in the meridional plane. The number of rays needed to be traced is defined by the aperture ratio. For optical systems with normal aperture ratio, say, i.e. with D/f' from 1:2.8 to 1:5.6 (relative aperture $f/2.8$ to $f/5.6$), the spherical aberration (an effect, to be discussed later, according to which the rays from an axial object point do not all cross the axis in image space at the paraxial focus) is to a sufficient accuracy defined by the third and fifth order terms, i.e., $\Delta s' \approx am^2 + bm^4$, therefore it is sufficient to compute the course of two rays: a marginal ray traversing the entrance pupil at a height of m_m and a zonal ray at a height of m_z. The height m_z is determined from the equation $\partial(\Delta s')/\partial m = 0$ on the condition that at the pupil edge $\Delta s'_m = 0$ and $m_m^2 = -a/b$. The height calculated in this way is $m_z = m_m\sqrt{1/2}$.

The circular zones of the entrance pupil, which are encircled by the

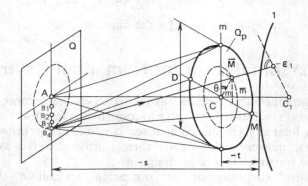

Fig. 8.10. Choice of initial data in raytracing

heights of the upper marginal ray and the zonal ray used as radii, turn out to be of the same area, consequently, through these areas the system collects identical fluxes of radiant energy.

High-speed optical systems and optical systems with aspherical surfaces exhibiting a complicated spherical aberration plot call sometimes for a greater number of rays to be traced, for example, three rays for aperture ratios from 1:1.5 to 1:2.8 (f-numbers $f/1.5$ to $f/2.8$), and four rays for aperture ratios from 1:1 to 1:1.5. Observing that the areas of the circular zones, in which the N rays break down the entrance pupil, are equal yields for the heights of the rays

$$m_i = m_N \sqrt{i/N} \tag{8.13}$$

where m_N is the height of the marginal ray m_m. To illustrate, for $N = 4$ $m_m = m_4$, $m_3 = m_4\sqrt{3/4}$, $m_2 = m_4\sqrt{1/2}$, and $m_1 = m_4\sqrt{1/4}$.

In catadioptric and mirror optical systems, the entrance pupil has an annular shape, as that shown in Fig. 8.11, because the central portion of the ray pencil is screened by one of the mirrors (shaded area in the figure). If we denote the height of the upper ray in the entrance pupil by $m_u = m_N$, the height of the lower ray by $m_1 = m_1$, and observe that the areas of the $N - 1$ annular zones for the N rays in the annular pupil are identical, then for the height of ray i we get

$$m_i = \sqrt{\frac{(i-1)m_u^2 + (N-i)m_1^2}{N-1}} \tag{8.14}$$

or

$$m_i = \sqrt{\frac{(i-1)m_N^2 + (N-i)m_1^2}{N-1}} \tag{8.15}$$

The tolerable amount of central screening is normally estimated by the

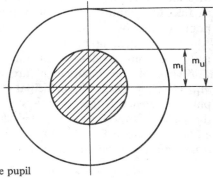

Fig. 8.11. An annular entrance pupil

ratio $k_A = m_1^2/m_u^2$ or the linear ratio $k = m_1/m_u$. Incorporating these ratios in (8.14) carries it to

$$m_i = m_u\sqrt{[i - 1 + (N - i)k_A]/(N - 1)}$$
$$= m_u\sqrt{[i - 1 + (N - i)k^2]/(N - 1)}$$
$$= m_N\sqrt{[i - 1 + (N - i)k^2]/(N - 1)}$$

A frequent choice is with a quarter of the pupil area being screened, i.e., with $k_A = 1/4$ or $k = 1/2$. In this case $m_1 = m_u/2$ ($m_1 = m_N/2$) and

$$m_i = (m_N/2)\sqrt{3(i - 1)/(N - 1) + 1}$$

For oblique pencils, in the meridional plane the rays are traced as a rule through the same heights at the entrance pupil as in the axial pencil, and the rays in the fan are symmetrically spaced on either side, i.e. upwards and downwards, of the principal ray so that the heights look like this: $+m_3$, $+m_2$, $+m_1$, $m_{pr} = 0$, $-m_1$, $-m_2$, $-m_3$. If the vignetting of the system is specified by the coefficient k_1, then for an oblique ray pencil in the meridional plane, $m_u = k_1 m_N$ and so on.

Experience collected with numerical raytracing of this type [2] indicates that reliable estimates of aberrations of off-axis points require the numerical raytracing for at least 15-30 rays (depending on the relative aperture of the lens and its aberrations). For camera lenses with narrow angular fields (20° to 30°) it suffices to compute one oblique ray fan, whereas for normal lenses (50° to 60° coverage) the number of slopes doubles, and for wide-angle lenses (90° to 120° coverage) triples.

The rays of the sagittal fan are traced at heights M numerically equal to the heights of rays in the meridional plane, for one half of the pupil symmetrical about the meridional plane, namely, $M_3 = m_3$, $M_2 = m_2$, etc.

Skew rays are traced in planes oblique to the meridional plane at angles θ. There can be two, four, six, etc. such planes, depending on the number of sectors into which these planes divide the entrance pupil. Here, it also suffices to compute the course of the rays through one half of the entrance pupil divided by the meridional plane.

By way of example, in Fig. 8.12, the entrance pupil of an optical system is divided into twelve sectors and constituted by three annular zones. The rays of the axial pencil are numbered by Roman numbers, viz., $m_{III} = m_m = D/2$. The principal ray of the oblique fan in the meridional plane is labelled 1, the other rays of the fan are labelled 2 to 7. The rays of the sagittal fan to be computed are labelled 8-10. The skew rays to be traced are numbered from 11 to 22. The coordinates of these rays at the entrance pupil can be determined as

$$m_{i,\theta} = m_i \cos\theta, \qquad M_{i,\theta} = m_i \sin\theta$$

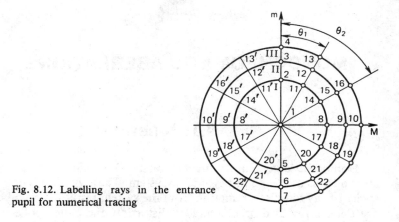

Fig. 8.12. Labelling rays in the entrance pupil for numerical tracing

The primed rays in Fig. 8.12 are symmetrical about the meridional plane to the evaluated rays and therefore are not computed.

MONOCHROMATIC ABERRATIONS

9.1 General

Homocentric ray bundles incident on a real optical system are no longer able to focus at one point beyond the system. These violations of homocentricity of ray bundles emanating from the system result in a number of image defects revealed as blurring or distortion and called collectively ray aberrations.

For the purpose of optical system design, aberrations are broadly divided into two groups. The first covers monochromatic aberrations which occur when the system is illuminated by monochromatic light. The other group deals with chromatic aberrations which occur owing to variations in system properties with wavelength of the incident polychromatic radiation. More often than not, however, aberrations of both groups occur simultaneously.

Normally aberrations are measured by the amount by which rays miss the paraxial image point. We shall measure aberrations by the linear displacement of the points at which real rays, traced by the expressions (8.2), intersect the image surface from the image point for the perfect system calculated by the equations of quasi-paraxial (zero-ray) optics. Approximate mathematical techniques of aberration analysis also exist.

It should be kept in mind that aberrations are unavoidable in actual optical systems and one of the aims of optical design is to correct the system being designed for aberrations. Some aberrations, however, inevitably remain and the purpose of image quality evaluations is to determine how strong the residual aberrations are. In general a modern lens design takes aberrations into account by solving the following problems:

— determine the residual aberrations for a system with specified structural parameters r, d, and n, or

— evaluate the structural parameters of the system which would keep the residual aberrations within the specified tolerable values.

Assume that the structural parameters of the system, namely, r, d, and n, are specified along with the distance s_1 from the front surface to the object plane A, and the distance t from the front surface to the entrance pupil — the relevant nomenclature is illustrated in Fig. 9.1. The object point B is

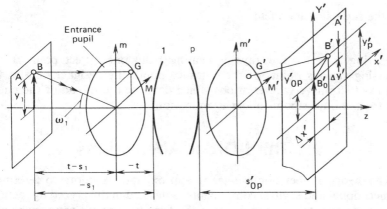

Fig. 9.1. Aberrations for a skew ray

elevated at a distance y_1 above the optical axis. Let us trace a skew ray BG that is one not in a meridional plane. If the coordinates y_1, s_1, and t of the ray are known, then the position of such a ray in space is defined by specifying the coordinates of a point G where it pierces the entrance pupil, namely m in the y axis and M in the x axis. For an infinitely distant object plane the coordinate y_1 gives way to the angle ω_1 at which the principal ray passing through the centre of the entrance pupil crosses the optical axis

$$\omega_1 = \arctan \left[y_1/(t - s_1) \right] \qquad (9.1)$$

Given the initial data for raytracing, then the coordinates where the ray meets surface p, i.e., x_p, y_p, and z_p, and the direction cosines λ_{p+1}, μ_{p+1}, ν_{p+1} of the ray leaving the system can be determined with the set (8.2). The coordinates of a point $B'(y', x')$ at which this ray pierces the image plane (more often than not the ideal image plane) can be derived then with Eq. (8.3).

The distance from the last surface to the plane of ideal image, s'_{0p}, can be determined by tracing a quasi-paraxial (zero) ray. This trace also results in the size of the ideal image

$$y'_{0p} = y_1 \beta_0$$

where β_0 is the transverse magnification of the perfect system.

With reference to Fig. 9.1 the transverse aberration is characterized by the line segment $B'B'_0$. In numerical raytracing this aberration is represented in terms of its projections on the axes of the image plane, namely, the meridional component

$$\Delta y' = y'_p - y'_{0p}$$

and the sagittal component

$$\Delta x' = x'$$

Having traced a few rays emanating from one object point B and traversing various points of the entrance pupil, the designer obtains for each ray its transverse aberrations which describe the spread of the image spot for the object point under examination.

9.2 Third-Order Aberrations

The theory of aberrations gives an approximate technique to determine the meridional and sagittal components as defined in the preceding section. These components $\Delta y'$ and $\Delta x'$ are the functions of the ray coordinates $y_1(\omega_1)$, m, M, the structural parameters of the system, position of the object plane and the plane of the entrance pupil. The aberration theory establishes the relation between the aberration components $\Delta y'$, $\Delta x'$ and the ray coordinates y_1, m, and M as

$$\Delta y' = f(y_1, m, M)$$
$$\Delta x' = f(y_1, m, M) \tag{9.2}$$

Because the system is symmetrical about the optical axis the functions (9.2) contain no terms of even order. Therefore, if such a function is developed into a power series, it will contain only odd-order terms, i.e. third-, fifth-, seventh-, and higher-order terms, in y_1, m, and M:

$$\Delta y' = \Delta y'_{III} + \Delta y'_V + \Delta y'_{VII} + \dots$$
$$\Delta x' = \Delta x'_{III} + \Delta x'_V + \Delta x'_{VII} + \dots \tag{9.3}$$

The terms appearing on the right-hand sides of these expansions are called respectively third-order, fifth-order, etc. meridional and sagittal components of aberrations. Note that third and fifth order are frequently referred to as *primary* and *secondary aberrations*. The components of order higher than three are referred to as *higher-order aberrations*.

Analytical expressions describing higher-order aberrations are extremely cumbersome and therefore inconvenient for practical computations. Therefore, optical engineers confine themselves to third order terms in the design of optical systems with specified residual aberrations. The theory of third-order (primary) aberrations yields approximate values for system's structural parameters and provides a tool for analysis of aberrations in the system at hand. More specifically, the theory of primary aberrations gives approximate components of aberrations as power series whose coefficients

$A, B, C, D,$ and E depend only on the system's structural parameters and on the positions of the object plane and the entrance pupil and are independent of the ray coordinates. Five coefficients are involved in the third-order aberration design. The ray coordinates y_1, m, and M enter as the factors y^α, m^β, M^γ in powers summing to $\alpha + \beta + \gamma = 3$ in each term.

Thus for the meridional and sagittal components of third-order aberrations we have, respectively

$$\Delta y'_{III} = Am(m^2 + M^2) + By_1(3m^2 + M^2) + Cy_1^2 m + Ey_1^3$$
$$\Delta x'_{III} = AM(m^2 + M^2) + 2By_1 mM + Dy_1^2 M$$

The five coefficients in these expansions are expressed in terms of the parameters of two auxiliary rays. One of these rays, labelled I in Fig. 9.2, crosses an axial point A_1 of the object plane at an angle α_1 to the axis and meets the principal plane of the front surface at a height h_1. The trace of this ray is computed by the expressions

$$\alpha_{k+1} = \frac{n_k}{n_{k+1}} \alpha_k + \frac{n_{k+1} - n_k}{n_{k+1}} \frac{h_k}{r_k}$$

$$h_{k+1} = h_k - d_k \alpha_{k+1}$$

(9.4)

where the symbols α denote tangents of the respective angles.

The other ray, labelled II in Fig. 9.2, intersects the central point of the entrance pupil making an angle β_1 with the axis and meets the principal plane of the first surface at a height H_1. This ray is traced with the expressions

$$\beta_{k+1} = \frac{n_k}{n_{k+1}} \beta_k + \frac{n_{k+1} - n_k}{n_{k+1}} \frac{H_k}{r_k}$$

$$H_{k+1} = H_k - d_k \beta_{k+1}$$

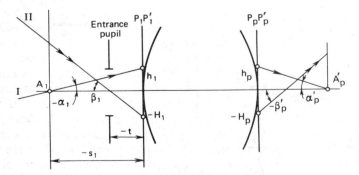

Fig. 9.2. Tracing of auxiliary rays in an optical system

In system design the types of glasses are subject to the designer's choice so that the indices of refraction in (9.4) are known values. Once he wishes to correct the system for aberrations the designer determines the parameters α and h of ray I so as to reduce aberrations to the tolerable level specified, which results in the system structural parameters

$$r_k = \frac{n_{k+1} - n_k}{\alpha_{k+1}n_{k+1} - \alpha_k n_k} h_k$$

$$d_k = \frac{h_k - h_{k+1}}{\alpha_{k+1}}$$

(9.5)

Expressing the coefficients A through E in terms of the ray parameters leads to the following formulae for the third-order aberration components

$$\Delta y'_{\mathrm{III}} = -\frac{m(m^2 + M^2)}{2n'_p(s_1 - t)^3\alpha_1^3\alpha'_p} S_{\mathrm{I}} + \frac{y_1(3m^2 + M^2)}{2n'_p(s_1 - t)^3\alpha_1^2\alpha'_p\beta_1} S_{\mathrm{II}}$$
$$- \frac{y_1^2 m}{2n'_p(s_1 - t)^3\alpha_1\alpha'_p\beta_1^2}(3S_{\mathrm{III}} + I^2 S_{\mathrm{IV}}) + \frac{y_1^3}{2n'_p(s_1 - t)^3\alpha'_p\beta_1^3} S_{\mathrm{V}}$$

$$\Delta x'_{\mathrm{III}} = -\frac{M(m^2 + M^2)}{2n'_p(s_1 - t)^3\alpha_1^3\alpha'_p} S_{\mathrm{I}} + \frac{2y_1 mM}{2n'_p(s_1 - t)^3\alpha_1^2\alpha'_p\beta_1} S_{\mathrm{II}}$$

(9.6)

$$- \frac{y_1^2 M}{2n'_p(s_1 - t)^3\alpha_1\alpha'_p\beta_1^2}(S_{\mathrm{III}} + I^2 S_{\mathrm{IV}})$$

The symbols S_{I}, S_{II}, S_{III}, S_{IV}, and S_{V} denote in these expressions the *Seidel sums*

$$S_{\mathrm{I}} = \sum_{k=1}^{p} h_k P_k \qquad \text{spherical}$$

$$S_{\mathrm{II}} = \sum_{k=1}^{p} h_k P_k \frac{\delta\beta_k}{\delta\alpha_k} \qquad \text{meridional coma}$$

$$S_{\mathrm{III}} = \sum_{k=1}^{p} h_k P_k \left(\frac{\delta\beta_k}{\delta\alpha_k}\right)^2$$

(9.7)

$$S_{\mathrm{IV}} = \sum_{k=1}^{p} \frac{\delta(\alpha_k n_k)}{h_k n_{k+1} n_k}$$

distortion

$$S_V = \sum_{k=1}^{p} \left[h_k P_k \left(\frac{\delta\beta_k}{\delta\alpha_k} \right)^2 + I^2 \frac{\delta(\alpha_k n_k)}{h_k n_{k+1} n_k} \right] \frac{\delta\beta_k}{\delta\alpha_k}$$

where

$$P_k = \left(\frac{\delta\alpha_k}{\delta\mu_k} \right)^2 \delta(\alpha_k \mu_k)$$

$$\mu_k = \frac{1}{n_k}$$

$$I = -n_1 \alpha_1 (s_1 - t)\beta_1$$

$$\delta\alpha_k = \alpha_{k+1} - \alpha_k$$

$$\delta(\alpha_k \mu_k) = \alpha_{k+1}\mu_{k+1} - \alpha_k \mu_k$$

The expressions appearing under the summation operators in (9.7) are known as the *Seidel sum coefficients*.

9.3 Normalization of Auxiliary Rays

Because the initial data for auxiliary ray tracing is a matter of designer's choice, Seidel's sums corresponding to different initial data will not be the same. This arbitrary choice of ray parameters does not, however, influence the third-order aberrations proper as follows from the expressions preceding the sums in (9.6). In order to be able to compare system designs in terms of Seidel's sums, these latter may be derived under certain normalizing conditions for the rays.

For the object at a finite distance from the system, the rays are chosen such that

$$\alpha_p' = 1$$

$$\alpha_1 = (n_p'/n_1)\beta$$

$$h_1 = s_1 \alpha_1$$

$$\beta_1 = 1 \qquad\qquad (9.8)$$

$$H_1 = t$$

$$I = -n_p'(s_1 - t)\beta$$

where β is the lateral magnification of the system.

Now the expressions (9.6) become

$$\Delta y'_{\text{III}} = -\frac{m(m^2 + M^2)}{2n'_p(s_1 - t)^3\alpha_1^3} S_{\text{I}} + \frac{y_1(3m^2 + M^2)}{2n'_p(s_1 - t)^3\alpha_1^2} S_{\text{II}}$$

$$- \frac{y_1^2 m}{2n'_p(s_1 - t)^3\alpha_1}(3S_{\text{III}} + I^2 S_{\text{IV}}) + \frac{y_1^3}{2n'_p(s_1 - t)^3} S_{\text{V}}$$

$$\Delta x'_{\text{III}} = -\frac{M(m^2 + M^2)}{2n'_p(s_1 - t)^3\alpha_1^3} S_{\text{I}} + \frac{2y_1 mM}{2n'_p(s_1 - t)^3\alpha_1^2} S_{\text{II}}$$

$$- \frac{y_1^2 M}{2n'_p(s_1 - t)^3\alpha_1}(S_{\text{III}} + I^2 S_{\text{IV}})$$

(9.9)

If the object is at infinity ($s_1 = -\infty$ and $\alpha_1 = 0$), the uncertainty arising in (9.6) is eliminated as illustrated in Fig. 9.3, namely,

$$|(s_1 - t)\alpha_1|_{\substack{s_1 \to -\infty \\ \alpha_1 \to 0}} = h_1$$

(9.10)

In addition, for a distant object it is more convenient to specify its angular dimension. Then from (9.1) we have

$$y_1/(t - s_1) = \tan \omega_1 \approx \omega_1$$

(9.11)

The relevant set of parameters is as follows

$$\alpha_1 = 0 \qquad h_1 = f' \qquad \beta_1 = 1$$

$$\alpha'_p = 1 \qquad H_1 = t \qquad I = -n_1 f'$$

(9.12)

With this choice of parameters and in view of (9.10) and (9.11) equations (9.6) become

$$\Delta y'_{\text{III}} = -\frac{m(m^2 + M^2)}{2n'_p f'^3} S_{\text{I}} - \frac{(3m^2 + M^2)\omega_1}{2n'_p f'^2} S_{\text{II}} - \frac{m\omega_1^2}{2n'_p f'}$$

$$\times (3S_{\text{III}} + I^2 S_{\text{IV}}) - \frac{\omega_1^3}{2n'_p} S_{\text{V}}$$

(9.13)

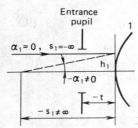

Fig. 9.3. Normalizing oblique rays at $s_1 = -\infty$

$$\Delta x'_{III} = - \frac{M(m^2 + M^2)}{2n'_p f'^3} S_I - \frac{2mM\omega_1}{2n'_p f'^2} S_{II} - \frac{M\omega_1^2}{2n'_p f'}$$
$$\times (S_{III} + I^2 S_{IV})$$

According to the normalizing conditions (9.12), Seidel's sums will now depend on the focal length of the system as $h_1 = f'$. To avoid the effect of the focal length on Seidel's sums it would be convenient to calculate them at $f' = 1$. Of course, this converts all linear dimensions of the system into fractions of the focal length. Such a system will be referred to as a *reduced optical system*. The set of parameters now looks as follows

$$\alpha_1 = 0 \qquad h_1 = 1 \qquad \beta_1 = 1$$
$$\alpha'_p = 1 \qquad H_1 = t/f' \qquad I = -n_1 \tag{9.14}$$

With this choice of parameters equations (9.13) become

$$\Delta y'_{III} = - \frac{m(m^2 + M^2)}{2n'_p f'^2} S_I - \frac{(3m^2 + M^2)\omega_1}{2n'_p f'} S_{II}$$
$$- \frac{m\omega_1^2}{2n'_p} (3S_{III} + I^2 S_{IV}) - \frac{\omega_1^3}{2n'_p} f' S_V$$

$$\Delta x'_{III} = - \frac{M(m^2 + M^2)}{2n'_p f'^2} S_I - \frac{2mM\omega_1}{2n'_p f'} S_{II}$$
$$- \frac{M\omega_1^2}{2n'_p} (S_{III} + I^2 S_{IV}) \tag{9.15}$$

Analysis of equations (9.9), (9.13) and (9.15) indicates that the image will be free from primary aberrations at any values of m, M, and y_1 (or ω_1) if all the Seidel sums are simultaneously zero. This unfortunately does not mean that the system is aberration-free because higher-order aberrations can also be noticeable. Design experience, though, indicates that a significant condition for the residual aberrations to be small is to keep the primary aberrations as small as possible.

The equations for aberration components also reveal that an optical system having small values of $\Delta y'_{III}$ for any m, M, and y_1 (or ω_1) has small values of Seidel's sums. This implies that the sagittal components of third-order aberrations, $\Delta x'_{III}$, will also be small for any values of m, M, and y_1 (or ω_1). Accordingly, at the initial stage of computing optical system aberrations the designer tends to correct the system for aberrations by tracing meridional ray fans, i.e. rays at $M = 0$. In this case the meridional component of the transverse aberration will in view of (9.9), i.e. for the object at a finite distance, have the form

$$\Delta y'_{III} = -\frac{m^3}{2n'_p(s_1-t)^3\alpha_1^3}S_I + \frac{3m^2 y_1}{2n'_p(s_1-t)^3\alpha_1^2}S_{II}$$

$$-\frac{my_1^2}{2n'_p(s_1-t)^3\alpha_1}(3S_{III}+I^2 S_{IV}) + \frac{y_1^3}{2n'_p(s_1-t)^3}S_V \qquad (9.16)$$

and for the object at infinity in terms of the reduced system (9.15) it will have the form

$$\Delta y'_{III} = -\frac{m^3}{2n'_p f'^2}S_I - \frac{3m^2\omega_1}{2n'_p f'}S_{II}$$

$$-\frac{m\omega_1^2}{2n'_p}(3S_{III}+I^2 S_{IV}) - \frac{\omega_1^3}{2n'_p}f'S_V \qquad (9.17)$$

Each Seidel sum appearing in the equations of third-order aberration defines different types of image defects: S_I, spherical aberration; S_{II}, coma; S_{III} and S_{IV}, astigmatism and field (image) curvature; and S_V, distortion.

To close this section we note that higher aperture ratio, angular field or object size covered at a given distance, all entail stronger aberrations of higher orders. Methods of system correction for these aberrations involve incorporation of additional lens elements and groups so that the corrected system becomes much more complicated.

9.4 Spherical Aberration

Consider the image of a point lying on the optical axis. The optical system possesses rotational symmetry about the optical axis, therefore in our examination it is sufficient to trace the rays lying in the meridional plane. Fig. 9.4 illustrates the situation for a single positive lens. The position of the ideal image A'_0 of the object point A is located by a paraxial ray which intersects the optical axis a distance s'_0 from the last surface. Rays

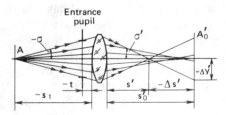

Fig. 9.4. Spherical aberration in imaging an axial object point with a positive lens

making with the optical axis finite angles σ exceeding the Gaussian angle domain fail to reunite at A_0', the ideal image point. For a single positive lens, an increase in the absolute value of the angle σ brings the ray to intersect the optical axis closer to the lens in image space. This phenomenon is attributed to unequal power of different zones of the lens, which increases with the distance from the optical axis.

This breakdown of homocentricity of the ray bundle emanating from the lens is measured as the distance

$$\Delta s' = s' - s_0'$$

by which the rays passing through the entrance pupil at nonparaxial heights and uniting at s' miss the convergence point for paraxial rays, s_0'. This difference is known as the *longitudinal spherical aberration*.

Spherical aberration is responsible for a spread of the perfect image point in the ideal image plane where in place of a sharp point we observe a spot of radius $\Delta y'$. In the image plane this radius measures the distance of an actual ray intersection with the plane from the optical axis. This distance is called the *transverse spherical aberration*. It is related to the longitudinal spherical aberration as

$$\Delta y' = \Delta s' \tan \sigma' \qquad (9.18)$$

Spherical aberration does not affect the symmetry of rays in the ray bundle emanating from the lens. As contrasted to other types of monochromatic aberration, spherical aberration takes place for all points of the optical system field. In the presence of only spherical aberration, a ray bundle issued from an off-axial point remains symmetric about the principal ray in image space, as illustrated in Fig. 9.5.

The equations of third-order aberration enable a spherical aberration to be estimated as a function of the Seidel sum S_I. To demonstrate, for an object at a finite distance from the lens, as in Fig. 9.4, we have $\tan \sigma = m/(s_1 - t)$. Within the validity limits of third-order aberration theory we may safely let $\sigma = m/(s_1 - t)$. For the system in air when $n_1 = n_p' = 1$ the normalization conditions (9.8) give $\alpha_1 = \beta$ and we have

$$\sigma' = m/(s_1 - t)\beta$$

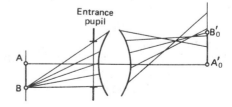

Fig. 9.5. Spherical aberration for an extra-axial point

In view of (9.3) we find that the transverse primary aberration for an object point at a finite distance from the system is

$$\Delta y'_{III} = -0.5\sigma'^3 S_I \qquad (9.19)$$

Accordingly, for the longitudinal primary spherical aberration we get by virtue of (9.18) and (9.19) under the assumption of $\tan\sigma' \approx \sigma'$

$$\Delta s'_{III} = -0.5\sigma'^2 S_I \qquad (9.20)$$

Equations (9.19) and (9.20) hold true for the object at infinity if S_I is computed under the normalization conditions (9.12), i.e. for a real focal length.

In practical design it is more convenient to use the equations of primary spherical aberration involving the ray coordinate at the entrance pupil. Assuming $n'_p = 1$ and S_I are derived under the normalization conditions (9.12) we obtain by virtue of (9.15) and (9.18)

$$\Delta y'_{III} = -(m^3/2f'^3)S_I$$
$$\Delta s'_{III} = -(m^2/2f'^2)S_I$$

For the reduced system with the normalizing set (9.14) we have in view of (9.15) and (9.18)

$$\Delta y'_{III} = -(m^3/2f'^2)S_I$$
$$\Delta s'_{III} = -(m^2/2f')S_I \qquad (9.21)$$

These equations indicate that for a given S_I the third-order spherical aberration is higher the greater the coordinate m at the entrance pupil.

Because spherical aberration takes place for all points of the field the primary attention in an optical system correction for aberration is paid to diminishing spherical aberration. The simplest optical system with spherical surfaces which can be freed from spherical aberration is a combination of a positive and a negative lens. The refracting zones closer to the edge in both lenses exhibit stronger refraction than zones closer to the optical axis. For a negative lens this phenomenon is illustrated in Fig. 9.6. The negative lens has a positive spherical aberration. Therefore if combined with a positive lens yields a system with diminished spherical aberration.

Unfortunately, spherical aberration can be completely eliminated for certain rays only rather than for rays piercing the entire entrance pupil. Therefore, for actual optical systems some residual spherical aberration is always present. Residual aberrations are normally tabulated or represented as graphs. For an axial object point the curves of $\Delta s'$ and $\Delta y'$ are plotted against m, σ' or $\tan\sigma'$. The curves of longitudinal and corresponding transverse spherical aberration are represented in Fig. 9.7.

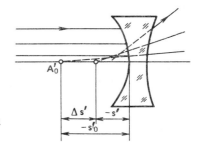

Fig. 9.6. Spherical aberration in the case of a negative lens

The plot in Fig. 9.7a corresponds to an optical system with undercorrected spherical aberration. If the spherical aberration of such a system is determined by the third-order aberrations only, then according to (9.20) the longitudinal spherical aberration curve assumes the form of a parabola, while the plot for the transverse aberration will look like a cubic parabola.

The plots in Fig. 9.7b correspond to an optical system corrected for the spherical aberration of marginal rays grazing the edge of the entrance pupil. The plots at (c) correspond to an optical system with overcorrected spherical aberration. A correction or overcorrection can be achieved by combining positive and negative lenses.

Transverse spherical aberration defines the transverse spread of an image which occurs in place of the ideally sharp point. The size of this spot is dependent on the location of the image plane, as can be seen in Fig. 9.8. If

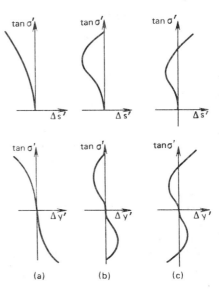

Fig. 9.7. Plots of longitudinal and transverse spherical aberration

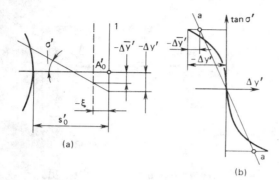

(a)

(b)

Fig. 9.8. Locating the image plane corresponding to the least spherical aberration

we shift this plane by a length ξ relative to the plane of ideal image (Gaussian plane), labelled 1 in Fig. 9.8a, then the transverse aberration $\overline{\Delta y'}$ will be related to the transverse aberration $\Delta y'$ in the Gaussian plane by

$$\overline{\Delta y'} = \Delta y' - \xi \tan \sigma' \qquad (9.22)$$

In the plot $\Delta y'$ versus $\tan \sigma'$ the term $\xi \tan \sigma'$ can be represented by a straight line through the origin. At $\xi = 0$ we obtain the plot of transverse spherical aberration for the Gaussian plane.

If we fit a straight line aa in the plot of transverse spherical aberration (Fig. 9.8b) so that the aberration curve deviates from this line less than from any other line through the origin, then this line will correspond to the image plane with the least transverse spread of the image. Displacement of any image plane from the Gaussian plane is given by

$$\xi = \Delta y'_1 / \tan \sigma'_1$$

where $\Delta y'_1$ and $\tan \sigma'_1$ are the coordinates of any point on the line aa.

The longitudinal spherical aberration can be represented by a polynomial in even powers of σ' or m, namely,

$$\Delta s' = am^2 + bm^4 + cm^6 + \dots \qquad (9.23)$$

where the coefficient a is expressed through S_1 and describes the aberration of third order, b defines the aberration of fifth order, c the aberration of seventh order, etc.

For many optical systems the longitudinal spherical aberration is rather accurately described by the two leading terms in (9.23), viz.,

$$\Delta s' = am^2 + bm^4 \qquad (9.24)$$

If the system is corrected for the spherical aberration at the margin of the pupil, where the marginal coordinate is denoted m_m, then

$$\Delta s'_m = am_m^2 + bm_m^4 = 0$$

whence

$$m_m^2 = -a/b \qquad (9.25)$$

We differentiate (9.24) with respect to m to determine the height of a zone, m_z, for which zonal spherical aberration is greatest. The differentiation yields $m_z^2 = -a/2b$, or in view of (9.25)

$$m_z = m_m/\sqrt{2} \approx 0.7m_m$$

Thus, to estimate the amount of correction of spherical aberration needed, it is evaluated for rays passing through the zones of m_m and $0.7m_m$ (see also Section 8.3).

9.5 Meridional Coma

In the preceding section we learned that the reduction of spherical aberration is a necessary condition of improving the quality of image for object points lying on the optical axis. The associated correction fails to improve the image quality for extra-axial points if the system is not corrected for coma. Coma aberration essentially spoils the symmetry of the ray bundle which on surviving the system is no longer symmetric about the principal ray. The corrupted symmetry in the emanating ray is explained by unequal refracting conditions for rays entering the system in different zones of the entrance pupil. This asymmetry for a meridional ray fan is known as the *meridional coma*.

Figure 9.9 shows the structure of a meridional ray fan emanating from an optical system which exhibits a spherical aberration and a meridional coma. The upper ray of coordinate $+m$ at the entrance pupil and the lower ray of coordinate $-m$ travel asymmetrically with respect to the principal ray in image space. A measure of meridional coma is the quantity

$$\Delta y' = (y_u' + y_l')/2 - y_{pr}'$$

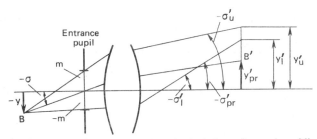

Fig. 9.9. The course of rays in defining the spherical aberration and meridional coma

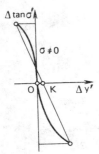

Fig. 9.10. Graphical definition of a meridional coma

In the absence of spherical aberration ($y_1' = y_u'$), the only aberration in view is coma. To estimate a meridional coma, the optical engineer is to trace the principal ray, and the upper and lower rays and determine their intercepts with the image plane. This raytracing is carried out by the formulae (8.2). The results are usually tabulated or represented in graphical form.

Figure 9.10 shows a plot of residual aberrations versus $\Delta \tan \sigma' = \tan \sigma' - \tan \sigma'_{pr}$ for a system suffering from spherical aberration and meridional coma. To determine a meridional coma from this plot one is to connect by a straight line two points on the curve with identical values of $\Delta \tan \sigma'$. The segment of abscissa OK gives the value of meridional coma.

A way to approximately compute a meridional coma is by the equations for third-order aberrations involving the Seidel sum S_{II}.

For an object at infinity, $n_p' = 1$, and the normalizing conditions (9.14) we get with (9.15) for the third-order meridional coma

$$\Delta y'_{III} = -(3m^2 \omega_1 / 2f')S_{II} \tag{9.26}$$

Thus, the meridional coma is proportional to the square of ray coordinate at the entrance pupil and the angular field of the optical system.

No coma takes place for a point on the axis ($\omega_1 = 0$).

9.6 The Sine Condition and Isoplanatism

If an optical system produces an aberration-free image of a point on the axis, then to produce an aberration-free image of a infinitesimal line segment perpendicular to the axis the system must satisfy the sine condition (see Section 7.7):

$$n' dy' \sin \sigma' = n dy \sin \sigma \tag{9.27}$$

where dy and dy' are the infinitesimal object and image line segments perpendicular to the axis, σ and σ' the slopes of the rays through the axial points of the object and image, and n and n' the refractive indices of the object and image media. This condition must be satisfied for any values of σ.

For an infinitely distant object point the sine condition becomes

$$m/\sin \sigma' = f' = f_0' \qquad (9.28)$$

where f_0' is the focal length for paraxial rays, m the incidence height at the entrance pupil for a ray which enters the system parallel to the optical axis and emanates from the system at an angle σ' with the optical axis. The condition (9.28) must be satisfied for all coordinates m. For rays grazing the edge of the entrance pupil where $m = D/2$, $\sin \sigma' = \sin \sigma'_A$.

Because the ultimate value of angular aperture in image space is $90°$, the maximum aperture ratio of the optical system satisfying the sine condition is confined by the inequality $D/f' < 1{:}0.5$ (i.e., the relative aperture $f'/D > f/0.5$).

Conjugate axial points for which spherical aberration is virtually absent and the sine condition is met are referred to as *aplanatic*. Optical systems capable of meeting these conditions include microlenses. In many cases, however, optical systems cannot produce a perfect image for an axial point. Large-pupil systems are corrected for spherical aberration for two, seldom for three, rays; the other rays of the axial ray bundle exhibit unremovable spherical aberration.

Systems with residual spherical aberration are often made isoplanatic i.e. such that ensure the image quality for points near the optical axis the same as for the axial point. For these points we say that the condition of *isoplanatism* is fulfilled. For an object at a finite distance from the system the quantity indicating the deviation from isoplanatism is defined as

$$\eta = \frac{\Delta\beta}{\beta_0} - \frac{\Delta s'}{s' - t'} \qquad (9.29)$$

where β_0 is the paraxial transverse magnification, $\Delta\beta = \beta - \beta_0$, $\beta = n \sin \sigma / n' \sin \sigma'$ is the transverse magnification for non-paraxial rays, $\Delta s'$ is the longitudinal spherical aberration, s' is the back focal length, and t' is the distance from the last surface to the exit pupil.

For an infinitely distant object plane, the deviation from isoplanatism is defind as

$$\eta = \frac{\Delta f'}{f_0'} - \frac{\Delta s'}{s' - t'} \qquad (9.30)$$

where $\Delta f' = f' - f_0'$, f' is computed with the formula (2.28), and f_0' is the paraxial focal length.

In order to make $\eta \approx 0$ so that the image formation for extra-axial object points may be the same as for axial points, i.e. the image will be coma-free, we see from Eqs. (9.29) and (9.30) that the quantities indicating the deviation from the sine condition, $\Delta\beta$ and $\Delta f'$, must be proportional to the residual spherical aberration $\Delta s'$ over the entire entrance pupil.

9.7 Astigmatism and Field Curvature

Examine the image formation for an extra-axial point by two ray fans lying in perpendicular planes, meridional and sagittal, as shown in Fig. 9.11. We suppose that both fans emanating from the point B consist of rays travelling very close to the principal ray, i.e. they are subtended by short diameters in the entrance pupil plane.

The curvature of the spherical surfaces met by these ray fans will be different for the fans, therefore the convergence points for the fans in image space will occur in different places. We denote the axial distances from the ideal image plane, containing point A_0', to the convergence points of meridional and sagittal ray fans by z_m' and z_s' respectively. These quantities can be computed with equations (8.12).

The aberration for an off-axis point in which the images formed by meridional and sagittal ray fans lie in different points is known as *astigmatism*. A measure of this aberration is the difference of z_s' and z_m', i.e.,

$$\Delta z_a' = z_s' - z_m'$$

With reference to Fig. 9.11, the meridional ray fan produces a horizontal line as it converges at B_m', while the sagittal fan produces a vertical line at B_s'. In the Gaussian plane, the image of a point in this case will be an ellipse with the vertical major axis. If we shift the image plane from B_m'

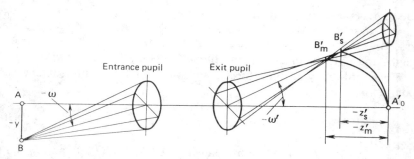

Fig. 9.11. Astigmatism

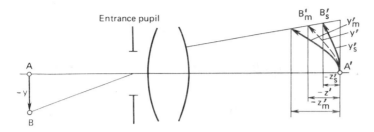

Fig. 9.12. Image surfaces formed by astigmatic ray bundles

toward B_s', then the image of the object point will appear at various locations of the plane as a horizontal line, horizontal ellipse, circle, vertical
ellipse, and vertical line. Quite appropriately, the ray bundles producing
this type of image are referred to as *astigmatic bundles*.

For an extended object, say a piece of plane, in a raytracing analysis we
have to examine a collection of points each of which is imaged by
astigmatic ray bundles. If the object is a line segment AB of length y lying
in the meridional plane, which is the plane of the page in Fig. 9.12, then to
each point of this line there correspond a meridional image, B_m', and a
sagittal image B_s'. Connecting the corresponding points yields the curves y_m'
and y_s' which are respectively the meridional and sagittal images of y. If we
rotate the curves y_m' and y_s' about the optical axis we obtain the astigmatic
surfaces of rotation tangent to the Gaussian plane at the axial point A'.

The locus of the mid-points between the curves y_m' and y_s' constitutes
the mid-curve y'. In the image surface resulted from the rotation of this
curve around the axis, each object point y will be imaged by a circle. All
these plane figures may be regarded as appearing in meridional sections of
the image field, thus revealing its curvature.

Thus when the system projects the image on a screen, astigmatism and
(image) field curvature blur the images of object points, the sharp definition of the image abading with the distance from the optical axis.

We emphasize the typical feature of imaging in the presence of
astigmatism for the case of a two-dimensional object shown in Fig. 9.13a.
The elementary meridional ray bundles imaging each point as lines perpendicular to variously oriented "meridional" planes yield a sharp image of
the circle because the elementary line segments of "meridional" images
overlap without sacrificing the sharpness of imagery. The points of the
radial object lines will be imaged as elementary lines perpendicular to the
radii of the image, the length of these line segments increasing farther from
the optical axis. This is illustrated with some exaggeration in Fig. 9.13b.
The elementary sagittal ray fans will image each object point as line

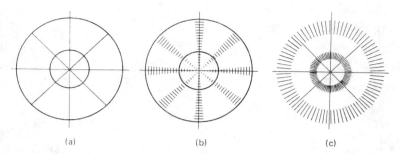

Fig. 9.13. Imaging a plane figure (*a*) by meridional (*b*) and sagittal (*c*) astigmatic ray fans

segments perpendicular to variously oriented "sagittal" planes. These lines will not distort the images of the radii, whereas the images of the circles will be constituted by elementary radial lines which become longer farther from the optical axis as shown in Fig. 9.13*c*.

Both astigmatism and field curvature are normally characterized by the familiar quantities z'_m and z'_s. These are tabulated or plotted as functions of slopes, for the principal rays emanating from various points of object, or object linear size y.

Figure 9.14 shows the plots of $z'_m = f(\sigma)$ and $z'_s = f(\sigma)$ for various cases of system correction for astigmatism and image curvature. The diagram at (*a*) represents these plots for the case of an astigmatism and image field curvature, this latter being represented by the mid-curve z' traced in between z'_s and z'_m. Owing to this aberration the image is not sharp over a plane surface even in the absence of astigmatism when $z'_m = z'_s$.

The diagram at (*b*) illustrates the aberrations for a system corrected for field curvature ($z'_m = -z'_s$), the astigmatism remaining uncorrected. To ensure sharp imagery over the entire field the system should be corrected for

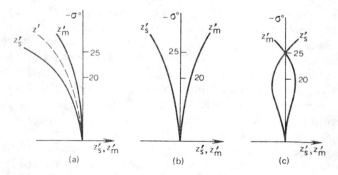

Fig. 9.14. Graphical representation of astigmatism and field curvature

both astigmatism and (image) field curvature. The systems corrected for both aberrations within a certain angular field, with the residual aberrations having tolerable values over the entire field of view, are called *anastigmats*. The diagram at (c) depicts the plots of the residual astigmatism and image curvature for an anastigmat practically freed from both aberrations up to an angular field of $2\sigma = 50°$, which are kept comparably small at the edge of the lens.

Approximate estimations of astigmatism and field curvature can be carried out with the third-order aberration formulae as functions of the Seidel sums S_{III} and S_{IV}. Let us derive these relationships for a meridional ray fan emanating from an extra-axial object point B at infinity. This situation is illustrated in Fig. 9.15. The respective meridional component of the third-order transverse aberration can be determined with the normalization set (9.14), subject to nonzero S_{III} and S_{IV}, with the help of (9.15) as

$$\Delta y'_{III} = -(m\omega_1^2/2)(3S_{III} + S_{IV}) \tag{9.31}$$

for the case of the object and image being in air, i.e. for $n'_p = n_1 = 1$.

From similar triangles in image space (see Fig. 9.15) we have

$$-\Delta y'_{III}/m = -z'_m/(f' + z'_m)$$

Observing that $f' \gg |z'_m|$ and substituting in the above proportion $\Delta y'_{III}$ from (9.31) we get

$$z'_m = -\frac{f'}{2}\,\omega_1^2(3S_{III} + S_{IV}) \tag{9.32}$$

Similar argument applied to (9.15) yields for the sagittal component of third-order transverse aberration

$$z'_s = -\frac{f'}{2}\,\omega_1^2(S_{III} + S_{IV}) \tag{9.33}$$

Now the amount of astigmatism

$$\Delta z'_a = z'_s - z'_m = f'\omega_1^2 S_{III}$$

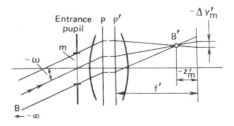

Fig. 9.15. Construction to deduce expressions for astigmatism and field curvature

Thus, the astigmatism is proportional to the squared angular field of the optical system. To correct the system for astigmatism in the area of third-order aberrations there must be $S_{III} = 0$. Then both astigmatic surfaces coalesce and in view of (9.32) and (9.33)

$$z_m' = z_s' = -(f'\omega_1^2/2)S_{IV}$$

that is, the coefficient S_{IV} defines the aberration of field curvature for a system corrected for astigmatism.

9.8 Distortion

Distortion is an aberration appearing as the curvature of straight lines in the object, i.e. as a breakdown of the geometric similarity between object and image. This aberration is independent of the ray coordinates at the entrance pupil and all the rays passing through a given object point form a homocentric bundle converging in the Gaussian plane at a point other than the ideal image. Distortion does not deteriorate the sharp definition of the image but distorts the shape of an object pattern.

For a given point of the field, distortion is determined as the difference of the ordinates of the principal ray y' and ideal image y_0':

$$\Delta y' = y' - y_0' \tag{9.34}$$

It is often specified as a percentage

$$\Delta' = \frac{y' - y_0'}{y_0'} 100 = (y'/y_0' - 1)100$$

With reference to Fig. 9.16, the lateral magnification of an optical system can be defined for a given conjugate object-image pair as

$$\beta = \frac{y'}{y} = \frac{(s' - t')\tan \omega'}{(s - t)\tan \omega} \tag{9.35}$$

If this quantity remains constant for any y and equal to the magnifica-

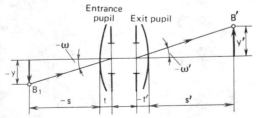

Fig. 9.16. The course of principal rays through a system with distortion

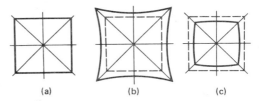

Fig. 9.17. Distortion of a pattern (*a*), positive or pincushion distortion (*b*), negative or barrel distortion (*c*)

(a)   (b)   (c)

tion β_0 of the perfect system, then distortion is absent and the system free from distortion is termed *ortoscopic*.

Magnification of a real optical system defined by (9.35) does not remain constant for various y because there exists spherical aberration in the pupils of the system and the angular magnification in the pupils fails to remain constant.

From Eqs. (9.34) and (9.35) it can be seen that if $|\beta|$ increases as the object point moves away from the optical axis, then the distortion of the system $\Delta y'$ also increases, i.e. $|\beta| > |\beta_0|$. This type of distortion is called "positive" or *pin-cushion distortion* (Fig. 9.17a) as a square is imaged into a figure shown in the diagram at (*b*). If $|\beta|$ decreases the distortion $\Delta y'$ also decreases, i.e. $|\beta| < |\beta_0|$. This is the case of "negative" or *barrel distortion* where in place of an object square there occurs a figure shown at (*c*).

Approximate estimations of distortion can be carried out by the third-order aberration formulae involving the Seidel sum S_V. According to (9.16) for an object plane at a finite distance, the distortion is as follows

$$\Delta y'_{III} = -[y_1^3/2n_p'(s_1 - t)^3]S_V$$

For an infinitely distant object plane we have from (9.17)

$$\Delta y'_{III} = -(\omega_1^3 f'/2n_p')S_V$$

More popular presentation of distortion is in the form of plots as a function of linear or angular position of the object point (Fig. 9.18).

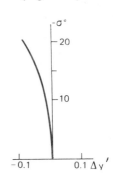

Fig. 9.18. Graphical representation of distortion

CHROMATIC ABERRATIONS

10.1 Axial Chromatic Aberration

Chromatic aberration appears as a result of imbalanced dispersion: a beam of white light on refraction at system surfaces appears dispersed into its monochromatic components. This phenomenon is observed already in the paraxial (Gaussian) domain: the paraxial images of an object formed by rays of different wavelengths will differ both in position and size depending on the optical characteristics of lens materials (see Section 5.1).

The aberration of optical systems as the result of which the images of an object point produced by rays of different wavelengths occur in different locations along the optical axis is known as *axial* or *longitudinal chromatic aberration*.

We consider this aberration with reference to Fig. 10.1 in which an optical system is hit by a polychromatic paraxial ray pencil emanating from an axial point A. We shall trace the rays of wavelength λ_1, λ_0, and λ_2. The zero labelled wavelength relates to maximum transmission or the detector sensitivity peak. The axial chromatic aberration of the system will cause the images A'_{λ_1}, A'_{λ_0}, and A'_{λ_2} to occur at distinct distances s'_{0,λ_1}, s'_{0,λ_0}, and s'_{0,λ_2} from the optical system. The axial chromatic aberration $\Delta s'_{\lambda_1,\lambda_2}$ for two wavelengths is measured as the difference of the respective distances

$$\Delta s'_{\lambda_1,\lambda_2} = s'_{0,\lambda_1} - s'_{0,\lambda_2}$$

When this type of aberration is present in the system, the image of a point in the paraxial plane for λ_0 will be blurred and coloured.

An accurate amount of axial chromatic aberration can be computed by numerical tracing of two paraxial rays of λ_1 and λ_2 from an axial point. An approximate estimation can be obtained by the formula

$$\Delta s'_{\lambda_1,\lambda_2} = (1/n'_p \alpha'^2_p) \sum_{k=1}^{p} h_k C_k \tag{10.1}$$

where h_k is the height of the first auxiliary ray,

$$C_k = \frac{\delta \alpha_k}{\delta \mu_k} \, \delta \left(\frac{\Delta n_k}{n_k} \right) = \frac{\delta \alpha_k}{\delta \mu_k} \, \delta \left(\frac{1 - \mu_k}{V_k} \right)$$

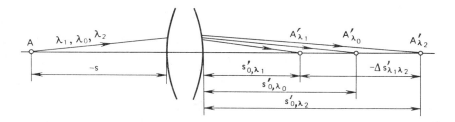

Fig. 10.1. Axial chromatic aberration

V_k is the V value, $\mu_k = 1/n_k$. Note that for air $\Delta n/n = (1 - \mu)/V = 0$.

We derive an expression for axial chromatic aberration of a single thin lens in air; the respective nomenclature is presented in Fig. 10.2. Recalling the conjugate distance relation (3.7), we have for $s = a$ and $s' = a'$

$$1/s' - 1/s = 1/f'$$

For various rays of the polychromatic beam the quantities s' and f' are variable. Differentiating the above equation yields $\partial s' \approx \Delta s'_{\lambda_1, \lambda_2}$

$$-\partial s'/s'^2 = -\partial f'/f'^2$$

whence

$$\partial s' = \Delta s'_{\lambda_1, \lambda_2} = s'^2(\partial f'/f'^2) \tag{10.2}$$

To determine $\partial f'$ we use the thin lens formula (5.9) for $d = 0$, namely,

$$-\partial f'/f'^2 = \partial n(1/r_1 - 1/r_2)$$

Then

$$-\partial f'/f'^2 = \partial n/f'(n - 1) \tag{10.3}$$

where $\partial n = \Delta n = n_{\lambda_1} - n_{\lambda_2}$, and $n - 1 = n_{\lambda_0} - 1$.

From (10.2) and (10.3) it follows that

$$\Delta s'_{\lambda_1, \lambda_2} = -s'^2/f' V \tag{10.4}$$

If the object is at infinity, $s = -\infty$, and for a thin lens $s' = f'$ then the

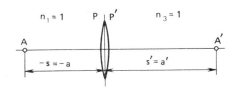

Fig. 10.2. Nomenclature to determine the axial chromatic aberration of a single thin lens

axial chromatic aberration is

$$\Delta s'_{\lambda_1, \lambda_2} = -f' / V \qquad (10.5)$$

This expression indicates that positive lenses exhibit negative axial chromatic aberration, and negative lenses exhibit positive axial chromatic aberration.

Example 10.1. Determine the axial chromatic aberration for a thin lens of $f = 100$ mm, $n_{\lambda_0} = n_e = 1.5$, $V = 60$, for (1) an infinitely distant object point, and (2) an object point at a distance $s = -2f'$ from the lens.

Solution. (1) From Eq. (10.5) we determine $\Delta s'_{\lambda_1, \lambda_2} = -1.67$ mm which implies that if the relative aperture of the lens were $f/10$, then in place of the ideal image point there would appear a spot 0.167 mm in diameter.

(2) The image is formed also at a double focal length $s' = 2f'$. Eq. (10.4) yields

$$\Delta s'_{\lambda_1, \lambda_2} = -4f'^2/f' V = -4f'/V = -6.68 \text{ mm}$$

that is, this axial chromatic aberration is four times that of the first case. This example indicates that as the object approaches the optical system the axial chromatic aberration increases.

Axial chromatic aberration for single lenses, both positive and negative, is illustrated by curves *1* and *2*, respectively, in Fig. 10.3. A suitable choice of lens materials and focal lengths of positive and negative lenses can balance their axial chromatic aberrations so that for the system $\Delta s'_{\lambda_1, \lambda_2} = 0$, and the system will be an *achromat* whose aberration is described by curve *3* in Fig. 10.3.

Let us consider the condition that the axial chromatic aberration may vanish for a thin cemented doublet in air for the case of infinitely distant

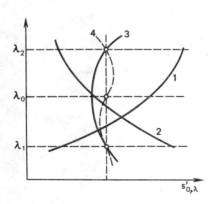

Fig. 10.3. Plots of longitudinal chromatic aberration

object. In view of (3.32) the power of such a cemented lens is

$$\phi = \phi_1 + \phi_2 \tag{10.6}$$

therefore

$$d\phi = d\phi_1 + d\phi_2 \tag{10.7}$$

where $d\phi_1 = \phi_1/V_1$, and $d\phi_2 = \phi_2/V_2$.

The condition that the lens becomes free from axial chromatic aberration is $d\phi = 0$. Substituting in (10.7) for $d\phi_1$ and $d\phi_2$ their expressions yields

$$\phi_1/V_1 = -\phi_2/V_2 \tag{10.8}$$

Solving (10.6) and (10.8) together leads to the expressions for the power of lenses constituting the achromatized doublet

$$\phi_1 = V_1\phi/(V_1 - V_2), \quad \phi_2 = -V_2\phi/(V_1 - V_2)$$

Analysis of the above equations leads to a number of conclusions as follows.

(i) To correct a system for axial chromatic aberration, a combination of lenses with opposite focal lengths is required (see Eq. (10.8)).

(ii) The positive lens of a doublet to be achromatized must be made of a material with a higher V value in the case of a positive objective, and of a material with a lower V value in the case of a negative objective. Hence the glass with higher V value (crown in most applications) is used to make the lens whose focal length sign defines the sign of the doublet.

(iii) The condition of an achromatic system, $d\phi = 0$, can be satisfied for a negative meniscus. This was first demonstrated by D.D. Maksutov in 1941. If we differentiate the lens formula (5.9) and put $\partial\phi = d\phi = 0$, then the condition that the axial chromatic aberration in the Maksutov meniscus may vanish is as follows

$$r_2 - r_1 = (1 - n^2)d/n^2$$

Referring to Fig. 10.3 it will be seen that for a two-lens objective the coloured paraxial images of axial points can be made to coincide only for the rays of wavelengths λ_1 and λ_2. The image in the major (related to maximum detector sensitivity or system transmission) colour occurs at a distance s'_{0,λ_0} other than the distance $s'_{0,\lambda_1} = s'_{0,\lambda_2}$. Therefore, the achromatized doublet exhibits a residual chromatic aberration called the *secondary (residual) spectrum*. It can be estimated by

$$\Delta s'_{\lambda_0,\lambda_1} = s'_{0,\lambda_0} - s'_{0,\lambda_1} = s'_{0,\lambda_0} - s'_{0,\lambda_2}$$

For a cemented doublet, the secondary spectrum can be estimated accurate to first order aberrations (on the condition that $\Delta s'_{\lambda_1, \lambda_2} = 0$) by the formula

$$\Delta s'_{\lambda_0, \lambda_1} = -f'(\gamma_1 - \gamma_2)/(V_1 - V_2) \qquad (10.9)$$

where γ_1 and γ_2 are the relative partial dispersions of the component lenses.

It is quite obvious that to reduce the secondary spectrum, the designer has to choose pairs of glasses for which the relative partial dispersions are as close as possible, and the V values differ widely.

We use the expression (10.9) to calculate the secondary spectrum for a number of glass pairs.

Case 1. Ordinary glasses of types K8 and F1 with $V_1 = 64.05$ and $V_2 = 36.93$.

$$\Delta s'_{F, D} = f'/2080$$

Case 2. Glasses of strongly different V values: LK3 of $V_1 = 70.02$ and TF10 of $V_2 = 25.36$.

$$\Delta s'_{F, D} = f'/1940$$

Case 3. Glasses with close relative partial dispersions: STK7 of $V_1 = 53.58$ and $\gamma_{F, D} = 0.707$, and BF23 of $V_2 = 52.41$ and $\gamma_{F, D} = 0.708$.

$$\Delta s'_{F, D} = f'/1170$$

Case 4. Glasses with close relative partial dispersions: BK12 of $V_1 = 58.33$ and $\gamma_{F, D} = 0.705$, and OF1 of $V_2 = 51.80$ and $\gamma_{F, D} = 0.706$.

$$\Delta s'_{F, D} = f'/6530$$

Cases 1 and 2 are seen to exhibit a secondary spectrum of about $f'/2000$ which has to be corrected in long-focus lenses, objectives for aerial photography, high-magnification lenses, colour camera lenses, and colour projection lenses. Small differences between γ_1 and γ_2, as in Cases 3 and 4, result also in small differences between V_1 and V_2, thus entailing a small radius of the cemented doublet preventing the spherical aberration from being corrected.

Curve *4* in Fig. 10.3 represents the chromatic aberration in an *apochromat*, i.e. a system corrected so that the images of three axial points coalesce, and $s'_{0, \lambda_1} = s'_{0, \lambda_0} = s'_{0, \lambda_2}$.

10.2 Transverse Chromatic Aberration

Another type of chromatic aberration of first order arising already in the paraxial region is *transverse* (or *lateral*) *chromatic aberration* also call-

ed *lateral colour*, or *chromatic difference of magnification*. It causes the
images of extra-axial points produced by rays of various wavelengths to appear at different distances from the optical axis, as shown in Fig. 10.4. The
quantities y' are derived by tracing the principal ray. The numerical
measure of transverse chromatic aberration is the difference of y' for the
extreme wavelengths involved, namely,

$$\Delta y'_{\lambda_1, \lambda_2} = y'_{\lambda_1} - y'_{\lambda_2}$$

This can also be given as a relative quantity with respect to the basic
wavelength as

$$\frac{\Delta y'_{\lambda_1, \lambda_2}}{y'_{\lambda_0}} = \frac{y'_{\lambda_1} - y'_{\lambda_2}}{y'_{\lambda_0}}$$

An approximate estimation of $\Delta y'_{\lambda_1, \lambda_2}$ may be derived with the formula

$$\Delta y'_{\lambda_1, \lambda_2} = (y'_{\lambda_0}/I) \sum_{k=1}^{k=p} H_k C_k$$

where H_k is the incidence height of the second auxiliary ray, $I = n_1 \alpha_1 (t_p - s_1)\beta_1$ is an invariant, and

$$C_k = \frac{\alpha_{k+1} - \alpha_k}{\mu_{k+1} - \mu_k} \left(\frac{1 - \mu_{k+1}}{V_{k+1}} - \frac{1 - \mu_k}{V_k} \right) = \frac{\delta \alpha_k}{\delta \mu_k} \delta \frac{1 - \mu_k}{V_k}$$

Transverse chromatic aberration is determined in the paraxial image
plane for the basic colour. The image receives undesirable coloured haloes
which deteriorate the sharp image. This chromatic aberration increases for
greater angular fields of optical systems and depends not only on the design
parameters and lens materials but also on the degree of correction for the
axial chromatic aberration.

For a thin system and the object at infinity, the transverse chromatic
aberration may be defined as

$$\Delta y'_{\lambda_1, \lambda_2}/y'_{\lambda_0} = -H_1 \sum_{k=1}^{k=p} C_k$$

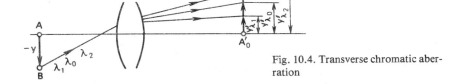

Fig. 10.4. Transverse chromatic aberration

It will be seen that this aberration depends on the position of the pupil ($H_1 = t$ at $\beta_1 = 1$) and the degree of system correction for axial chromatic aberration. For $s_1 = -\infty$, $n' = 1$, and $\alpha' = 1$, the axial chromatic aberration of a thin system is

$$\Delta s'_{\lambda_1, \lambda_2} = h_1 \sum_{k=1}^{k=p} C_k$$

Therefore, if a thin system is corrected for axial chromatic aberration, which is achieved only at $\sum C_k = 0$, then it is also corrected for transverse chromatic aberration. Also, if the entrance pupil of the system coincides with the first surface ($t = 0$) then the transverse chromatic aberration also vanishes.

10.3 Spherochromatism

Spherochromatism or *chromatic variation of spherical aberration* is an error in the image of an axial point which arises because of the different spherical aberrations for rays of wavelengths λ_1, λ_0, and λ_2.

Figure 10.5 shows the spherical aberrations for three wavelengths, plotted as functions of height of passage through the entrance pupil, m. This system is amendable to the correction for the spherical aberration at the basic colour for marginal rays, to make $\Delta s'_{m, \lambda_0} = 0$, and for the axial chromatic aberration, to make $\Delta s'_{\lambda_1, \lambda_2} = 0$. However, the image quality may be unsatisfactory due to a large difference of the spherical aberrations for λ_1 and λ_2. Moreover, the higher the rays pass through the entrance pupil the stronger the effect of this aberration.

Spherical aberration (see Section 9.4) may be represented as

$$\Delta s' = s' - s'_0 = a_1 \sigma^2 + a_2 \sigma^4 + \dots \tag{10.10}$$

where a_1, a_2, etc. are some constants independent of m, and σ is the en-

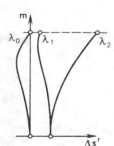

Fig. 10.5. Chromatic variation of spherical aberration

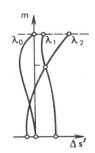

Fig. 10.6. Plots of spherochromatism after optimal correction

trance angular aperture (current value). For two wavelengths, λ_1 and λ_2, this representation becomes

$$\Delta s'_{\lambda_1} = s'_{\lambda_1} - s'_{0\lambda_1}$$
$$\Delta s'_{\lambda_2} = s'_{\lambda_2} - s'_{0\lambda_2}$$

Accordingly, the spherochromatic aberration is

$$\Delta_{\lambda_1 - \lambda_2} = \Delta s'_{\lambda_1} - \Delta s'_{\lambda_2} = (s'_{\lambda_1} - s'_{\lambda_2}) - (s'_{\lambda_1} - s'_{\lambda_2})_0 \qquad (10.11)$$

Here, $(s'_{\lambda_1} - s'_{\lambda_2})_0 = s'_{\lambda_1, \lambda_2}$ is the axial chromatic aberration. Because spherical aberration is determined with respect to the paraxial image plane for λ_0, Eq. (10.11) rewrites

$$\Delta_{\lambda_1 - \lambda_2} = s'_{\lambda_1} - s'_{\lambda_2}$$

Therefore, in a system corrected for spherochromatism $s'_{\lambda_1} = s'_{\lambda_2}$ or $a_{1\lambda_1} = a_{1\lambda_2}$, $a_{2\lambda_1} = a_{2\lambda_2}$, etc. If one confines himself to third-order aberration, then $S_{I, \lambda_1} = S_{I, \lambda_2}$.

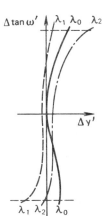

Fig. 10.7. Chromatic variation of
oblique aberration

Equation (10.11) suggests that to ensure freedom from spherochromatism there should be $s'_{\lambda_1} - s'_{\lambda_2} = \Delta s'_{\lambda_1, \lambda_2}$. In this case, however, one has to reconcile with some residual axial chromatic aberration.

An optimally corrected system will be the one free from spherochromatic aberrations for zonal rays with $(s'_{\lambda_1})_z = (s'_{\lambda_2})_z$ as can be seen in Fig. 10.6.

To close the above consideration we conclude that spherochromatism relates to the aberrations of a wide axial ray bundle examined over the entire range of operable wavelengths. By analogy we may also conclude for a wide oblique bundle of real rays that extra-axial aberrations computed for a polychromatic light may also appear different even for a meridional ray fan. Plots of such oblique chromatic aberrations are presented in Fig. 10.7. The system can be freed from the transverse chromatic aberration to make $y'_{pr, \lambda_1} = y'_{pr, \lambda_2}$, but the presence of oblique chromatic aberration would deteriorate the image quality for extra-axial points by adding undesirable colouring.

THE EYE AS AN OPTICAL SYSTEM

The eye is a living optical system transducing the incident radiant energy in the visible perceptory signals transmitted into the brain. Most optical instruments utilize the eye as a final element in one way or another. It is vital therefore that the designers of optical systems recognize the relevant possibilities of the eye.

11.1. The Structure of the Eye

From outside the eyeball is covered by a white and opaque shell, the *sclera*, labelled *1* in Fig. 11.1, exept for the convex front portion (*cornea*) which is transparent. The *cornea* (*10*) supplies most of the refractive power of the eye.

The *iris* (*7*) which gives the eye its colour, plays the part of a diaphragm in the eye. Light enters the eye through the iris aperture, the *pupil*, and propagates through the *aqueous humour* (*9*), which is a watery fluid, to the biconvex elastic *lens* (*8*). The refraction of the lens is varied by a multitude of fibres or *ligaments* (*11*) which buldge or flatten the lens, altering the curvature of this flexible capsule. The lens separates the inside of the eyeball into two chambers — that of the *aqueous humour* (*9*) and of the *vitreous humour* (*12*), a gelly-like material.

The inner back surface of the eyeball is covered with the *retina* (*6*) which contains blood vessels (*2*), nerve fibres, the light-sensitive rod and cone cells, and a pigment layer, met in this order by the light. The *optic nerve* (*5*) leaves the eyeball for the brain at a place called the *blind spot* (*4*). The light-sensitive receptors of the retina, rods and cones, are terminations of the optical nerve fibres. Rod cells are cylinders about 2 μm in diameter, while cones are the shape of an incandescent lamp with 4.5 to 6.5 μm diameter in the largest cross section.

The blind spot, as the name implies, has no light sensitive elements.

Slightly to the outer side of the *optical axis* (*14*) there is an area of the *macula* (*3*) shaped as an oval of 1-mm horizontal and 0.8 mm vertical axes, subtending about 6° of the field of vision. The central portion of the macula is the *fovea* where the sensitive layer of the retina consists entirely of cones each of which terminates in its own nerve fibre. The 0.3 mm diameter of the fovea is responsible for the most sharp vision. It subtends about 2° 30′ of the field of view.

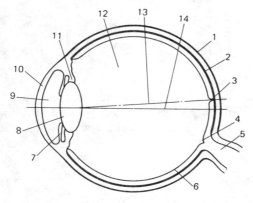

Fig. 11.1. The structure of the eye

Constant parameters

Refractive indices	
cornea	1.376
aqueous humour and vitreous humour	1.336
lens	1.386
Distance (mm) from the vertex of cornea	
to back surface of the cornea	0.5
to back surface of the lens	7.2
Radii (mm)	
cornea to air	7.7
cornea to aqueous	6.8

Variable parameters

Distance (mm) from the vertex of cornea to	at rest	max. stress
front lens surface	3.6	3.2
first principal point	1.348	1.772
second principal point	1.602	2.086
first nodal point	7.078	6.533
second nodal point	7.332	6.847
first focal point	−15.707	−12.397
second focal point	24.387	21.016
entrance pupil	3.047	2.668
exit pupil	3.667	3.212
Lens surface radii (mm)		
front	10	5.33
back	−6	−5.33
Focal length (mm)		
first	−17.055	−14.169
second	22.785	18.930
Power (dioptres)	58.64	70.57
Magnification between the pupils	0.909	0.941

The line through the centre of the fovea and the second nodal point of the lens will be called the *axis of vision* (*13*). It makes about 5° with the optical axis of the eye.

There are about seven million cones in the retina, about 100 million rods and only about one million nerve fibres. In the outer portions of the retina farther away from the macula the number of cones connected multiply to one nerve fibre increases (several hundred to a fibre), and the sensitive cells are more widely spaced, thus accounting for the less distinct vision in this area of the retina.

The pupil diameter varies from individual to individual and depending on illumination from 1.5 to 8 mm. It contracts at bright light and expands at dark.

The suspending muscles turn the eyeball from side to side within 45-50° in the tracking of moving objects and in scanning large patterns.

The interpupillary distance (interocular distance) of adults varies from 58 to 72 mm, the average spacing is taken to be 65 mm. In viewing nearby objects the eyes roll inside so that their visual axes make a convergence angle as large as 32°.

11.2 Characteristics of the Eye

The optical system of the eye forms at the retina the real images of objects being viewed. These images are sensed by the light-sensitive receptors of the eye — cone and rod cells. This system may be regarded as constituted by two lenses, the cornea and the lens, separated by the aqueous humour. The front surface of the cornea faces air, while the back surface of the lens is immersed in the vitreous humour.

Some parameters of the eye as an optical system derived by statistical analysis are given below.

Figure 11.2 schematizing the optical system of the eye gives some rounded data for the eye parameters. It will be seen that the second focal length, defining the power of the eye, can vary by about 20 per cent. This ability of the eye called *accommodation* is effected by the fibres of the ligaments changing the curvature of the lens faces. Accommodation helps the lens to focus the images of objects situated at various distances from the viewer onto the retina.

When the eye is fixed at infinity, its second (posterior) focal length is the greatest (22.785 mm), the second principal focus being at the retina. This case corresponds to the eye at rest and is pictured in Fig. 11.2a. The condition of accommodation of the eye is conveniently specified by giving the position of the accommodation point on which the eye is focused. For

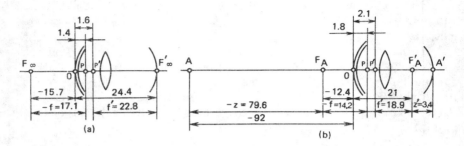

Fig. 11.2. The optical system of an eye (*a*) accommodated at an object at infinity, and (*b*) at a strong accommodation

the eye at rest, fixating a point at infinity, the position of the accommodation point will be characterized by the 'distance of distinct vision' equal to -15.7 mm.

When the eye is in a state of the greatest stress of accommodation muscles, the posterior focal length decreases to 18.93 mm, which corresponds to the retinal image of an axial point *A* (in Fig. 11.2*b*) 92 mm apart from the vertex of the first surface of the cornea. The distance of the point of accommodation for this case is called the least distance of distinct vision.

The inverted difference of these extreme distance of distinct vision in dioptres is sometimes used as a measure of accommodation capacity of the eye. For the eye schematized in Fig. 11.2 this total power of the eye equals 11 dioptres. This characteristic varies with the age of individual as the least distance of distinct vision grows. For example, at the age of fifty this distance is 400 mm so that the power of the eye is 2.5 D.

For a normal sighted adult, the most convenient distance for reading (at a good illumination of about 50 lx) and working with small objects measures at 250 mm. This distance is taken to be the distance of most acute vision.

The field of vision of an eye approximates an ellipse about 125° high and 150° wide, however, only a small portion of the field is the area of acute vision. The portion is governed by the size of the macula usually subtending about 6 to 8° of the field. The outside portion of the field of view is used mostly for orientation. High mobility of an eye aids in swiftly bringing the images of viewed objects in the area of the macula. Within one minute the eye can fixate as many as 120 observation points allowing 0.2 to 0.3 second to fixate each of them.

We shall learn later that the resolving power of a system is the least angular separation of two point sources that are just detectably separated by the instrument. For a perfect optical system, the resolving power is given by

$$\psi = 140''/D \qquad (11.1)$$

where D is the pupil diameter in mm. If we regard the eye as an ideal optical system, then for $D = 1.5$ to 2.5 mm the resolving power is about one minute of arc. The same value of resolution will be obtained by evaluating the angular subtense of a sensitive receptor within the fovea from the back nodal point, i.e. for 5 μm of cone cell diameter, and posterior focal length of about 20 mm. The average resolving power of an eye may thus be concluded to be 1 minute of arc.

The aforestated resolving power does not remain the same at different viewing conditions. For example, in viewing images on a screen $\psi = 2'$ to 3′, in viewing through common optical instruments it is 1′, with high quality instrumentation $\psi = 30''$, and with range finders it is as small as 10 seconds of arc. The high resolution in operating with the latter instruments may be attributed to the fact that the eye exhibits better sensitivity with respect to the bending of lines, say, with respect to a lateral shift of lines. This sensitivity with respect to lines is due to a staggered arrangement of sensitive cone cells of the retina and consistent small motions of the eye bringing the shifted portions of a line onto various cones.

As can be seen from Eq. (11.1), the resolving power of the eye must improve for larger diameters of the pupil, D. This statement, however, holds only for $D \leqslant 3$ to 4 mm. A further increase in pupil diameter fails to improve the resolving power of the eye as it is determined by the diameter of a cone cell in the retina. Moreover, aberrations of the optical system of the eye also increase with D. Hence a further increase in D can only result in decreasing resolution.

The best resolving power of an eye can be achieved at the most favourable conditions for the eye operation, i. e. with an object illumination of 50 lx and a wavelength of 0.55 μm at which the eye has the highest sensitivity.

In general, owing to visual adaptation the eye is capable of sensing and reacting to a rather wide range of luminance, from 10^{-7} to 10^5 candelas per square metre. The eye illuminated after a darkness takes about 20 to 30 minutes to regain its light sensitivity. When returned to a dark room after light the eye looses its sensitivity at first, but recovers after a few minutes. This process is known as *dark adaptation*. The complete dark adaptation takes about an hour.

At low luminance levels (up to 10^{-3} cd m^{-2}) the pupil of the eye expands and admits more light so that the retina becomes more sensitive and switches from cone vision to rod vision. Rod cells are very sensitive receptors for scotopic vision but they fail to discern colours. As the level of object luminance increases, the pupil contracts and the eye switches more cones into action so that colours become discernible, the situation which takes place in the twilight. With a further increase in object luminance (above 1 cd m^{-2}) the light is brought to the fovea and the eye switches completely to cone vision. Note that as the pupil diameter increases from 2 to 8 mm during the dark adaptation process, the light flux admitted to the eye increases 16 times.

There is also an electrochemical mechanism involving rhodopsin, the visual purple pigment, that aids dark adaptation. On the other hand, the transfer of dark pigment in the retina protects the receptors from high levels of irradiance.

The lowest level of illuminance of the object under the given viewing conditions is called the *threshold of vision*.

Visual sensitivity of the eye gives the least amount of energy capable of arising an ocular response. For a pupil diameter of about 8 mm, the lowest light flux to which rod cells respond amounts to 2×10^{-14} lm. The sensitivity of the eye to light is a function of the wavelength of the light and the level of illumination to which the eye is adapted. Fig. 11.3 shows the relative sensitivity of the eye, $V(\lambda)$, to different wavelengths for normal levels of illumination (photopic vision) and under conditions of dark adaptation (scotopic vision). At low levels of illumination, when the eye operates in the rod vision mode, the maximum of sensitivity for the dark adapted eye shifts towards the blue end of the spectrum to peak at 0.51 μm. This *Purkinje shift* owes its existence to the different chromatic sensitivities of the rods and cones of the retina.

After removing the illumination of the eye, the visual images presented to the brain do not disappear at once. They persist for about 0.05 to 0.2 s, the time delay depends on the luminance, the spectral composition of the flux and on the degree of adaptation of the eye.

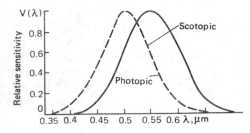

Fig. 11.3. Relative sensitivity of the eye

The contrast sensitivity of the eye is a function of field luminance. The threshold luminance is measured as a fraction of the smallest perceptible difference in luminance between two adjacent fields (ΔL) over the background field luminance (L_b), also called the *reciprocal contrast sensitivity*. The *contrast sensitivity* of the eye $L_b/\Delta L$ increases with the field luminance to reach a maximum of about 60 for $L_b = 130$ to 6400 cd m^{-2}.

In periodic illumination there exists a critical frequency of light pulses at which the viewed field seems to be of constant luminance. This frequency is a function of the background field illumination, being 10 Hz at illuminations up to 0.1 lx, 30 Hz at 10 lx, and 40 Hz at 100 lx.

Stereoscopy. If we would observe an object with one eye only, we could be robbed of some information on the object. In estimating the distances to nearby objects (about 5 m), accommodation functions, head turning, and eye rolling take part. Larger distances are judged by their image size at the retina. Both types of estimation are subject to large subjective errors.

In viewing with two eyes, the images in both eyes are combined and a single visual fused impression results. The fixation to an object point is maintained by the motor muscle of the two eyes which are largely yoked in their action, so that fusion of the images is maintained without conscious effort. Corresponding areas of the two retinae are supposed to be responsible when simultaneously simulated, for a single image point in the binocular field. The convergence of the visual axes brings the retinal images into the corresponding points. As soon as the images move away from these points a doubled perception occurs. High mobility of the eyes expands the area of the object field capable of producing a fused impression.

Viewing with two eyes gives an impression of the depth of field, i.e. produces a stereoscopic effect responsible for the three-dimensional vision. The stereoscopic impression occurs by comparing the retinal images in both eyes and results in the judgement of relative distances to the objects being viewed.

Figure 11.4 schematizes the convergence of the visual axes giving rise to the stereoscopic effect. The difference $\Delta\varepsilon$ of two angles ε_B and ε_C subtended by the interocular distance b at point B and C, respectively, is known as the *binocular parallax*. When the eyes fixate a point at a large distance R compared with the interocular distance b we may safely set $\varepsilon_B = b/R$ whence

$$\Delta\varepsilon = b\Delta R/R^2$$

An experienced observer is able to notice the difference between the images of an object point in the eyes ($C_1'B_1' - C_2'B_2'$) proportional to the

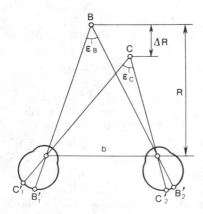

Fig. 11.4. Construction to illustrate stereo-scopic vision

binocular parallax of at least 10 seconds of arc. This value may be termed the *limit of stereoscopic perception*. Accordingly, the smallest distance difference which the unaided eyes can notice for the object distance R may be estimated by putting $\Delta\varepsilon_{min}$ equal to the stereoscopic parallax so that

$$\Delta R_{min} = \frac{R^2}{b}\Delta\varepsilon_{min}$$

where $b = 65$ mm and $\Delta\varepsilon_{min} = 10'' = 4.9 \times 10^{-5}$ rad. This distance difference can no longer be noted starting with the distance $R = b{:}10'' = 1320$ m. This distance will be called the *radius of stereoscopic vision*.

11.3 Defects of the Eye

The normal condition of muscular balance in which the visual axes of both eyes are parallel, when the eyes are relaxed and accommodated at infinitiy, is known as orthophoria. In a healthy pair of eyes at rest the images are at the retinae. Two defects of the eye called nearsightedness (myopia) and farsightedness (hyperopia) result when the image of a distant object is formed by the relaxed eye ahead and behind the retina, respectively. These defects are schematized in Fig. 11.5 in the diagrams at (*a*) and (*c*), respectively.

Since myopia results from an excessive amount of positive power, it is corrected by placing a negative lens before the eye, as illustrated in Fig. 11.5*b*, the power of the negative lens being chosen so that the image is brought to the retina for the most distant point on which the myopic eye can focus. For this purpose the posterior focus of the lens, F_v'', is brought to this distant point of acute vision, V, of the myopic eye. Denote the distance from the lens to the vertex of the cornea by d. Then the second (posterior) focal length of the lens is

$$f_1' = a_v + d \tag{11.2}$$

Recall that for a_v in millimetres both myopia and hyperopia are given in dioptres as

$$A = 1000/a_v \tag{11.3}$$

where a_v is positive as for a hyperopic eye the distant point of acute vision lies behind the retina. Conversion of a_v in (11.2) into dioptric form yields

$$D = 1000/f_1' = \frac{1000}{d + 1000/A} = \frac{A}{1 + dA/1000} \tag{11.4}$$

The dioptric power of a positive lens for correction of hyperopia (Fig. 11.5d) is determined in the same manner.

Equation (11.4) indicates that the dioptric power of a spectacle lens, D, does not coincide with the ametropia A of the eye. This fact is accounted for in prescribing spectacle lenses. For contact lenses, d is very nearly zero, so that $D = A$.

Another defect of the eye is astigmatism which is a difference in the power of the eye from one meridian to another. Astigmatism usually results from an imperfectly formed cornea, which has a stronger radius in one direction than in another. As a result the images of mutually perpendicular lines are not identically sharp. It is quite obvious that to correct an eye for astigmatism the spectacle lens must exhibit different power in

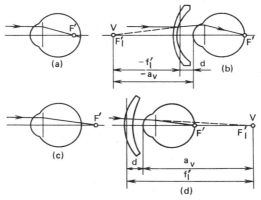

Fig. 11.5. Correction of myopia (a) and hyperopia (c) by spectacle lenses

perpendicular meridional sections, that is, must have toroidal or cylindrical surfaces.

An opaque or cloudy lens (cataract) is still another defect of the eye. Such a lens is frequently removed surgically to restore vision (and replaced with a synthetic lens in modern surgery). Such an aphakic eye lacking the lens, cannot accommodate. The resultant loss of power is made up by positive spectacle lenses of extremely strong power. Normally these are combination lenses of 10 to 11 dioptres for distant vision and of 13 to 14 dioptres for near objects.

ILLUMINATION SYSTEMS

A special class of optical systems is formed by those designed to illuminate an object by a directed beam of light. Illumination systems made for the purpose of flooding large areas with light and operating as search lights and naval beakons relate to illumination engineering and will not be discussed in this book. We shall focus our attention on the optical systems which solve the problem of maximal use of a light flux intercepted by the system, and the problem of uniform illumination of the object.

12.1 Purpose and Types of Illumination Systems

In solving the problem of illumination of finite objects, three system arrangements are in use as follows from the diagrams of Fig. 12.1.

(a) The object y to be illuminated is at infinity. The source of light (1) is at the first focal point of the optical system (2) which is referred to in this case as a *collimator*.

(b) The optical system (2) images the source of light (1) on to the illuminated object y. Such an optical system is referred to as a *condenser*.

(c) The illuminated object y is located in the beam of light passing through the condenser (2) which projects the source of light (1) into the entrance pupil D of the following optical system. The object is normally located near the condenser as this allows for smaller condenser lens diameter.

Choosing one of the two last arrangements is decided by the distribution of radiance over the emitting surface of the light source. If its radiance is uniform and the heating of the object is not objectionable — this should be always taken care of in projecting slides — then the most popular arrangement is the system imaging the emitting body of the source into the plane of the object to be illuminated. In this case to every illuminated point of the object there will correspond a conjugate point in the light emitting surface. Unless the radiance of the source is uniform, the problem will be solved with the system imaging the source filament into the entrance pupil of the following (projection) system as in this arrangement every point of the object receives beams emitted from all the points of the luminous surface.

As with other classes of optical instrumentation, illumination systems

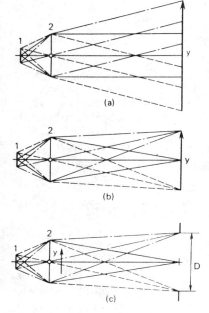

Fig. 12.1. Illumination of object y by the
light from source *1 passing through system 2*

can be dioptric, i. e. constituted by lenses only, catadioptric, i. e. constituted by lenses and mirrors, and purely mirror systems.

12.2 Searchlight and Collimator Systems

Searchlights have optical system which concentrates a portion of the source light into a narrow beam for illuminating distant objects and transmitting signals over large distances.

The Soviet classification of searchlights divides them into devices of long range having the exit pupil diameters D' in the range from 800 to 2100 mm, close range of D' from 500 to 650 mm, and signalling devices of D' from 105 to 250 mm. Collimators are qualified as systems in which the illuminated object is placed in the immediate vicinity to the device.

The principal optical characteristics of a searchlight or collimator, as outlined in this book, include the luminous intensity, gain in illumination, critical distance of beam formation, angle of divergence, and angle of interception.

The illumination in the image can be given in view of Eq. (7.62) as

$$E' = \tau\pi L' \sin^2\sigma'_A.$$

where $\tau = \tau_s\tau_a$, τ_s is the transmission factor of the system, τ_a the transmission factor of the atmosphere or some other medium traversed by the beam behind the searchlight, $L' = (n'/n)^2 L$ the radiance of the image determined by Eq. (7.49), L the radiance (brightness) of the source of light, and n and n' are the refractive indices of the medium embracing the source and the medium of image space, respectively; most common situations have $n = n' = 1$.

In order to determine $\sin \sigma'_A$, we refer to Fig. 12.2. It pictures a source of light of a luminous surface, measured $c \times b$, placed at the first focal plane of the optical system represented as an indefinitely thin lens. The object to be illuminated is at a large distance p' from the optical system, therefore

$$\sin\sigma'_{A'} \approx D/2p'$$

where D is the entrance pupil diameter of the system, which for many practical situations may be taken equal to the diameter D' of the exit pupil. Thus,

$$E' = \tau\pi(n'/n)^2 L(D^2/4p'^2) \qquad (12.1)$$

Comparison of this formula with Eq. (7.12) at $\varepsilon = 0$

$$E' = I_s/p'^2$$

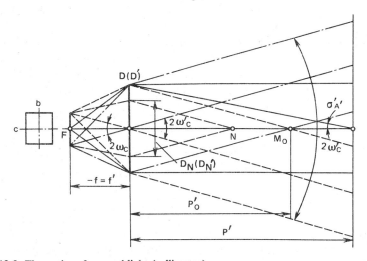

Fig. 12.2. The optics of a searchlight (collimator)

where I_s is the radiant intensity of the searchlight (or collimator), yields

$$I_s = \tau\pi(n'/n)^2 L(D^2/4) \tag{12.2}$$

or at $n' = n = 1$

$$I_s = \tau L A_p'$$

where A_p' is the area of the exit pupil (or entrance pupil at $D = D'$).

It will be seen that the radiant intensity of a searchlight grows with the area of the exit (entrance) pupil for the same radiance of the source of light.

The formulae (12.1) and (12.2) are valid for objects to be illuminated being at a distance $p' \geqslant p_0'$ from the searchlight. The critical distance p_0' may be called the distance of beam formation; it is located by the point M_0 (see Fig. 12.2) which is the first point beyond the system formed as an intersection of the rays passing at the edge of the entrance pupil of diameter D. For a point N, the effective diameter of the entrance (exit) pupil diminishes to $D_N(D_N')$.

For a searchlight (collimator),

$$p_0' \approx Df'/c$$

The parameter which may be called the searchlight gain is the ratio of the radiant intensities of searchlight and source in the normal direction

$$k_s = I_s/I_0 = \tau L A_p'/L A_0 = \tau(D/d)^2$$

where D is the diameter of the entrance pupil $(D = D')$, and d the diameter of the source of light. In well fabricated searchlights this gain can attain values as high as 10 000.

The convergence angle $2\omega'$ of a searchlight (see Fig. 12.2) is a function of the dimensions c and b of the luminous body of the source and the spherical aberration of the optical system. From Fig. 12.2, the angle of beam convergence, shown as $2\omega_c' = 2\omega_c$ in the meridional plane coinciding with the page, can be defined as

$$\tan \omega_c' = c/2f' \tag{12.3}$$

In the other meridional plane perpendicular to the page, this angle can be determined from the expression

$$\tan \omega_b' = b/2f' \tag{12.4}$$

Since usually f' by far exceeds both c and b, we may take

$$2\omega_c' \approx c/f' \quad \text{and} \quad 2\omega_b' \approx b/f' \tag{12.5}$$

In the case of a point source, the angle of divergence is due to diffraction

$$2\omega' \approx \lambda/D$$

where λ is the wavelength, O the diameter of the entrance pupil of the searchlight system equal to the diameter D' of the exit pupil (we assume the thin lens system). At a wavelength of $\lambda = 6 \times 10^{-4}$ mm

$$2\omega' \approx 6 \times 10^{-4}/D \text{ [radians]} \approx 120''/D \qquad (12.6)$$

Because the optical system of a searchlight or collimator usually suffers from spherical aberration, the actual angle of divergence of the beam will be somewhat larger than the angle derived with the above equations.

At a distance p' the diameter of illuminated spot on the screen is

$$2y' = D + 2\omega'p'$$

where again the diameter D of the entrance pupil is taken equal to that of the exit pupil D'.

We wish to emphasize an important point concerned with choosing the focal length of the collimator lens system. From Eqs. (12.3)-(12.5) it follows that the greater the focal length f', the smaller the angle of beam divergence defined by the dimensions c and b of the source. System designers determine the focal length by the given tolerable angle of divergence and the known dimensions of the luminous surface with account of the spherical aberration and diffraction.

With reference to Fig. 12.3 the angle of 'interception' $2\sigma_A$ is seen to be the double aperture angle (reflector's angular subtense) in object space. It characterizes the efficiency of utilization of the light radiated from the source.

The searchlight and collimator systems in use are both dioptric and catadioptric. Let us look at mirror systems. More often than not such a system is built around a spherical or paraboloidal reflector made as a first surface mirror. Fig. 12.3 shows a spherical first surface mirror of radius r. Here, D stands for the diameter of the entrance (exit) pupil aperture. If the

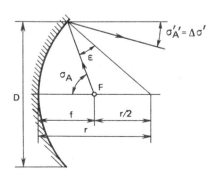

Fig. 12.3. Use of a spherical mirror

point source is situated in the focus of the mirror reflector, it emits a collimated beam in the paraxial domain. As the incidence height of the rays on the reflector increases, the aperture angle $\sigma_A' = \Delta\sigma'$ also increases, that is, the reflected rays will cross the optical axis at a finite distance from the mirror, the distance decreasing for greater aperture angles σ_A. This variation of the exit aperture angle will be naturally recognized as the angular spherical aberration of the mirror which deteriorates the homogeneity of image illumination.

Referring to Fig. 12.3, we have by the law of sines

$$r/2\sin\varepsilon = r/\sin\sigma_A$$

or

$$2\sin\varepsilon = \sin\sigma_A$$

The angular spherical aberration of the mirror reflector is then

$$\Delta\sigma' = \sigma_A - 2\varepsilon$$

This aberration limits the aperture ratio of a spherical mirror reflector.

For a paraboloidal mirror reflector with a point source at the focus, the beam divergence is only controlled by diffraction estimated by Eq. (12.6).

The catadioptric (i. e. combined reflecting and refracting) system of a searchlight may be represented in its simplest form as a system with one refracting surface used twice and one reflecting surface. The *Mangin mirror* is perhaps the simplest of catadioptric systems (Fig. 12.4). It consists of a second surface spherical mirror with the power of the first refracting surface chosen to correct the spherical aberration of the reflecting surface. Notice a rather wide angle of interception for this reflector. For a glass of $n = 1.5$, a correct design for this system dictates $r_1 = f$ and $r_2 = 1.5f$.

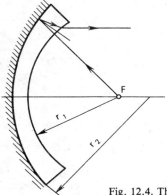

Fig. 12.4. The Mangin mirror

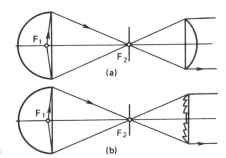

Fig. 12.5. Optical systems of searchlights

Figure 12.5 schematizes the optics of searchlights consisting of ellipsoidal reflectors with point sources situated at their first foci, diaphragms placed at the second focal points of the ellipsoids and (*a*) a lens whose spherical aberration is kept to a minimum or (*b*) a Fresnel lens. The first foci of the lenses coincide with the second foci, F_2, of the ellipsoids.

12.3 Catadioptric Systems

In this section we examine mirror illumination system designed to illuminate objects at a finite distance from the system with the aim to observe these objects or to image them with some optical devices.

These catadioptric systems are devoid of chromatic aberrations, intercept light from the source at embracing angles of 180° and even larger, weigh lesser than dioptric systems of the same relative aperture, and transmit more light than the latter. In many cases these advantages decide the choice for this type of illumination system.

The concave spherical reflector is perhaps a simplest catadioptric illumination system. Its application, however, is limited owing to a large spherical aberration, high losses of light and nonuniform illumination of the screen. Spherical mirrors subtend the angle up to 110° at the point source and produce a magnification of at most five times. Often this mirror is used as a concentric reflector with a light source placed at the centre of mirror curvature as this arrangement provides a more effective utilization of the light flux.

The optics of an ellipsoidal mirror is suggested by Fig. 12.6*a*. The point filament of an incandescent lamp placed at the focus F_1 of the mirror is imaged at the focus F_2 which coincides with the centre, *C*, of the entrance pupil of the following optical system, say, projecting. The object *y*, for example a slide or a negative film, is placed near the mirror at a distance e. The maximum size of the object is obviously D', the diameter of the exit pupil of the reflector.

The convergence angle $2\sigma'_A$. (doubled angular aperture of the mirror) must be equal to or must somewhat exceed the angular field 2ω of the following lens. The distance s from the vertex of the ellipsoid to the focal point F_1 is selected so that to have enough room for the lamp with a socket. The placement of the transparency y with respect to the entrance pupil of the following lens system, given by the distance p in Fig. 12.6a, is selected from the required magnification of the lens system.

With reference to Fig. 12.6a we observe

$$\tan \sigma'_A. = y/2p \tag{12.7}$$

The doubled aperture angle of the mirror $2\sigma'_A$. must exceed the angular field of the lens system.

Given a distance e between the mirror exit section and the transparency to be illuminated, the clear diameter (exit pupil diameter) of the mirror is

$$D' \approx D = 2(e - p) \tan \sigma'_A. \geq 2(e - p) \tan \omega \tag{12.8}$$

Let us determine the semiaxes a and b, the sag q, and the angle of interceptance $2\sigma_A$ of the mirror. The distance F_1F_2 between the foci F_1 and F_2 of the ellipsoid

$$F_1F_2 = 2\sqrt{a^2 - b^2} \tag{12.9}$$

The focus F_1 is a distance

$$s = a - \sqrt{a^2 - b^2} \tag{12.10}$$

apart from the pole of the ellipsoid.

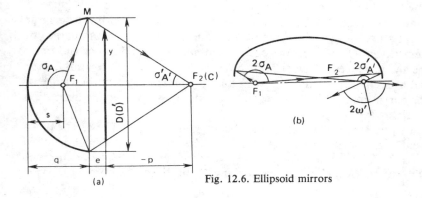

Fig. 12.6. Ellipsoid mirrors

Assume a point M be at the edge of the exit pupil and in a meridional section of the ellipsoid (ellipse), then

$$F_1M + MF_2 = 2a \tag{12.11}$$

Referring to Fig. 12.6a we obtain from Eqs. (12.9)-(12.11), with some trigonometric relationships, the ellipse parameters as follows

$$a = \frac{s(s + e - p)}{2s + e - p - D/2\sin \sigma'_A} \tag{12.12}$$

$$b = \sqrt{2as - s^2} \tag{12.13}$$

$$q = 2a - s - e + p \tag{12.14}$$

The interception angle $2\sigma_A$ is obtained from

$$\tan \sigma_A = -D/2(q - s) \tag{12.15}$$

Example 12.1. Given the angular field of the lens system $2\omega = 55°$, $-p = 108.9$ mm, $y = 113$ mm (diagonal of a 6 × 6 cm slide), $s = 40$ mm, and $e = 50$ mm. Determine the ellipsoidal reflector characteristics.

Substitution into Eqs. (12.7), (12.8), (12.12)-(12.15) yields $2\sigma'_A = 55°10'$, $D = 165$ mm, $a = 132$ mm, $b = 107.8$ mm, $q = 65.7$ mm, and $2\sigma_A = 207°$.

The angle of interception $2\sigma_A$ at which ellipsoid mirrors embrace the light source often exceeds 180° thus providing a high efficiency of light flux utilization.

The ellipsoidal reflector projects the radiant cavity of the light source into the entrance pupil of the lens filling with the image the entire area of the pupil. If this were the pupil of an eye, it would observe a "complete flash" in the projector. Normally the diameter of the entrance pupil of the following lens system should exceed the diameter of the light emitting member of the source. In order to fulfill this condition, the magnification of the mirror system (see Fig. 12.6a) must be

$$\beta = -(q + e - p)/s$$

and simultaneously

$$\beta = -D_1/D_s$$

where D_1 is the diameter of the entrance pupil of the following lens system, and D_s the diameter of the radiant cavity of the light source.

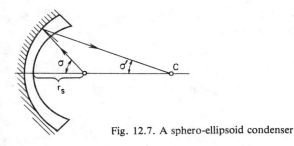

Fig. 12.7. A sphero-ellipsoid condenser

The ellipsoid mirror reflector shown in Fig. 12.6*b* has not only a wide embracing angle $2\sigma_A$, but also a large convergence angle $2\sigma_A'$, which is also an important characteristic of illumination systems (to meet the condition $2\sigma_A' > 2\omega$, where 2ω is the angular field of the following optical system, for instance, a wide-angle projection lens).

The sphero-ellipsoidal reflector presented in Fig. 12.7 is a popular piece of motion picture projection systems. It is made as a second surface ellipsoidal mirror with a spherical refracting surface. The embracing angle of such reflectors can be as high as 140°, and the magnification is −6 to −8.

An obvious advantage of spherical mirrors is due to the simplicity of their manufacture. However, spherical aberration present already in the axial beams often sets a limit to their application.

Ellipsoidal and paraboloidal mirror reflectors are free from aberration for axial bundles of rays, but their aberration for oblique bundles exceeds that of spherical mirrors, to say nothing of a more complicated technology for these types of reflectors.

To free a spherical mirror from spherical aberration the reflected rays are passed through an aspheric corrector plate as, for example, in the Schmidt system shown in Fig. 12.8.

Many setbacks of mirror systems can be eliminated in lens systems even though these systems are liable to exhibit chromatic aberrations. Lens il-

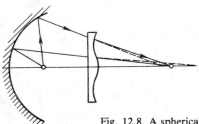

Fig. 12.8. A spherical mirror with a correction plate

lumination systems, called condenser systems, find numerous applications in different types of optical devices.

12.4 Condenser Systems

Condensers are essentially optical devices producing real images of light sources at a finite distance from the lens system. If a lens system images the source of light at infinity then such a system is termed a collimator. The number of lenses in a condenser system is determined by the sum of the angular subtense $2\sigma_A$ and convergence $2\sigma'_A$ angles.

The optical characteristics of a condensing system include the focal length f', lateral magnification β, relative aperture (f-number), angular subtense $2\sigma_A$, and convergence angle $2\sigma'_A$..

The entrance aperture angle σ_A (one half of the angular field) and the relative aperture $F = f'/D$ are related by an expression which can be derived from examining Fig. 12.9. We have

$$\tan \sigma_A = D/2a = D\beta/2a\beta = D\beta/2a'$$

or by observing that, from (3.10), $a' = (1 - \beta)f'$,

$$\tan \sigma_A = D\beta/2(1 - \beta)f'$$
$$= \beta/2(1 - \beta)F$$

Condenser lens. A single lens can be used as a condenser if the sum of the embracing subtense and the convergence angle does not exceed 45°. The shape of the lens depends on the desired lateral magnification. If the light source is situated at a distance from the condenser exceeding its focal length more than 20 times (or the image of the source is to be produced at more than 20 focal lengths from the condenser lens) then the condenser may be a plano-convex lens turned by its spherical surface toward the source (or its image).

If the condenser is to project the filament of the source (or arc) with a 1:1 magnification ($\beta = -1$), the problem can be solved by a double convex lens with faces of equal curvature.

Fig. 12.9. Schematic to deduce a relation between angle σ_A and the f-number

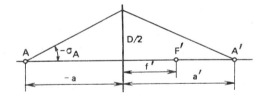

If a single lens condenser is to be used with magnification other than unity, its shape is determined from the condition that the design will be lowest in spherical aberration.

A two-lens design is suitable for situations with the sum of embracing angular subtense and the convergence angle being within 60°.

Because a plano-convex lens has the least spherical aberration at an infinite (rather large) distance from the image, it is apparent that the optimal two-lens design will be the one illustrated in Fig. 12.10. For the lenses *1* and *2* in contact, the power of this lens is $\phi = \phi_1 + \phi_2$ or $2\phi_1$ if the lens components are identical. These condenser designs find their use for $\beta = -1$ (it is allowed also $\beta = -3$). However, if $|\beta|$ is other than unity then the system should be such that $f_2'/f_1' = |\beta|$.

Triple lens condensers enable the sum of the angles of interception and convergence to be increased up to 100°. When still higher angular sums are desired, the condensing system, to solve this problem, involves 4, 5, and 6 lenses. The respective lens system design is then carried out with account of the spherical aberration introduced by these lenses. It should be noted that for condensing doublets with magnification $|\beta| > 3$ or $|\beta| < 1/3$ the lenses are figured from the condition of minimal spherical aberration.

The number of lenses in designs for large angles of interception and convergence can be cut down by incorporating aspheric surfaces. A condensing system of a multi-chambered projector (multiplex) consists of two plano-convex lenses with ellipsoidal surfaces ensuring both the aperture angle and convergence angle equal to 122°.

A very large aperture angle can be achieved with Fresnel lenses (see Section 5.8) at practically no (for a point source) spherical aberration with any desired magnification.

Most condensing systems can be improved by the addition of a spherical reflector behind the source as in searchlights (Fig. 12.5).

Microscope condensers where a large convergence angle is required must be achromatized. The resultant achromat may become rather sophisticated, as the design shown in Fig. 12.11.

Some designs of microscope illumination system incorporate a collector lens with the purpose of imaging the source of light into the aperture stop

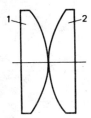

Fig. 12.10. A two-lens condenser

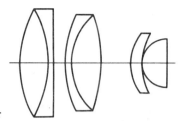

Fig. 12.11. An achromatic condenser

of the following condenser. This arrangement excludes the heating effect of the source on the condenser system and the object being viewed. The requirements imposed on the collector are essentially the same as those on the condenser. The illumination system involving a collector lens and a condenser is essentially a multi-stage arrangement.

MICROSCOPES

A microscope is an optical system which conveys to the eye an enlarged image of a near object. We say that the image is enlarged meaning that it subtends at the eye a greater angle than the object does when viewed with the unaided eye at normal viewing distance. This normal viewing distance is conventionally assumed to be 250 mm, being as will be recalled the average distance at which most people see small features most clearly. The magnification or the magnifying power of a microscope is defined as the ratio of the visual subtense of the image to the angle subtended by the object at the unaided eye at a distance of 250 mm.

13.1 The Simple Microscope

The simple microscope, magnifying glass, or loupe consists of a lens used to view the object located at its first focal point. We shall discuss the simple microscope in terms of its magnifying power Γ, object size (linear field coverage) $2y$, and exit pupil diameter D'.

With the object at the first focal point of the simple microscope the eye receives from each point of the object bundles of parallel rays. In other words, the relaxed eye can view this object without accommodation.

We define the magnification of a simple microscope as the ratio of the tangent of the image subtense when the object is viewed through the lens to the tangent of the object subtense when this is viewed with the unaided eye. With reference to Fig. 13.1a it will be seen that for an object in the first focal plane of the lens, the tangent of the object angular subtense is

$$\tan \omega' = y/f'$$

whereas in viewing with the unaided eye (Fig. 13.1b) at normal viewing distance, the tangent of the object subtense is

$$\tan \omega = y/250 \qquad (13.1)$$

Therefore, the magnifying power of a simple microscope in the absence of accommodation is as follows

$$\Gamma = \tan \omega' /\tan \omega = 250/f' \qquad (13.2)$$

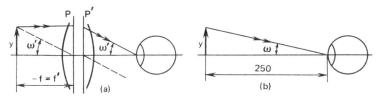

Fig. 13.1. Construction to derive the expression for the magnifying power of a simple microscope in the absence of accommodation

In the general case, the object viewed through the magnifying glass may be situated at some distance z from the first focal plane (for a normal eye, $z \geqslant 0$). The image y' is sighted by the eye accommodated at some distance p', as illustrated in Fig. 13.2. The tangent of the angle subtended by the image is

$$\tan \omega' = y' / - p' \qquad (13.3)$$

From this figure we have account of the formulae of the perfect lens system

$$p' = z' - z_p' \quad \text{and} \quad y' = -yz'/f'$$

Substituting into (13.3) yields

$$\tan \omega' = \frac{y}{f'} \left(1 + \frac{z_p'}{z' - z_p'} \right) \qquad (13.4)$$

Now, in agreement with (13.1) and (13.4) we get

$$\Gamma_{ac} = \frac{250}{f'} \left(1 + \frac{z_p'}{z' - z_p'} \right) \qquad (13.5)$$

At $z' = \infty$ ($z = 0$) we arrive at Eq. (13.2).

From Eq. (13.5) it follows that when the eye is at the second focal point of the magnifying glass ($z_p' = 0$), $\Gamma_{ac} = \Gamma$.

Analysis of ray bundle limitation and of the aperture and field characteristics of a simple microscope should be carried out for the system

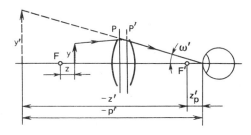

Fig. 13.2. Construction to derive the expression for the magnifying power of a simple microscope for an accommodated eye

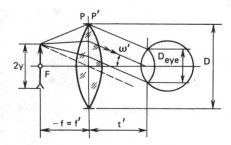

Fig. 13.3. Construction to deduce
the linear coverage of a magnifier

of the microscope and the eye. Fig. 13.3 shows a magnifying glass made as
a single lens of diameter D. The pupil of the eye of diameter D_{eye} is at a
distance t' from the lens. Normally $D > D_{eye}$, therefore, the role of the exit
pupil of the lens-eye system is performed by the pupil of the eye
($D' = D_{eye}$).

In most applications the first focal plane of the magnifying glass has no
field stop and the field of view of the lens is not confined. The rim of the
lens is simultaneously a vignetting diaphragm and the exit window. In the
absence of vignetting the angular field $2\omega'$ of the magnifying glass in image
space is defined by a ray just grazing the edge of the exit window and the
upper edge of the exit pupil (Fig. 13.3), so that

$$\tan \omega' = (D - D_{eye})/2t'$$

and the respective linear field coverage in the object space will be

$$2y = 2f' \tan \omega' = f'(D - D_{eye})/t'$$

This equation suggests that for a given focal length and diameter of a
simple microscope the linear field coverage can be increased by placing the
eye as close to the lens as possible. The oblique ray bundles beyond the circle of diameter $2y$ are vignetted. As can be seen from Fig. 13.4, the angular

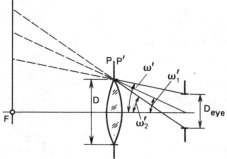

Fig. 13.4. Angular field of a magnifying glass at various vignetting

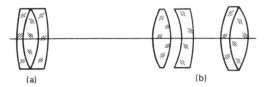

Fig. 13.5. Steinheil magnifier (a), and a four-lens anastigmat (b)

field $2\omega_1'$ corresponding to 50 per cent vignetting can be determined from the formula

$$\tan \omega_1' = D/2t'$$

while that corresponding to a vignetting of 100 per cent from the formula

$$\tan \omega_2' = (D + D_{eye})/2t'$$

With the magnifying power up to $7\times$ the magnifier is made as a single lens. The diameter of the linear field $2y$ within which the lens is capable of producing a satisfactory imagery does not exceed $0.2f'$.

A magnifier may be formed of two lenses almost touching each other. The object may be viewed either through one of the lenses or through both of them. Accordingly, such a magnifier has three magnifying powers namely, Γ_1, Γ_2, and $\Gamma_3 = \Gamma_1 + \Gamma_2$, where Γ_1 and Γ_2 are the magnifying powers of the two component lenses.

The performance of magnifiers can be improved at the expence of system simplicity. More complex designs cope, specifically, with aberrations in the system. The Steinheil aplanatic lens shown in Fig. 13.5a is constituted by a biconvex crown lens and two negative flint menisci. This magnifier can have a magnifying power of $6\times$ to $15\times$ and an angular field of up to $20°$. Four lens anastigmats (Fig. 13.5b) of large magnification ($10\times$ to $40\times$) are the more perfected magnifiers with a high degree of correction for both axial and oblique ray bundles.

13.2 The Compound Microscope

The compound microscope as a magnifying glass is also designed to view small nearby objects. As illustrated in Fig. 13.6, a compound microscope consists of an objective lens and an eyelens. We shall discuss the compound microscope in terms of its magnifying power Γ, linear field coverage in object space $2y$, and exit pupil diameter D'.

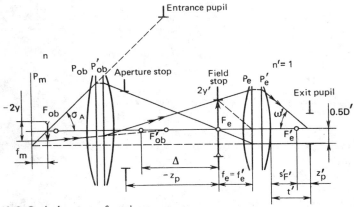

Fig. 13.6. Optical system of a microscope

The objective lens produces a real, enlarged, inverted image. The magnification of the objective lens is described as

$$\beta_o = -\Delta/f_o'$$

where f_o' is the focal length of the objective lens, and Δ known as the tube length is the distance between the second focus of the objective and the first focus of the eyepiece.

The objective lens images the object into the first focal plane of the eyepiece. This image is viewed through the eyepiece which magnifies it still further, thus operating as a magnifying glass of the power

$$\Gamma_e = 250/f_e' \tag{13.6}$$

Hence, the magnifying power of the compound microscope is

$$\Gamma = \beta_o \Gamma_e \tag{13.7}$$

The object being viewed is placed in the first focal plane with respect to the microscope as a whole, and the magnification of the compound microscope can be defined in the same manner as for a magnifying glass, namely, denoting by f_m' the second focal length of the microscope,

$$\Gamma = 250/f_m' \tag{13.8}$$

The field of view of a microscope is limited by a field stop placed in the first focal plane of the eyepiece. The diameter D_{fs} of this stop depends on the angular field $2\omega'$ of the eyepiece confining an image of satisfactory quality. From Fig. 13.6 it follows

$$D_{fs} = 2y' = 2f_e' \tan \omega'$$

or in view of (13.6)

$$D_{fs} = 500 \tan \omega' / \Gamma_e \qquad (13.9)$$

For a given aperture of field stop and magnification of the objective, the object size (linear field coverage) is given as

$$2y = D_{fs}/\beta_o \qquad (13.10)$$

Combining the two last expressions we get

$$2y = 500 \tan \omega' / \beta_e \Gamma_e$$

or taking into account the magnifying power of the microscope defined by (13.7)

$$2y = 500 \tan \omega' / \Gamma$$

As can be seen from this expression, for a given angular field of the eyepiece $2\omega'$, the field of view of the microscope $2y$ is smaller the larger the microscope magnifying power Γ.

The exit pupil of the microscope can be the image of the mount of the last objective lens produced through the eyepiece, which is the aperture stop for the microscope, or this can be the image of the aperture stop proper placed between the objective and its second focal point (see Fig. 13.6). Sometimes the aperture stop is placed at the second focal point of the objective, thus shifting the entrance pupil of the microscope to infinity. This arrangement renders the course of the principal ray in object space telecentric, which is a desirable feature for many measuring microscopes.

Referring to Fig. 13.6,

$$0.5D' = -f_m \tan \sigma_A \qquad (13.11)$$

where σ_A is the aperture angle of the microscope in object space.

If the object being viewed is immersed in a medium of refractive index higher than unity (immersion fluid), then according to Eq. (3.3) the first focal length of the microscope is $f_m = -f_m' n$. Then from (13.11) the exit pupil diameter of the microscope becomes

$$D' = 2nf_m' \tan \sigma_A \qquad (13.12)$$

Because in aberration analysis of the objective lens the designer tends to make it aplanatic, by the sine condition Eq. (13.12) gives way to

$$D' = 2f_m' n \sin \sigma_A \qquad (13.13)$$

Recalling that $n \sin \sigma_A = NA$ is the numerical aperture of the microscope, we obtain from (13.8) and (13.13)

$$D' = 500NA/\Gamma \qquad (13.14)$$

With reference to Fig. 13.6 the distance t' of the exit pupil from the last surface of the eyepiece is

$$t' = s_{F'}' + z_p'$$

where $s_{F'}'$ is the back focal of the eyepiece which depends on the eyepiece design, and z_p' is determined by the Newton formula (3.4) as

$$z_p' = -f_e'^2/z_p$$

In viewing objects with a compound microscope the eye is placed at the exit pupil of the microscope to see the full field of view; then t' is called the *eye relief*. It should be observed that the exit pupil of a microscope is smaller than the pupil of the eye in most cases.

13.3 The Resolution of a Microscope

When the separation of two points is such that it is just possible to determine that there are two points and not one, the points are said to be resolved. The resolution of a microscope is limited by both diffraction and the resolution of the eye. From diffraction theory, the smallest separation between two object points that will allow them to be resolved is given by Sparrow's criterion

$$\delta = \lambda/2NA \qquad\qquad (13.15)$$

where λ is the wavelength of light in which observation is carried out, and NA is the numerical aperture of the microscope. Note that the refractive index n and the slope of the marginal ray σ_A in $NA = n\sin\sigma_A$ are those at the object.

From the above criterion for resolution we see that the resolving power of a microscope can be improved by using shorter wavelengths and greater numerical apertures. The first way is realized by photographing the object in the ultraviolet. To realize the second way we should look at the components of $NA = n\sin\sigma_A$. In modern high-performance microscopes the value of the aperture angle σ_A is practically increased to the limit. Therefore the only possibility to increase the numerical aperture is by increasing the index n. This can be done by immersing the object in a fluid of high n. This can be water of $n \approx 1.33$, bromonaphthalene of $n \approx 1.7$, and the like.

In order that the viewer's eye can completely utilize the resolving power of a microscope, the latter must possess a respective magnifying power. If two points of the first focal plane of an optical system are separated by a distance δ, then the angular spacing between these points in image space is

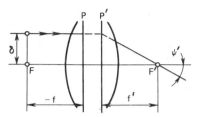

Fig. 13.7. Construction to determine the magnifying power of a microscope

$\psi' = \delta/f'$. The respective geometry is suggested by Fig. 13.7. The eye is able to determine that these are two points and not one if the angular separation between these points is not less than the visual resolution limit ψ_{eye}, that is,

$$\psi' = \delta/f' \geqslant \psi_{eye} \qquad (13.16)$$

Substituting Eqs. (13.16) and (13.15) into (13.8) yields for the necessary magnifying power of the microscope

$$\Gamma \geqslant 500NA\ \psi_{eye}/\lambda \qquad (13.17)$$

This is the smallest magnification at which the eye will use the resolution of the microscope completely, i. e., the eye can resolve all the details of the image. Hence, any magnification beyond this "useful" value is "empty magnification".

In using the formula (13.17) it should be kept in mind that in many cases the diameter of the microscope exit pupil is 1-0.5 mm. This increases the visual resolution from one minute of arc to two or four minutes. For the mean wavelength in the visible range $\lambda = 0.55 \times 10^{-3}$ mm, assuming the visual resolution ψ_{eye} in the range 0.0006 to 0.0012 radians (2-4 minutes of arc), by Eq. (13.17) the limits of the useful magnifying power of the microscope are

$$500NA < \Gamma < 1000NA$$

A magnification under 500 NA is not enough to distinguish all details of the object imaged as separate features by the objective of a given numerical aperture. Magnification in excess of 1000 NA is undesirable because features smaller than those revealed at the "useful" magnification cannot be discerned whatever the magnifying power above this level.

13.4. Depth of Field

An object viewed through a microscope is placed at the first focal point. A sufficiently sharp (for the eye) image, however, will be produced also for close object points lying ahead and behind this plane. This portion of object space measured along the optical axis is called the *depth of field*. The depth of field for a microscope may be considered as consisting of three contributions, namely, accommodation, system geometry, and diffraction.

Depth of field due to accommodation. In viewing an object the eye is in turn accommodated at points of various depth. Owing to the subjective perception the result of this process is an impression that the entire depth of space is seen simultaneously in focus. The process of perception of the depth of object space through an optical system is very much the same.

If the eye can accommodate in the range from 250 mm to infinity, then it will see through a microscope a sharp image in case the object plane is located within a positive distance z from the first focal plane. For the exit pupil of the microscope situated near the second focal point, the distance z defining the accommodation depth of field is derived by the Newton formula (3.4) as follows

$$T_{ac} = z = f_m''^2/250$$

or by virtue of (13.8)

$$T_{ac} = 250/\Gamma^2 \qquad\qquad (13.18)$$

Depth of field due to system geometry. The object set at the first focal point of a microscope sends to the eye parallel bundles of ray from every object point. In this case a sharply defined image occurs on the retina without accommodation. From the points of the object planes A_1 and A_2 shifted with respect to the first focal point as shown in Fig. 13.8, the eye will receive divergent and convergent ray bundles, respectively, and the retinal image of a point will be a blur spot. If the diameter of the blur spot does not exceed the ultimate value consistent with the visual resolution, then such blur spots will be perceived as sharply defined.

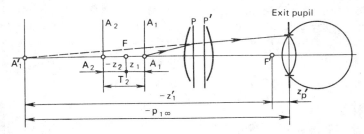

Fig. 13.8. Derivation of the depth of field for a microscope

Let a point A_1 lie a distance z_1 from the first focal point. Then by Newtonian relation its image A_1' will be produced by the microscope at a distance of

$$z_1' = -f_m'^2/z_1 \qquad (13.19)$$

If the image point A_1' lies farther from the eye than a distance $p_{1\infty}$, set to be the close end of infinity, then this image will seem to be sharp. If the distance from the second focal point of the microscope and the eye is denoted by z_p' (see Fig. 13.8), then

$$p_{1\infty} = z' - z_p'$$

Observing that usually $|z_1'| \gg z_p'$, we get $p_{1\infty} = z_1'$ and from (13.19)

$$z_1 = -f_m'^2/p_{1\infty} \qquad (13.20)$$

The distance $p_{1\infty}$ depends on the visual resolution ψ_{eye} and can be determined either as

$$p_{1\infty} = -D_{eye}/\psi_{eye}$$

if the eye pupilar diameter is smaller than the exit pupil size of the microscope, or as

$$p_{1\infty} = -D'/\psi_{eye} \qquad (13.21)$$

if the exit pupil diameter of the microscope is smaller than the eye pupil diameter.

Assuming that the exit pupil diameter of the microscope is smaller than the eye pupil diameter we have from (13.20) with account of (13.8), (13.14), and (13.21)

$$z_1 = 125 \, \psi_{eye}/\Gamma NA$$

A similar dependence with a minus sign results for the outside shift z_2 from the point of focus. The geometric depth of field $T_g = z_1 - z_2$ is therefore

$$T_g = 250 \, \psi_{eye}/\Gamma NA \qquad (13.22)$$

Diffraction effects of apertures. The presence of diffraction phenomena in the microscope increases the depth of field by another amount

$$T_d = n\lambda/2(NA)^2$$

where n is the refractive index of the immersion liquid.

Thus, the overall depth of field for a microscope is the sum of the three contributions defined by (13.18), (13.22), and (13.23)

$$\begin{aligned} T &= T_{ac} + T_g + T_d \\ &= 250/\Gamma^2 + 250\psi_{eye}/\Gamma NA + n\lambda/2(NA)^2 \end{aligned} \qquad (13.24)$$

This expression indicates that the contribution to the depth of field due to accommodation depends only on the magnifying power of the microscope, the 'geometric' contribution depends on the magnifying power and the numerical aperture, and the diffractional component depends only on the numerical aperture. We should note here that in microscopes using an eyepiece with a reticle $T_{ac} = 0$ because the eye is accommodated on the image of the reticle.

13.5 Objectives and Eyepieces

The wide variety of problems solved by optical microscopy means a large family of microscopes covering a large range of characteristics. This variability is achieved by various combinations of objectives and eyepieces.

The existent designs of microscope objectives may be broadly classified as follows:

— by the amount of correction of residual aberrations (achromats, apochromats, planachromats, etc.);

— as immersion and immersionless systems;

— as dioptric, reflection, or catadioptric systems; and

— by the length of tube of the microscope.

The tube of a microscope is a cylinder tapped at both ends to accept an objective screwed at end *3* and an eyepiece screwed at end *1* (Fig. 13.9). Most microscopes using transmitted light illumination of the object have a 160-mm long tube, whereas the microscopes for reflected-light work have 190-mm tubes.

In microscopes of 160-mm tube length, the object plane *4* is 33 mm from the tube, and the distance from the image plane of the objective, *2*, to the upper end of the tube is 13 mm. Accordingly, the object plane is at a standard distance of 180 mm from the objective image plane.

Microscopes usually receive from the manufacturer a set of replaceable objectives and eyelenses. The objectives are selected such that the image they produce can be adjusted into the aperture of any eyepiece from the set. Microscope objectives are normally specified by power and numerical aperture. Modern objectives are characterized by magnifying powers from $1\times$ to $120\times$ and numerical apertures from 0.01 to 1.4.

The optical design of microscope objective grows in complexity for higher apertures, magnifications and stronger control of residual aberrations. Achromatic lenses of power $5\times$ to $10\times$ and numerical aperture up to 0.2 consist of two cemented double-lens components. When it is required to increase the numerical aperture up to 0.3 this system includes a third component — a front plano-convex lens. An immersion achromatic

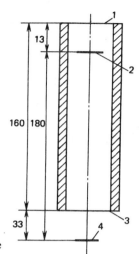

Fig. 13.9. Drawtube of a microscope

objective of power 90 × and numerical aperture 1.25 (specified as a 90 ×
NA 1.25), for example, consists of four groups, a front plano-convex lens,
a positive meniscus, and two cemented doublets. Apochromatic objectives
utilize fluorite (CaF_2) crowns to reduce or eliminate secondary spectrum. A
salient feature of objectives with corrected curvature of field
(planachromats and planapocchromats) is the use of a negative component
or a meniscus of considerable thickness.

By way of example, Fig. 13.10 presents a schematic and specification
for an achromatic 10× *NA* 0.25.

The resolution criterion (13.15) indicates that the resolving power of a
microscope can be improved by using light of shorter wavelengths.
However, optical glass absorbs strongly in the ultraviolet and becomes
practically of no use for wavelengths shorter than 350 nm. Instead, objectives can be made of quartz. Such designs must of course be exploited at a
certain wavelength, and being monochromats need not be achromatized. A
tight control of the spherical aberration is achieved by using aplanatic
menisci and lenses designed for minimal spherical aberration.
Monochromats have magnifying powers up to 100× and numerical apertures up to 1.30 with glycerine immersion, thus enabling the resolution of
details 0.1 μm across at a wavelength of 0.276 μm.

Objectives for use in the ultraviolet or infrared spectral regions (high-
temperature metallographic investigations, say) are frequently made in
reflecting form. An attractive feature of these objectives is that they can be
used over a wide spectral range, from the ultraviolet to the infrared,

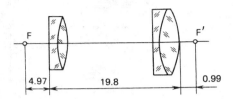

Fig. 13.10. Objective of a microscope

Magnifying power	10×
Numerical aperture	0.25
Focal length	15.3 mm
Object to image distance	180.3 mm
Working distance (front vertex to object)	− 6.5 mm
Rear vertex to image distance	154.0 mm

$r_1 = \infty$ $\qquad\qquad\qquad$ $n_1 = 1$
$r_2 = 6.40$ $\qquad d_1 = 1.0$ $n_2 = 1.6242$
$r_3 = -6.40$ $\qquad d_2 = 2.5$ $n_3 = 1.5100$
$r_4 = \infty$ $\qquad d_3 = 12.8$ $n_4 = 1$
$r_5 = 12.3$ $\qquad d_4 = 1.0$ $n_5 = 1.6242$
$r_6 = -12.13$ $\qquad d_5 = 2.5$ $n_6 = 1.5100$
$\qquad\qquad\qquad\qquad\qquad\qquad\qquad$ $n_7 = 1$

avoiding the necessity for refocusing the microscope. Catadioptric objectives can possess a magnifying power of up to 125 × and a numerical aperture of up to 1.1 (glycerine immersion).

Figure 13.11 shows the optical scheme of a Maksutov objective (60× *NA* 0.85). The object is placed at the centre of curvature of surface *1*. The parameters of surfaces *2* and *3*, which are close to concentric, are designed so that on reflection from them the rays traverse surface *4* without refraction. This objective is practically free from chromatic aberrations and can be operated without refocusing for observation and photography in the range from 200 nm to 600 nm.

In addition to the aforementioned objectives there exist objectives for interference and polarization microscopes, epidiascopes operating in

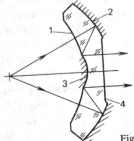

Fig. 13.11. The Maksutov reflecting objective

reflected light and a number of other designs for which we refer the reader reading in Russian to [19].

Eyepieces. The family of eyepieces used in microscopes includes the Huygenian, Kellner, compensating, symmetrical, orthoscopic, and negative eyepieces. The magnifying power of these eyepieces ranges from 4× to 30×, the angular field from 40° to 70° which corresponds to the linear field coverage 24 mm to 16 mm.

The Huygenian eyepiece consists of two plano-convex elements, an *eyelens* and a *field lens* with the convex surface of each toward the objective (Fig. 13.12). The first focal plane of this eyepiece is between the lenses. The real image of an object point produced by the objective at the first focal point F of the eyepiece is the imaginary object point A_1 for the collecting field lens *1*. This lens forms a real image A_1' at the first focal point F_2 of the eyelens *2* which images A_1' to a point at infinity. The field stop of the eyelens is placed at the first focal point of the eyelens. The Huygenian eyepiece has a magnifying power in the range from 4× to 15× at an angular field coverage of 30° to 40°.

The orthoscopic eyepiece consists of a single element eyelens (usually plano-convex) and a cemented triplet. This type of eyepiece is used in conjunction with achromatic objectives of moderate numerical apertures at appreciable magnifications of the eyepiece and an angular field coverage of up to 50°. It is well corrected for lateral colour, astigmatism and distortion.

The compensating eyepiece is used in conjunction with apochromatic objectives, planobjectives, and apochromatic objectives of large magnifying power. It compensates the lateral colour of the objective. The optics of this eyepiece is essentially a complicated Huygenian design or similar to the orthoscopic form.

Negative optical systems are inserted in microscopes instead of eyepieces to project the enlarged image onto the photographic film. The aberrations of these systems are designed such as to offset the field curvature and lateral colour of the objective.

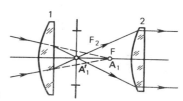

Fig. 13.12. The Huygenian eyepiece

13.6 Illumination Systems

Most objects investigated with the use of a microscope are not self-luminous, therefore illumination of the object from an external source of light is necessary. The illumination system is expected to provide contrast and uniformly illuminated images. This arrangement implies that the estimation of resolving power of the microscope must modify the criterion (13.15) so that it takes into account the numerical aperture of the condenser system. This may be done as follows

$$\delta = \frac{\lambda}{NA_o + NA_c}$$

where NA_o is the numerical aperture of the objective, and NA_c the numerical aperture of the condenser.

In order to analyze transparent and opaque objects, the microscope is supplied with illumination systems for transmitted and reflected light work.

Two illumination techniques are in wide use, for bright field and dark field work. In the method of bright field, the rays from the illumination system arrive at the objective after having passed through a transparent ob-

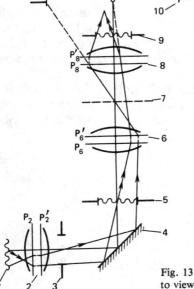

Fig. 13.13. The Köhler illumination system to view by transmitted light

ject or having reflected from an opaque object. The background field on which the object is observed will be bright.

The dark field illumination technique uses the rays scattered from the object in a diffuse manner. When the object is absent, no ray from the illumination system reaches the objective.

The Köhler illumination system shown diagrammatically in Fig. 13.13 is the most popular in optical microscopy. The source of light *1* is projected by the collector *2* into the plane of the iris diaphragm *5* of the condenser *6*. This condenser projects the diaphragm *5* into the aperture of the entrance pupil *10* of the objective *8*. After the objective, the image of the source is formed in the plane of the aperture stop *9* of the microscope. In the immediate vicinity of the collector lens there is an iris field stop *3* which is projected by the condenser into the object plane *7* of the microscope. The plane mirror *4* changes the direction of the optical axis.

Altering the size of the stop *3* changes the size of the illuminated area in the object plane *7*, the numerical aperture of the condenser remaining the same. If we change the diameter of the diaphragm *5*, then it will change the numerical aperture only. These properties of the illumination system make it a useful component for microscopes of various numerical apertures.

Figure 13.14 shows the method of illumination of a nontransparent object with the Köhler system. This illumination system is termed an opaque illuminator. The collector lens *2* projects the source of light *1* into the plane of the iris aperture stop *3*. The condensers *4* and *6* project this stop into the plane of the aperture stop *8* of the objective *9*. The clear aperture of the collector lens *2* is projected by the condenser *4* into the plane of the iris field stop *5* and then by the condenser *6* and objective *9* into the object plane of the microscope. On reflection from the opaque object *10* the rays pass through the objective *9*, and a semitransparent plate *7* and strike the

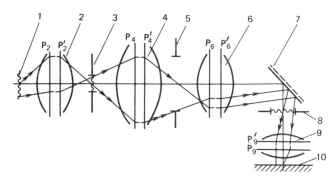

Fig. 13.14. The Köhler illumination system to view by reflected light

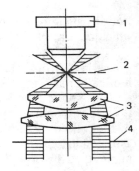

Fig. 13.15. Condenser for dark-field work

eyepiece. In place of the semitransparent plate *7* a 90° reflecting prism filling one half of the objective aperture can be used.

To illuminate objects by the dark-field method, the numerical aperture of the condenser must exceed that of the objective. Illumination can be from one side only or from all sides. Fig. 13.15 shows the optics of a dark-field condenser. It incorporates an annular diaphragm *4* whose central disc stops the light beam corresponding to the aperture of the objective. If the object plane *2* has no object, the observer sees in the eyepiece a dark field as the rays emanated from the condenser *3* fail to reach the objective *1*. When the object is in the object plane, its details diffuse the light and appear bright on the dark background.

14

TELESCOPES

A large group of optical instruments designed with a purpose of viewing distant objects includes binoculars, terrestrial telescopes, astronomical telescopes, periscopes, rangefinders, rifle scopes, gun sights, and some other military and geodetic surveying instruments. The optics of these systems is built around a telescopic scheme. Indeed, the primary function of a telescope is to enlarge the apparent size of a distant object. This is accomplished by presenting to the eye an image which subtends a larger angle (at the eye) than does the object. The magnification, or magnifying power, of a telescope is simply the ratio of the angle subtended by the image to the angle subtended by the object. For large angles, the angular subtenses of this definition are replaced with their tangents. It is assumed that a telescope works with both its object and image located at infinity: a bundle of parallel rays entering the entrance pupil of a telescope leaves the exit pupil also as a parallel bundle. Therefore the telescope is referred to as an *afocal* instrument because it has no focal length. In the following material, a number of basic relationships for telescopes will be presented, all based on systems with both object and image located at infinity.

14.1 The Fundamentals of the Telescope

The ray bundles incident on a telescope are thought to be parallel since the entrance pupil of a telescope is very much smaller than the distances to the object being viewed. Thus, the rays from points on the optical axis are parallel to the axis of the system, and the rays in extra-axial (oblique) bundles have the same, within the bundle, inclination to the optical axis, ω. The farther from the axis an extra-axial object point is located, the larger the slope of the ray bundle emanating from this point. In order that the viewer may see the image formed by a telescope without an undue fatigue on the eye, i. e. with the eye in a relaxed state, the ray bundles emerging from the telescope must be parallel too. The bundles from off-axial points make an angle ω' with the axis behind the telescope.

Optical telescopes consist essentially of two lens systems each of which can be an optical surface, as shown in Fig. 14.1 (see also Section 8.1), or a combination of optical components, as illustrated in Fig. 14.2. The compo-

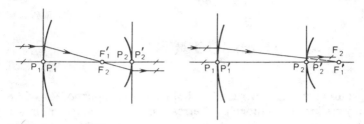

Fig. 14.1. The optics of the simplest telescope

nent nearer the object is called the *objective*, and the other, nearer the eye, the *eyepiece* or *ocular*.

The objective and eyepiece are mounted so that the second focal point F_1' of the objective coincides with the first focal point F_2 of the eyepiece. The eyepiece thus delivers parallel rays out of the system and the power (bending of rays in the instrument) of such an instrument is zero. Note that this arrangement of lens systems follows from Eq. (3.27) for a two-lens system of zero power in a uniform medium

$$\phi = \phi_1 + \phi_2 - \phi_1\phi_2 d = 0$$

which is satisfied if the distance $L = d$ between the principal planes of the objective and eyepiece equals the sum of their focal lengths,

$$d = f_1' + f_2'$$

The objective of a telescope forms a real inverted image in its second focal plane and therefore is a positive component, whereas the eyepiece is used as a magnifier to examine this image. Fig. 14.3 shows that the eyepiece of a telescope may be a positive or negative lens. The instrument constituted by positive objective and eyepiece is known as the *astronomical* or *Kepler telescope*, whereas the system of a positive objective and a negative eyepiece is known as the *Galilean telescope*.

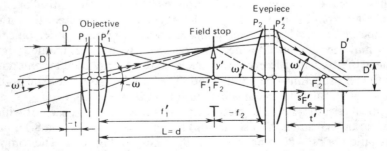

Fig. 14.2. Ray diagram of a telescope

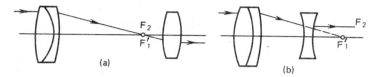

Fig. 14.3. The two basic types of refracting telescopes: (*a*) astronomical or Kepler, (*b*) Galilean

The principal optical characteristics of a telescope include the magnifying power Γ_t (denoted by M. P. in other texts), angular field of view 2ω, exit pupil diameter D', least angular separation for visual resolution ψ, barrel length of the system along the axis L, and the positions of the entrance and exit pupils, t and t' respectively.

The magnifying power of a telescope, Γ_t, equals its angular magnification γ_t, and for large angles

$$\Gamma_t = \tan \omega' / \tan \omega = \gamma_t \qquad (14.1)$$

With reference to Fig. 14.2 we see that

$$\Gamma_t = -f_1' / f_2' \qquad (14.2)$$

and

$$\Gamma_t = D/D' \qquad (14.3)$$

The sign convention here is that a positive magnification indicates an erect image. Thus, if objective and eyelens both have positive focal lengths, the magnification is negative and the telescope is inverting. The Galilean scope with objective and eyelens of opposite sign produces a positive magnification and an erect image.

The telescope images an erect object y of size $D/2$ into a length segment $D'/2$ perpendicular to the axis. Accordingly, the transverse or linear magnification is $\beta = n/n'\Gamma_t$, while the longitudinal magnification will be

$$\alpha = n/n'\Gamma_t^2 \qquad (14.4)$$

whence it follows that the magnifications are constant over the field of the system, and that the instrument produces a distorted perspective. Indeed the apparent size of the object appears enlarged Γ_t times because it is presented to the eye at an angle ω' which is Γ_t times the angle ω, whereas the image space is compressed, as it were, along the optical axis inversely as the square of the magnifying power. This statement is mathematically expressed by Eq. (14.4). Accordingly all object being viewed appear closer to the viewer and the image space seems to be compressed along the line of sight.

15*

The angular field of view of a telescope, 2ω, depends on the angular field $2\omega'$ of the eyepiece and the magnifying power

$$\tan \omega = \tan \omega' / \Gamma_t$$

The angular field of eyepiece varies within a comparatively narrow range of 50 or 70° (specific designs are known with $2\omega = 100°$); the magnifying power of most telescopes does not exceed 10 × to 30 ×, therefore the field of view of telescopes (angular field of the objective) never exceeds 10°. It is limited by the field stop of diameter D_{fs} placed in the plane of the internal virtual object image,

$$\tan \omega = -D_{fs}/2f_1' \tag{14.5}$$

where $D_{fs} = 2f_2' \tan \omega'$.

The diameter of the exit pupil controls the amount of light energy delivered out by the instrument, i. e., serves as the key parameter in estimating the transmitting power of the telescope, $H = E'/L$. Recalling Eq. (7.69) we have for the transmitting power of a telescope whose exit pupil diameter D' does not exceed the pupil size of the eye, D_{eye},

$$H = gD'^2 \quad \text{or} \quad H = g(D/\Gamma_t)^2$$

where $g = (n'/n)^2 \tau \pi / 4 f_{eye}'^2$.

If the pupil size of the eye is less than the exit pupil diameter of the telescope, then

$$H = gD_{eye}^2$$

In this case the retinal image formed by the telescope has a luminance which differs from the luminance of the image seen with the unaided eye by the losses of light in the instrument. If the diameter of the eye pupil exceeds that of the exit pupil the subjective luminance produced by the instrument on the retina will be less than that in the unaided eye. Therefore, the observer's eye is placed at the exit pupil of the telescope. The optimal case, of course, is when these two pupils are of the same size.

Nominally the exit pupil of a telescope is the image of the entrance pupil. The exit pupil data specified in the design of telescopes include its diameter and *eye relief*, the distance from the last surface, denoted by t' in Fig. 14.2. The entrance pupil is often the rim of the objective which takes the part of aperture stop. Recall that the eye pupil varies in size, increasing at dark. Therefore, to match this variation with the exit pupil size, telescope systems for day-time are designed with exit pupil diameters of 2 to 5 mm, whereas those for work at low level of illumination, as night glasses, with a diameter of 5 to 7 mm.

14.2 Resolving Power and Useful Magnification

The resolving power is a measure of the ability of a telescope to produce detectably separate images of object details that are close together. It is defined for image space and measured as the angular separation ψ of two point sources that are just detectably separated by the instrument.

The resolving power of a telescope is determined by the resolution of the objective. When the aberrations of the objective are kept rather low the resolving power of the telescope will be controlled by diffraction, and determined by Eq. (11.1). For objectives (diameter D) of astronomical and surveying systems, the resolution can be estimated with the expression

$$\psi = 120''/D \tag{14.6}$$

The *Rayleigh criterion* for resolution based on diffraction states that two point sources are resolved by a telescope objective of diameter D provided their angular separation is not less than 1.22 λ/D, λ being the wavelength. For instruments working in the infrared at a wavelength $\lambda = 1.1$ μm, the least angular separation becomes $\psi = 280''/D$.

In general, the angular resolution limit of the objective of a telescope depends on the size of the entrance pupil. For a reflecting telescope, or reflector for short, which is an optical telescope with a large-aperture concave mirror for gathering and focusing light from astronomical bodies, the resolving power is controlled by the diameter of the mirror. To illustrate, a 6-m diameter reflecting telescope has a theoretical resolution in the visible range equal to 0.02 seconds of arc.

The resolving power of a telescopic system used with the eye will be limited by the visual resolution of the eye ψ_{eye} which, as will be recalled, is one minute of arc (60''').

If two objects to be resolved are separated by an angle ψ then after magnification by a telescope their images will be separated by

$$\psi' = \psi\Gamma_t$$

If $\psi\Gamma_t$ exceeds one minute of arc, the eye will be able to separate the two images; if ψ' is less than one minute, the two objects will not be seen as separate and distinct. In order that the eye may completely utilize the resolving power of a telescope, its magnifying power must not be less than the useful magnification

$$\Gamma_u = 60''/\psi \tag{14.7}$$

or in view of (14.6)

$$\Gamma_u \approx 0.5D \tag{14.8}$$

Hence, for a constant diameter D of the entrance pupil, the resolving power of a telescope will not improve if we would increase the magnification of the instrument beyond Γ_u.

Equation (14.8) is not universal in the sense that it is derived for an average eye. Some individuals with acute vision can resolve objects with angular separations as low as 30 seconds of arc. On the other hand, in surveying instruments exit pupils of 1.0 to 1.5 mm are common, since size and weight are at a premium and resolution is the most desired characteristic; for critical work a value considerably larger than indicated in Eq. (14.8) is often selected in order to minimize the visual fatigue of the viewer. Therefore, it is not unusual to utilize magnifications within the following limits

$$0.2D \leqslant \Gamma_u \leqslant 0.75D$$

14.3 Objective Systems and Eyepieces

The principal characteristics of telescopic objectives are the focal length f', aperture ratio D/f' (or relative aperture, F-number, f'/D), and angular field of view 2ω in object space, called the *real field* of a telescope to distinct from the *apparent field* which is the angular field in image (i. e. eye) space.

As we noted in Section 14.1, the real field of most telescope systems, especially systems for visual work, is confined to within values of 6 to 10°. These angular fields avoid the necessity for correcting the system for the aberrations of narrow oblique ray pencils (astigmatism, curvature of field, and distortion). A rather high quality of image can be achieved by the control of chromatic aberration, spherical aberration and coma, therefore fairly simple objective designs can solve the problem.

The most popular design of the telescope objective is a two-lens system (doublet) of a positive and negative lens, either cemented or air separated.

The cemented doublets come out in two modifications as shown in Fig. 14.4, namely (*a*) a crown-in-front combination with a positive crown lens *1* facing the object, and (*b*) a flint-in-front combination with a negative flint lens *2* facing the object. It should be noted that crown glasses are less sensitive to atmospheric and mechanical impacts. Cemented doublets of real field 1° to 2° and relative aperture $f/10$ provide perfect imagery and are utilized in astronomical instruments and devices for testing other optical systems.

Experience collected with cemented doublets indicates that they ensure good performance of telescopes (longitudinal spherical aberration is kept below 0.1% to 0.2% focal length) at relative apertures not higher than $f/4$

Fig. 14.4. Cemented doublets (*a*) crown-in-front, (*b*) flint-in-front

(a)   (b)

and real fields $2\omega \leqslant 6°$, the focal length f' (in mm) at specific F-numbers not exceeding the shown values.

$$f/4 \quad f/5 \quad f/6 \quad f/8 \quad f/12$$
$$150 \quad 300 \quad 500 \quad 1000 \quad 1000$$

If the aberrations of the eyepiece or other components of a telescope offset, at least partially, the aberrations of the objective, then the relative aperture of the objective may be increased to a value of $f/2$ and the real field may be increased up to 8° or 11° for a crown-in-front design and up to 15° for a flint-in-front design.

The secondary spectrum of achromatic doublets amounts to about $f'/2000$, and at large magnifying powers (above 10 ×) in long-focus objectives this aberration can markedly deteriorate the quality of image. This situation can be coped with by using apochromatic objectives.

The entrance pupil of objective is often made to coincide with the rim or situated ahead of the objective at a distance up to $0.7f'$. Because the eye cannot notice a 50% fall in illumination, the width of oblique bundles in telescopes for visual work may be diminished to $2m_1 \approx 0.5D$, so that the following components may be also of smaller size.

Objective lenses larger than 60 to 70 mm are used noncemented. An airspaced doublet shown in Fig. 14.5 is a somewhat more versatile form in the sense of improving the image quality and attaining a specified focal length. It should be kept in mind, however, that airspaced objectives exhibit higher reflection losses than their cemented counterparts, apt for glare, and these forms are more critical in mounting and centring.

A three-lens design consisting of an air separated lens and a cemented doublet (Fig. 14.6) is a popular design for surveying instruments. Its secondary spectrum is kept rather small so that the relative aperture may be increased to $f/2$.

When a terrestrial telescope has a low magnifying power, its angular field of view may be enlarged appreciably and the objective will be a wide-angle design. Eyepieces of large focal length are occasionally used in place of such an objective. The entrance pupil of such objectives is always ahead of the lens system.

Fig. 14.5. An air spaced doublet

A further increase in the real field and aperture ratio may be achieved by way of adding lens components, introducing nonspherical surfaces, and the use of new types of glass.

Catadioptric and reflecting objectives are attractive propositions because of freedom from chromatic aberrations at remarkably low size and weight.

Eyepieces. The key characteristics of an eyepiece are its focal length f' which defines the magnification of this component $\Gamma_e = 250/f'$, apparent field $2\omega'$ (angular field in image space), and diameter D' of the exit pupil.

The position of the exit pupil $t' = s'_{F'} + z'_p$ (see Fig. 13.6), called the eye relief, depends on the back focal length $s'_{F'}$, while the position of the field stop depends on the front focal length s_F which determines whether or not a reticle can be inserted and how far the eyepiece can be shifted in focusing.

The apparent field $2\omega'$ of eyepiece is Γ'_t times the angular field of the objective, therefore in correcting the aberrations of eyepieces the major emphasis is placed on field aberrations. It is common to evaluate the aberrations of an eyepiece by tracing rays in reverse direction from infinity, i. e. from the eye side, to analyze their focusing in the first focal plane. The focal length of eyepieces is normally a multiple of 5 and ranges from 5 to 80 mm. For $f' < 5$ mm the eye relief t' defining the position of the exit pupil and the viewer's eye, will be too small to make the two pupils to coincide. On the other hand, a long f' entails large lenses in the eyepiece, which means larger dimensions of the instrument. The popular designs of eyepiece have their focal length in the range of 20 to 30 mm.

The simplest eyepiece may be a single lens, but more often than not the eyepiece system consists of at least two lenses, an *eyelens* nearer the eye and a *field lens* facing the objective. The function of a field lens is discussed at

Fig. 14.6. A triplet

length in Section 14.5. Thus far we confine ourselves to note only that if this lens is placed exactly at the internal image of the telescope objective, it has no effect on the power of the telescope, but it bends the ray bundles (which would otherwise miss the eyelens) back toward the axis. In this way the field of view may be increased without having to increase the diameter of the eyelens. Usually field lenses are rarely located exactly at the internal image plane, but rather they are situated either ahead of or behind the image so that to make imperfections of the lens, say scratches and air bubbles, invisible and to give room for inserting a reticle.

In principle, microscopes and telescopes may utilize identical eyepieces. Fig. 14.7 presents the most popular formulations.

The Ramsden eyepiece (Fig. 14.7a) consists of two plano-convex elements of equal focal length, the curved surfaces facing one another. The apparent field $2\omega'$ of this eyepiece ranges within 30° to 40°. The front and back focal lengths are $-s_F = s'_{F'} \approx 0{,}3f'$, and the spacing of the lenses (overall length of the eyepiece) is $\Sigma d \approx f'$. The quality of image is not very high as the simple design of the eyepiece cannot be freed completely from chromatic and monochromatic aberrations. For the most part this design is used in instruments with small exit pupils, such as surveying and astronomical instruments.

The Kellner eyepiece is essentially a Ramsden eyepiece with an achromatized eyelens to reduce the lateral colour. The design shown in

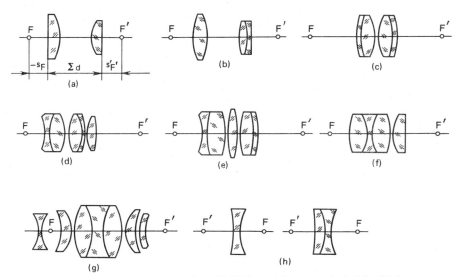

Fig. 14.7. Eyepiece designs: (a) Ramsden, (b) Kellner, (c) symmetrical, (d) with long eye relief, (e) Erfle, (f) orthoscopic, (g) wide-angle, (h) negative

Fig. 14.7b consists of a single field lens and a two-component eyelens. The apparent field is $2\omega' = 40°$ to $50°$; $s_F = -0.3f'$, $s'_{F'} \approx 0.4f'$. The overall length of this eyepiece is $\Sigma d \approx 1.25f'$. This system demonstrates a better performance than the Ramsden eyepiece and is frequently used in low-cost binoculars.

The symmetrical, or Plössl, eyepiece (Fig. 14.7c) is composed of two achromatic doublets, usually identical, with their crown elements facing each other. This formulation performs well in controlling the aberrations at short air spacing between the components. The apparent field is $2\omega' = 40°$ to $50°$. The front and back focal lengths are about equal to the overall length of the system, $-s_F \approx s'_{F'} \approx \Sigma d \approx 0.75f'$, thus allowing for a larger eye relief than in the Kellner eyepiece.

The eyepiece of long eye relief (Fig. 14.7d) has the eye relief $t' \approx f'$ owing to the front negative component. The apparent field is $2\omega' = 50°$; $-s_F = 0.3f'$, $s'_{F'} \approx f'$; $\Sigma d \approx 1.4f'$. The system is well corrected for aberrations.

The Erfle eyepiece (Fig. 14.7e) is probably the most widely used wide-angle eyepiece. It consists of five lenses and has the apparent field $2\omega' \approx 65°$ to $70°$. The overall length $\Sigma d \approx 1.6f'$; $-s_F \approx 0.35f'$, and $s'_{F'} \approx 0.7f'$.

The orthoscopic eyepiece (Fig. 14.7f) consists of a single element eyelens (usually plano-convex) and a cemented triplet (usually symmetrical). Distortion correction is quite good (4%, in eyepieces of other types it can be as high as 10%), therefore this eyepiece is used in telescopes with a reticle over the entire field. The apparent field is $2\omega' = 40°$, the overall length $\Sigma d \approx 0.75f'$; $-s_F \approx 0.6f'$, and $s'_{F'} \approx 0.75f'$ so that the eye relief is rather long. It should be noted that the cemented triplet of this formulation is hard to manufacture.

The wide-angle eyepiece shown in Fig. 14.7g has the apparent field $2\omega'$ as wide as $90°$ and consists of seven elements. An additional negative field lens placed ahead of the field stop aids in field curvature control. The overall length of the design is $\Sigma d \approx 2.5f'$; $-s_F \approx 0.4f'$, and $s'_{F'} \approx 0.45f'$.

The negative eyepiece presented in Fig. 14.7h consists of one element or two elements. This eyelens finds its use in Galilean terrestrial telescopes. Its apparent field $2\omega'$ never exceeds $20°$. The eye relief depends on the position of the observer's eye.

14.4 Focusing the Eyepiece

Since most telescopes are visual instruments, they must be designed to be compatible with the various focusing characteristics of the human eye.

This means that an adjustment is necessary to produce at the exit pupil of the system (a) parallel ray bundles for a normal-sighted eye, (b) diverging bundles for a short-sighted eye, and (c) converging ray bundles for a long-sighted eye.

The system can be adjusted to fit any type of eye by moving the eyepiece along the optical axis. Moving the eyepiece toward the objective will produce diverging ray bundles, and moving it back will render the bundles converging.

With reference to Fig. 14.8 examine a telescope, consisting of a positive focusing objective and a positive eyepiece, which is hit by a ray bundle from an infinitely distant axial point A. Suppose that the viewer is a short-sighted person, and the eyepiece is adjusted for a normal-sighted eye as shown in the diagram at (a), then the parallel bundle of rays emanating from the instrument will reunite at the focal point F'_{eye} which is ahead of the retina. The image of the point on the retina will be a blurred spot of size δ''. To bring the sharp image A'' on the retina of the short-sighted eye, the diverging bundle must come to the eye from the point of acute vision $A_{a.v}$. This bundle can be produced by shifting the eyepiece toward the objective a distance $-\Delta$; this adjustment is shown in the diagram at (b). In this adjustment, the first focus of the eyepiece is shifted to the left from the second focus of the objective, that is A' appears a distance $z = -\Delta$ behind the first focus of the eyepiece. The point of acute vision $A_{a.v}$ is at a distance $-z'$ from the second focal point of the eyepiece, and at a distance $-a_{a.v}$

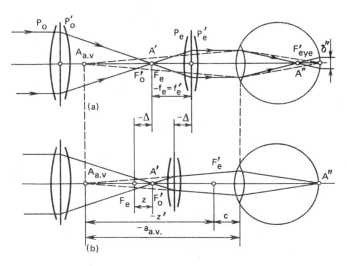

Fig. 14.8. Focusing of an eyepiece

from the eye. If we denote the distance from the second focus of the eyepiece to the eye by c, then $z' = a_{a.v} + c$. The distances z and z' are related by the Newtonian relation $zz' = -f_e'^2$. Substituting for z and z' their values as derived above we get

$$\Delta = f_e'^2/(a_{a.v} + c)$$

For $a_{a.v}$ in mm expressed in dioptres $a_d = 1000/a_{a.v}$, this expression becomes

$$\Delta = f_e'^2/(1000/a_d + c)$$

In view of $|c| \ll |a_{a.v}|$ for positive eyepiece this expression may be reduced to

$$\Delta = f_e'^2 a_d/1000$$

For the case of negative eyepieces, c is comparable with $a_{a.v}$ and should not be neglected if an accurate value of Δ is desirable.

The dioptric range taken into account in the design of visual instruments is usually from -5 to $+5$ dioptres. For example, the displacement resulted for an eyepiece of $f_e' = 25$ mm from account of $a_d = \pm 5$ dioptres is $\Delta = \pm 3.125$ mm.

14.5 The Function of a Field Lens

The field lens, as we learned in the earlier sections of this chapter, is the front lens of an eyepiece placed at the plane of internal image, or close to it with the purpose of bending oblique rays back toward the optical axis to prevent them from missing the eyelens.

Incorporating a field lens in the eyepiece shifts the position of the exit pupil and affects the size of the following elements of the system.

The function of a positive-focusing field lens in the system of a telescope is demonstrated in Fig. 14.9. In the absence of the field lens the diameter of the eyelens will be traced by the ray $B'M_1$ and the principal ray will locate the exit pupil at C_0', and the eye relief will be t_0'. The field lens bends the bundle of rays to the axis and the diameter of the eyelens, established now with the ray $B'M$, becomes smaller. The new eye relief equals t'.

A field lens placed exactly at the second focal plane of the preceding component, say objective, or at the first focal plane of the following component (eyelens) has no effect on the power of the telescope in that it does not alter the focal lengths of the adjacent components and has no effect on the course of an axial bundle.

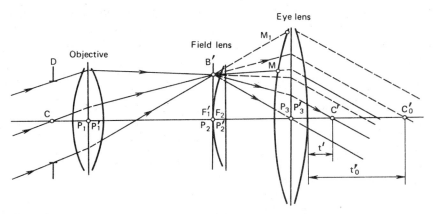

Fig. 14.9. The action of a field lens in increasing the field of view

Let us determine the power of a field lens (2) placed at the coincident focal planes of the objective (1) and eyelens (3) of a telescope shown in Fig. 14.10, given the focal length of the objective f_1', the distance of the entrance pupil from the front component t, the focal length of the eyelens f_3', and the eye relief specification t'.

The power of the field lens can be uniquely determined when a_2 and a_2' are known, namely,

$$\phi_2 = 1/a_2' - 1/a_2 \qquad (14.9)$$

where

$$a_2 = a_1' - f_1' = a_1' - 1/\phi_1 \qquad (14.10)$$

and

$$a_2' = a_3 + f_3' = a_3 + 1/\phi_3 \qquad (14.11)$$

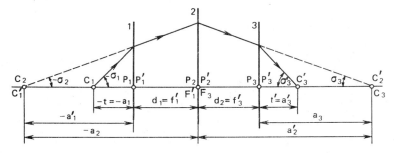

Fig. 14.10. Construction to determine the power of a field lens

In turn, the distances a_1' and a_3 are related to $a_1 = t$ and $a_3' = t'$ as

$$a_1' = t/(1 + t\phi_1), a_3 = t'/(1 - t'\phi_3) \qquad (14.12)$$

Substituting Eqs. (14.12) first in (14.10) and (14.11), and then in (14.9) yields

$$\phi_2 = \phi_1(1 + t\phi_1) + \phi_3(1 - t'\phi_3)$$

It is apparent that the field lens affects the quality of imagery for off-axial points and the distribution of illumination over the image field.

Now we turn to Fig. 14.11 to establish the relation between the eye relief t_0' in the telescope without the field lens and that in the presence of this lens, t'. Here P_2 and P_2' denote the principal planes of the field lens of focal length f_2'. The course of the principal ray in the system without this lens is traced by a dashed line. From inspection of the figure we have

$$t' = t_0' - \Delta h/\tan \omega' \qquad (14.13)$$

where Δh is the drop of the incidence height for the principal ray due to the field lens, ω' is the semiangle of the apparent field, $\tan \omega' = h_2/f_3'$, $h_2 = (\tan \sigma_3 - \tan \sigma_2)/\phi_2$ is the height of the principal ray between the principal planes of the field lens.

By Fig. 14.11, $\Delta h = f_3'(\tan \sigma_3 - \tan \sigma_2)$. Substitution of the ratio $\Delta h/\tan \omega'$ in (14.13) leads to

$$t' = t_0' - f_3'^2/f_2'$$

where $t_0' = s_{F'}' + z_{0p}'$, z_{0p}' is the distance of the exit pupil from the second focal point of the eyelens F_3' in the system without the field lens. By the formula for longitudinal magnification, $z_{0p}' = z_{0p}/\Gamma_t^2$ and $z_{0p} = t + f_1'$, so that, finally,

$$t' = s_{F_3'}' + (t + f_1')/\Gamma_t^2 - f_3'^2/f_2' \qquad (14.14)$$

When the design specifications include the position of the entrance pupil, eye relief, focal lengths of the objective and eyelens, Eq. (14.14) is useful in deriving the focal length of the necessary field lens.

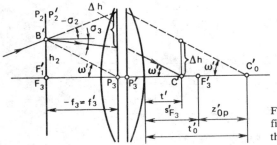

Fig. 14.11. The effect of a field lens on the position of the exit pupil

14.6 Designing a Kepler Telescope

In practical design it is often required to develop a telescope which would have a magnification Γ_t, angular field of view 2ω, exit pupil diameter D', fit in the length L, resolve an angular separation ψ, have the eye relief t' or the entrance pupil distance t, and vignetting ratio (relative effective aperture) k_1. The design is to determine all the longitudinal dimensions and sizes of the components constituting the system along with their optical characteristics.

Having adopted a type of telescope system, the designer usually begins with a first-order layout of the powers and spacings of the system components. We shall proceed along these lines and assume the objective and eyepiece to be thin lenses.

First of all we compute the focal lengths of the objective and eyepiece.

(*i*) From $\Gamma_t = -f_1'/f_2'$ and $L = f_1' + f_2'$,

$$f_1' = \Gamma_t L/(\Gamma_t - 1)$$

and

$$f_2' = L/(1 - \Gamma_t)$$

(*ii*) Given the diameter of the exit pupil and the telescope magnification we are in a position to determine the diameter of the entrance pupil of the telescope which also will be the entrance pupil of the objective. By Eq. (14.3), we have

$$D = D'\Gamma_t$$

By this we are in a possession of all the principal characteristics of the objective, namely, f_1', f_1'/D, and 2ω.

(*iii*) We learned in Section 14.3 that the eyepiece of a telescope is a rather complex optical system. At present there exists a wide selection of mass-produced eyepieces, therefore in practice it is often wise to choose an eyepiece from a catalogue rather than to design it.

The eyepiece is chosen by the desired, or specified, focal length and angular field. An important point of the choice concerns the compatibility of the eye relief. The fact is that the eyepiece is designed by ray tracing in reverse direction (from right to left) in which the oblique ray bundles are assumed to pass through the entrance (for this ray tracing) pupil of the eyepiece separated from the vertex of the lens by a certain distance t_e. The aberrations of these bundles are designed for this eye relief. The objective of the telescope is then designed (or selected) such as to offset the aberrations of the eyepiece. A necessary condition for this compensation is that the positions of the pupils be consistent (i. e., $t' = \overleftarrow{t_e}$), which implies that the entrance pupil of the telescope must be a certain distance from the instrument.

Figure 14.12 indicates that the points C and C', and F_1 and F_2' are conjugates, therefore the distances z_p and z_p' may be related by the longitudinal magnification which, in agreement with Eq. (14.4), is inversely proportional to the square of the magnifying power

$$z_p = z_p' \, \Gamma_t^2 \tag{14.15}$$

The quantities t' and $s_{F'}'$ are known from the condition for consistency of the eyepiece and objective, therefore,

$$z_p' = t' - s_{F'}' \tag{14.16}$$

Combining Eqs. (14.15), (14.16) and observing that $t = z_p - f_1'$ leads to

$$t = (t' - s_{F'}')\Gamma_t^2 - f_1' \tag{14.17}$$

A telescope may suffer from the spherical aberration of the principal ray $\Delta z_p'$ which must be included in our design too. Now

$$t = (t' - s_{F'}' - \Delta z_p')\Gamma_t^2 - f_1'$$

(iv) Let us compute the angular field of view of the eyepiece equal to the apparent field of the telescope, i. e. the field in image space. From (14.1),

$$\tan \omega' = \Gamma_t \tan \omega$$

Thus, we determined the principal characteristics of the eyepiece f_2' and $2\omega'$ essential in selecting, unless specified, the type of eyepiece, the quantities s_F, $s_{F'}'$, Σd, and the diameters of the field lens (D_2) and eye lens (D_3) from a catalogue.

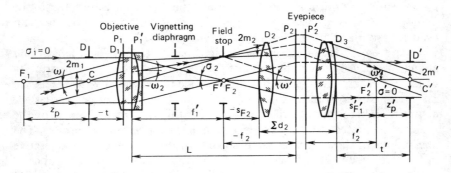

Fig. 14.12. Optical layout of the Kepler telescope

(v) The field stop ensures the given value of the angular field. By Eq. (14.5),

$$D_{fs} = -2f_1' \tan \omega$$

(vi) The diameter of objective D_1 can be determined by tracing either an axial or oblique marginal ray, that is,

$$D_1 = 2m_1 + 2t \tan \omega \qquad (14.18)$$

If $D_1 < D$, the diameter of the objective is assumed to be equal to the diameter of the entrance pupil, $D_1 = D$.

The meridional dimension $2m_1$ of the oblique ray fan is related to the entrance pupil diameter by the vignetting ratio (relative effective aperture) $k_1 = 2m_1/D$, whence $2m_1 = k_1D$. The given amount of vignetting can be achieved by computing D_2 or D_3 of the eyepiece using $2m_2$ of $2m'$ as the initial data. In Fig. 14.12 it can be seen that in this case one portion of the oblique bundle is vignetted out. In order to vignet out the other portion (in this case the upper portion in the entrance pupil) a vignetting diaphragm is placed, for example, between the field stop and the objective.

(vii) The diameter D_2 of the field lens is evaluated by tracing an oblique ray bundle. By examining the course of the principal ray and the lower (at the entrance) ray successive from the object space to the objective, between the objective and the field stop, and from the stop to the eyepiece we can write

$$D_2 = 2t \tan \omega - 2(f_1' - s_{F_2}) \tan \omega_2 + 2m_2 \qquad (14.19)$$

where ω_2 is the slope of the principal ray behind the objective determined by

$$\tan \omega_2 = \frac{f_1' + t}{f_1'} \tan \omega$$

m_2 is the spacing between the incidence heights of the principal and upper ray at the field lens, $m_2 = -m_1 s_{F_2}/f_1'$.

($viii$) The diameter D_3 of the eyelens is evaluated by tracing the upper ray through the exit pupil .

$$D_3 = 2t' \tan \omega' + 2m' \qquad (14.20)$$

where $2m' = k_\omega D'$ or $m' = m_1/\Gamma_1$.

As a rule the found values of D_2 and D_3 are compared with the nearest counterparts from the catalogue to check the feasibility of the design formulation.

It is important to observe that this design evaluated the clear apertures for the optical elements. The corresponding full diameters should exceed these values by the allowance necessary for the adopted method of lens mounting.

This type of telescope is mainly used in geodetic and astronomical instruments, and also in surveying instruments where an erect image is obtained with reflecting prisms.

For measurement purpose, the telescope receives a reticule which is applied on a plane-parallel plate placed exactly at the common focal points of the objective and eyepiece. The scale of such a reticule Δy (spacing between two adjacent hair lines) depends on the angular separation $\Delta\omega$ subtended by this spacing and the focal length of the objective,

$$\Delta y = f_1' \tan (\Delta\omega)$$

14.7 The Galilean Telescope

The Galilean telescope, as we learned in Section 14.1, consists of a positive objective and a negative eyelens and therefore produces an erect image of objects under observation. The optics of the telescope is presented in Fig. 14.13. The internal image formed at the coincident focal planes of the system is imaginary, rather than erect as in an astronomical telescope, therefore there is no place to insert a reticule in this type of instrument.

Consider the formula (14.17) with reference to a Galilean telescope. Assuming that the eyelens is indefinitely thin we get $s_{F'}' = f_2'$ and

$$t = (t' - f_2')\Gamma_t^2 - f_1'$$

This expression can be readily transformed to the form

$$t = t'\Gamma_t^2 + f_1'(\Gamma_t - 1)$$

or

$$t = \Gamma_t(t'\Gamma_t + L) \tag{14.21}$$

The positive separation of the entrance pupil in this telescope implies that the entrance pupil is imaginary and is far to the right behind the viewer's eye.

The position and size of the aperture stop and exit pupil in the Galilean telescope are all functions of the eye pupil. The field of view in this telescope is limited by a vignetting diaphragm (rather than a field stop which is absent) whose role is undertaken by the objective mount. For the most part, the objective is a two-lens system of aperture ratio of about $f/3$

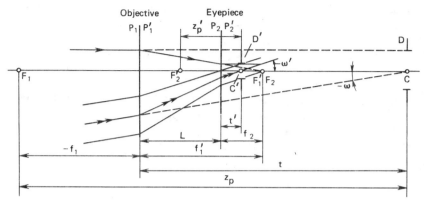

Fig. 14.13. Principle of the Galilean telescope

and real field of at most 6° to 8°. For angular field of this size at appreciable separation of the entrance pupil, objectives must be of large aperture.

The eyepiece is usually a single negative lens or a negative doublet ensuring an angular field of at most 30° to 40° on the condition that the field aberrations are compensated for by the objective. Hence, a Galilean telescope cannot be designed for high magnifying power (it does not exceed 6× to 8×, more often a figure between 2.5× and 4× is encountered). The dependence of the semifield ω on magnifying power for this type of telescope is presented in Fig. 14.14.

The major advantages of a Galilean telescope include an erect image, simple design, and short constructional length which for a similar astronomical telescope would be longer by two focal lengths of the eyepiece.

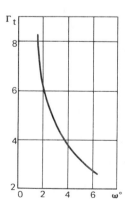

Fig. 14.14. Semiangular field of the Galilean telescope as a function of the magnifying power

Among the setbacks of Galilean telescopes one finds small field of view
and magnifying power, and the absence of a real image preventing the use
of a reticule and cross hairs.

In the following illustrative example we demonstrate the design of a
Galilean telescope with the expressions derived for astronomical telescopes
in the previous section.

Example 14.1. Design a Galilean telescope of magnifying power
$\Gamma_t = 3 \times$, angular field $2\omega = 4°$, exit pupil diameter $D' = 4\,mm$,
$t' = 12\,mm$, and $L = 40\,mm$.

(1) The focal lengths of objective and eyelens are

$$f_1' = 3 \times 40 \div 2 = 60\,mm$$

$$f_2' = 40/-2 = -20\,mm$$

(2) The entrance pupil diameter is $D = D'\Gamma_t = 12\,mm$. Hence, the objective has the focal length $f_1' = 60$ mm, aperture ratio $D/f_1' = 1:5$, and
real angle $2\omega = 4°$.

(3) By Eq. (14.21), $t = 228$ mm.

(4) The apparent field of the eyelens, $2\omega' = 12°$, is derived from

$$\tan \omega' = \Gamma_t \tan \omega = 0.105$$

(5) With reference to Fig. 14.13, the diameter of the objective
$D_o = 2t \tan \omega = 15.96$ mm. We take it to be 16 mm. For this diameter
and the angular field $2\omega = 4°$, the vignetting is 50 per cent, i. e. $k_\omega = 0.5$.
If vignetting is objectionable, that is k_ω is to be unity, the diameter of the
objective must be enlarged by the aperture of the entrance pupil and
becomes equal to 28 mm.

(6) The diameter of the eyelens will be determined by Eq. (14.20)

$$D_e = 2m' + 2t' \tan \omega'$$

For $k_\omega = 0.5$, $2m' = 2$ mm and $D_e = 4.52$ mm; whereas for $k_\omega = 1$,
$2m' = 4$ mm and $D_e = 6.52$ mm.

14.8 Erecting Prism Systems

In an ordinary astronomical telescope, the objective forms an inverted
image of the object, which is then viewed through the eyepiece. Since the
internal image is inverted, and the eyepiece does not re-invert the image,
the view presented to the eye is inverted top to bottom and reversed left to
right. To eliminate the inconvenience of viewing an inverted image in terrestrial conditions, an erecting system is often provided to re-invert the im-

age to its proper orientation. This may be a lens system (to be discussed in the next section) or a prism system which will be elucidated below as applied to prism monoculars.

A prism monocular is essentially a Kepler telescope with an inverting prism system. In addition to re-inverting the image, a prism system incorporated in a telescope is helpful in placing the optical axes of the objective and eyepiece at a desired angle to provide a comfortable position of the observer's head for example in viewing high objects. Note that if the inverting system is a single reflecting prism, this has to be a roof prism.

Figure 14.15 presents some popular systems of prism monoculars. The monocular with a Schmidt prism shown at (a) has an angular field of view of at most 8° and a deviation angle (between the axes of the objective and eyepiece) of 45°. The system with an Abbe prism shown at (b) is sometimes encountered in prism binoculars. The Pechan prism shown in the diagram at (c) is a space-saving device owing to a long optical path in the prism, an attracting feature if a compact system is desired. If a compact and flexible binocular system (constituted by two monoculars) is desired, a good choice would be the erecting system of a Leman (or Springer) prism shown at (d). The diagrams at (e) and (f) present the monoculars with Malafeyev prism systems of first and second type respectively. In these systems, the optical axes of the objective and eyepiece are non-coplanar. The diagram at (g) pictures the optical system of a stereoscope monocular: 1 cover glass, 2 leading prism, 3 objective, 4 roof prism, 5 wedge, 6 reticle on a substrate,

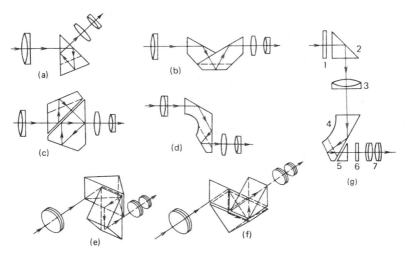

Fig. 14.15. Layouts of prism monoculars

and *7* eyepiece. This system is unique in its periscopic feature measured as the distance between the optical axes of the objective and eyepiece.

The first-order layout of a monocular is carried out on the same lines as the design of a simple telescope. An additional stage concerns itself with the design of an appropriate prism system. For convenience of design the optical system of a monocular is unfolded in the horizontal axis and the prism is replaced by a plane-parallel plate reduced to an air equivalent distance. Prisms can be placed both in converging ray bundles behind the objective (diagrams *a-f* in Fig. 14.15) and in parallel bundles, say ahead of the objective as in (*g*).

For prisms subjected to converging bundles, the formula (5.25) is invoked to account for the additional optical path in the prism. This distance is a measure of additional space allocated for the prism system in a real instrument.

The objective of a prism design is to determine the clear aperture of the ray bundle the prism is to propagate, and the locations of the prism system in the space between the objective and the eyepiece. As a rule all the other relevant dimensions for reflecting prisms may be looked up in optical manuals and handbooks. By way of example, *Appendix 1* illustrates how these data are given in Soviet manuals for a ray pencil of diameter D.

The size analysis for a reflecting prism placed in a parallel ray bundle is given in Section 5.5. We look at the design procedure for a prism located behind the objective. To evaluate the size of the prism Fig. 14.16 shows the course of rays surviving the objective. The clear aperture at the front face of the prism is determined either by tracing the upper extra-axial ray *2* (D_1 in the diagram at (*a*)), or by tracing the axial ray *1* (D_0 in the diagram at (*b*)). The largest clear aperture of the last face (D_2) can be traced out by the ray *2*.

The distance b_2 from the last surface O_2B_2 of the prism to the focal plane of the objective is selected such that to minimize the size of the prism

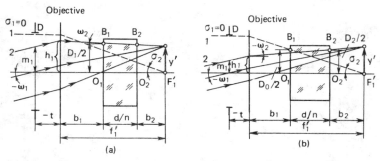

Fig. 14.16. Construction for a prism in a converging ray bundle

and to gain an allowance for the roof of the prism as wide as possible. These conditions can be met by placing the prism in the immediate vicinity of the focal plane. It should be observed, however, that the back face of the prism should not be placed near the focal plane as otherwise all the defects of the glass (air bubbles, scratches, deposits of particulate matter, etc.) will be clearly seen in the field of the eyepiece and will obscure the view. On the other hand, the farther the prism is situated from the focal plane, the stronger the effect of doubling the image due to errors in the manufacturing of the roof angle. Therefore, an optimal tradeoff is to place the prism so that its last surface placed in front of the focal plane of the eyepiece is imaged beyond the accommodation range of the eye. This condition can be satisfied by 10-20 dioptres of convergence decrement a_d behind the eyepiece. For f_2' and b_2 in millimetres,

$$b_2 = f_2'^2 \, a_d/1000$$

Now we derive a set of expressions for clear apertures D_i with reference to Fig. 14.16. Here, D is the diameter of the entrance pupil, t the distance to the entrance pupil, m_1 the semiwidth of the meridional oblique ray bundle ($m_1 = k_\omega D/2$), b_1 the distance from the principal plane of the objective (thin lens approximation) to the first face of the prism ($b_1 = f_1' - b_2 - d/n$). After some manipulations, observing $d = cD_i$ ($i = 0, 1, 2$), the above relationships lead to

$$D_0 = nb_2 D\phi_1/(n - cD\phi_1)$$

$$D_2 = 2y' + 2b_2[0.5k_\omega D\phi_1 + (1 + t\phi_1)\tan\omega_1]$$

$$D_1 = 2n \frac{b_2[0.5 \, k_\omega \, D\phi_1 + (1 + t\phi_1)\tan \omega_1] + y'}{n - 2c \,[0.5k_\omega \, D \, \phi_1 + (1 + t\phi_1)\tan \omega_1]}$$

or

$$D_0 = nb_2 D/(nf_1' - cD)$$

$$D_2 = 2y' + 2b_2[0.5 \, k_\omega \, D + (f_1' + t)\tan \omega_1]/f_1'$$

$$D_1 = 2n \frac{b_2[0.5 \, k_\omega \, D + (f_1' + t)\tan \omega_1] - f_1'^2 \tan \omega_1}{nf_1' - 2c \,[0.5 \, k_\omega \, D + (f_1' + t) \tan \omega_1]} \quad (14.22)$$

or

$$D_2 = 2(y' + Ab_2/f_1') \quad (14.23)$$

$$D_1 = 2n \frac{b_2 A - f_1'^2 \tan \omega_1}{nf_1' - 2cA} \quad (14.24)$$

where $A = 0.5 \, k_\omega \, D + (f_1' + t) \tan \omega_1$, $A = h_1 - y'$.

In view of $A/f_1' = \tan \omega_2$ for $\tan \omega_2 > 0$ the front face of the prism has the largest dimension calculated by Eqs. (14.22) or (14.24). If

$2y' > D$, then $\tan \omega_2$ is always under zero and the largest dimension belongs to the back (exit) face of the prism calculated by Eq. (14.23).

The largest clear aperture D_{max} found in this way should be increased somewhat to allow for a mount and the resultant value is used as the initial data in deriving the other dimensions of the prism. Note that the aperture D_{max} limits the course of rays in the upper part of the oblique ray bundle, and the lower part of this bundle is limited by the mount of the eyepiece lenses.

14.9 Lens Erecting Telescopes

In the preceding section we saw that prism erecting systems reduce the size of the instrument, but the designer has not to overlook that prism systems increase weight and entail problems due to prism manufacture and alignment. An alternative solution for a Kepler telescope is the *lens* erecting system. Even with a complicated, say five-lens, erecting system the resultant system weighs half as much as the telescope with a prism erecting system. A telescope with a lens erecting system is often called a *terrestrial telescope* because it presents to the eye an erect image convenient for terrestrial observations.

Erector systems come in all sizes and shapes. Occasionally a single element may serve as an erector, but two simple elements in the general form of a Huygenian eyepiece is a more popular construction.

Figure 14.17 shows the optics of a telescope with a single-lens erector *3*, objective *1*, field lens *2*, and eyelens *4*. We illustrate the design of such a system using the nomenclature presented in this figure.

As a rule, specifications used for the design of such a telescope include the magnifying power Γ_t, real angular field 2ω, exit (D') or entrance (D) pupil diameter, telescope length L, entrance pupil separation t or eye relief t', vignetting ratio (relative effective aperture) k_ω, and the lateral magnification of the erector β_{er}. The focal length of the eyepiece, f_4', may also be specified.

If we take the first three components to be the objective system, the magnifying power of the group is

$$\Gamma_t = -f_{1,2,3}'/f_4' \qquad (14.25)$$

where $f_{1,2,3}' = h_1/\tan \sigma_4$, $\tan \sigma_4 = \tan \sigma_3/\beta_{er}$, $\tan \sigma_3 = \tan \sigma_2$, and $f_1' = h_1/\tan \sigma_2$.

Incorporating this nomenclature in Eq. (14.25) yields

$$\Gamma_t = -(f_1'/f_4')\beta_{er}$$

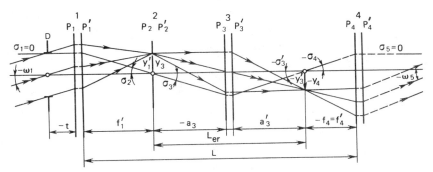

Fig. 14.17. A telescope with a single-lens erector system

The following design proceeds with the expressions

$$f_1' = -\Gamma_t f_4'/\beta_{er}$$

$$D = D'\Gamma_t$$

$$D = 2m_1 + 2t \tan \omega$$

If calculations give $D_1 < D$, the correction $D_1 = D$ must be made to prevent vignetting for the axial ray bundle.

Having determined the key characteristics of the objective (f_1', D/f_1', and 2ω), the designer may look up the respective component in catalogues or finalize its design.

We proceed next to determine

$$L_{er} = L - (f_1' + f_4')$$

$$a_3 = L_{er}/(\beta_{er} - 1)$$

$$a_3' = \beta_{er} L_{er}/(\beta_{er} - 1)$$

$$f_3' = -(L_{er} - \Delta_{P_3 P_3'})\beta_{er}/(1 - \beta_{er})^2$$

$$D_3 = -D a_3/f_1'$$

$$\tan \omega_3 = -y_1'/a_3$$

These quantities constitute the data set of the erector system: f_3', D_3/f_3', $2\omega_3$, and β_{er}. Notice that $D_2 = -2f_1' \tan \omega_1$, and because the field lens directs the principal ray into the first principal point of the erector system,

$$\frac{1}{f_2'} = \frac{1}{f_1'}\left(1 + \frac{t}{f_1'}\right) + \frac{1}{f_3'}$$

The parameters of eyepiece are computed in the same manner as the parameters of a Kepler telescope (see Eqs. (14.19) and (14.20)). As a rule, the magnification of the erector $\beta_{er} = -1$, therefore for a single-lens erector system $L_{er} = 4f_3'$, $a_3 = -2f_3'$, and the aperture ratio of the erector is twice as large as that of the eyepiece. Attempts to diminish the magnification of the erector (in absolute value) lead to greater values of its aperture ratio which is undesirable. When, on the contrary, an increase in the magnification is attempted (by absolute value) it would cause a rapid growth of the field $2\omega_3$ and expansion of the system length L, which is undesirable too.

In telescopes with two-lens erector systems (Fig. 14.18), rays travel between the erector components as parallel bundles, therefore the distance d_3 can be varied and thereby the erector length $L_{er} = f_3' + d_3 + f_4'$. The first three components of this telescope form a subsystem of magnification Γ_{t1}, the two last components a subsystem of magnification Γ_{t2} so that

$$\Gamma_t = \Gamma_{t1}\Gamma_{t2} = (-f_1'/f_3')(-f_4'/f_5') = (-f_1'/f_5')\beta_{er}$$

The principal ray in such erector systems is usually directed by the field lens to cross the optical axis midway between the erector lenses at $d_3/2$ from either component. With reference to Fig. 14.18 we can arrive at the following relationship between d_3 and f_3'

$$d_3 = -\frac{(1 - k_\omega) D}{\tan \omega_1}\left(\frac{f_3'}{f_1'}\right)^2 \tag{14.26}$$

Assuming that we are in a possession of the same characteristics we knew in the case of the single-component erector, we write the expressions for the two-element erector system

$$f_1' = -\Gamma_t f_5'/\beta_{er}$$
$$D_1 = 2m_1 + 2t \tan \omega_1$$

where $2m_1 = k_\omega D$,

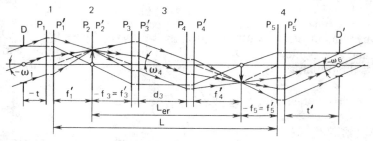

Fig. 14.18. A telescope with a two-lens erector system: 1 objective, 2 field lens, 3 erector, 4 eyepiece

$$f_3'^2(1 - k_1) D/f_1'^2 \tan \omega_1 + f_3'(\beta_{er} - 1) + [L - (f_1' + f_5')] = 0$$

to be solved for f_3',

$$f_4' = -\beta_{er} f_3'$$

d_3 is determined by Eq. (14.26),

$$D_2 = -2f_1' \tan \omega_1$$

$$D_3 = Df_3'/f_1', \quad D_4 = D_3$$

$$\frac{1}{f_2'} = \frac{1}{f_1'}\left(1 + \frac{t}{f_1'}\right) + \frac{1}{f_3'}\left(1 - \frac{d_3}{2f_3'}\right)$$

The aperture of the field stop inserted in the first focal plane of the eyepiece is determined as

$$D_{fs} = -2f_5' \tan \omega_6' \qquad (14.27)$$

The diameters of the eyepiece lenses can be evaluated in the same way as for the Kepler telescope.

14.10 Variable Power (Zoom) Systems

There exist telescopic systems with discrete and continuous variation of magnifying power. Discrete variation can be achieved in one of the following ways: by changing (1) eyepieces, (2) objectives, (3) certain components of the objective system, (4) erector system; (5) by reversing the placement of the erector in the telescope; (6) by moving the objective of the erector system along the optical axis; and (7) by inserting afocal attachments in the parallel beams inside the telescope system.

The simplest method of achieving discrete variation of power in telescopes is by replacing the eyepieces. It is widely used in geodetic and astronomical instruments and, rather often, for rifle sights and gunsight optical systems. In theodolites, for example, the replacement set of eyepieces can be constituted by units of $f' = 8, 9, 10, 13.5, 16.7$, and 20 mm. The Zeiss reflecting telescope with $f_0' = 1.1$ m is supplied with a kit of eyepieces of $f' = 6, 10, 16$, and 25 mm.

Replacing the eyepiece of a telescope with one having a shorter focal length expands the angular field and diminishes the exit pupil aperture. In some designs an increase in magnifying power is by necessity accompanied by contraction of the angular field. The extent of the field in this case will be determined by the field stop inserted in the first focal plane of the short-focus eyepiece (see Eq. (14.27)).

Change of the objective is a less popular approach. This technique is used in some periscopes and gunsight optical systems.

The third method — change of a component from the objective system — is a convenient approach for elaborate objective systems constituted by a few groups of lenses, as is the case, say, with a telephoto lens both refracting and reflecting. Replacing the front or back group in such an objective alters the focal length of the objective system. A modification of this method is achieved by inserting a new optical element, a lens group or a mirror, into the objective system.

In erecting telescopes, variations can be introduced in the erector system. One or both lenses may be the subject of replacement in two-lens erector systems, as indicated in Fig. 14.19. Recall that the lateral magnification of erector system is $\beta_{er} = -f_2'/f_1'$. It is worth emphasizing that this method leaves the image size $2y'$ and the relative aperture of the subsequent components unchanged, that is keeps unchanged also the exit pupil aperture D' of the entire telescope system, whereas the linear field coverage $2y$ of the preceding system and the relative apertures of the objective lenses ahead of the erector undergo variations. If the linear magnification of the erector system increases in absolute value, then the angular field coverage 2ω decreases while the entrance pupil aperture D increases. As a result we have the following invariants:

$$D_{min} \tan \omega_{max} = D_{max} \tan \omega_{min}$$

or

$$\Gamma_{t, min} \tan \omega_{max} = \Gamma_{t, max} \tan \omega_{min}$$

A discrete alteration in magnification can be achieved by reversing the erector system from the setting of Fig. 14.19b into the one indicated in Fig.

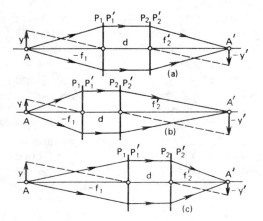

Fig. 14.19. Replaceable erector systems of (a) $\beta = -1$, (b) $\beta < -1$, (c) $\beta > -1$

14.19c. The prime lens of the erector becomes its back component and vice versa. Whereas in the first setting the image magnification is β_1, after resetting it is $1/\beta_1$, that is the overall magnifying power will change by β_1^2.

Exchange of erector lens or inversion of the entire erector system complicates the mechanics of the instrument and causes a larger cross section. Freedom from these limitations may be achieved with the sixth method in which the prime lens of the erector system is made to move along the optical axis. A modification of this method is to move a central lens of the three-lens erector, keeping the two extreme lenses intact. The simpler method, however, is by moving the prime lens PP' a distance d along the optical axis (Fig. 14.20). In the first setting, diagrammed at (a), the magnification is $\beta_1 = a_1'/a_1$, and in the setting (b) $\beta_2 = a_2'/a_2$. The conjugate distance equation (3.7) yields

$$a_1^2 + a_1(2f' - d) - f'd = 0$$

Observing that $a_1 - d = a_2$ and the fact that both a_1 and a_1' and a_2 and a_2' are conjugates, we solve the quadratic equation and obtain two pairs of fixed conjugate planes. The first pair is for the planes of the object A and the image A', while the other pair relates to the planes of the entrance and exit pupils centered on C and C'. If the length of the erector $L = -a_1 + a_1'$ is known, then for a given magnification β_1 we obtain for the focal length of the first lens

$$f' = -\beta_1 L/(1 - \beta_1)^2$$

In the second setting, the magnification will be $\beta_2 = 1/\beta_1$.

For a given linear magnification β_1 (or β_2) and f', the distances a_1 and a_1' (or a_2 and a_2') can be determined as $a_k' = (1 - \beta_k)f'$.

Variable magnification may be also attained by using a negative first lens and with a positive focusing lens set to move between two fixed lenses,

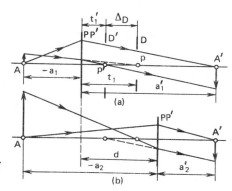

Fig. 14.20. Moving the prime lens of an erector along the optical axis

positive or negative, as has been discussed at considerable length by Churilovsky [21].

A number of systems achieve variable magnification by means of an additional Galilean telescope introduced in the parallel beam of the basic instrument. These inserts come either as attachments fitted on the first objective lens of the instrument, or as an expander incorporated within the instrument where the bundle of rays is parallel. Fig. 14.21 shows such a Galilean telescope which is made turnable as a system, around point C, to be located in three positions for three values of magnifying power ($\Gamma_{t, max}$, $\Gamma_{t, 0}$, and $\Gamma_{t, min}$). If the magnifying power of the basic instrument is Γ_t, and the power of the attachment in the settings indicated at (a), (b), and (c) is respectively Γ_{t1}, Γ_{t2}, and unity, then the overall magnifying power is, respectively,

$$\begin{aligned}
\Gamma_{t, max} &= \Gamma_t \Gamma_{t1} \\
\Gamma_{t, 0} &= \Gamma_t \\
\Gamma_{t, min} &= \Gamma_t \Gamma_{t2}
\end{aligned}$$

where, as will be recalled, $\Gamma_{t2} = 1/\Gamma_{t1}$.

Tiltable or replaceable attachments of this type may be used not only with telescope systems, but with other instruments where there is a parallel course of rays, with the purpose of upscaling or downscaling the image size.

An additional telescope system is used with the entire basic telescope or its part for which the position of the entrance pupil is defined by the placement of the observer's eye. Consequently, the exit pupil D' of the attachment must always coincide with the entrance pupil of the following portion of the basic system, as indicated in Fig. 14.22.

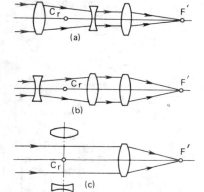

Fig. 14.21. The effect of a tiltable attachment in the form of a Galilean telescope

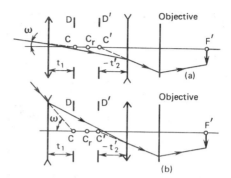

Fig. 14.22. Positions of entrance and exit pupils of Galilean telescopic attachments at magnifying power of (a) $\Gamma = \Gamma_t$ and (b) $\Gamma = 1/\Gamma_t$

With reference to Fig. 14.22, an afocal Galilean telescope attachment can be made to satisfy the condition $t_1 = -t_2$, then the axis of rotation C_r of the attachment is midway between the entrance pupil, D, and the exit pupil, D'. From Eq. (14.21) upon substitution $t = t_1$ and $t' = t_2'$ we get

$$t_1 = f_1'(\Gamma - 1)/(\Gamma^2 + 1)$$

where Γ is the magnifying power of the attachment, and f_1' is the focal length of its prime lens. Given $\Gamma = 4$ and $f_1' = 100$ mm, we have $t_1 = 17.6$ mm. Consequently, we can now determine the distance between the second principal plane of the second component of the attachment and the first principal plane of the objective lens of the basic telescope, because the position of the entrance pupil in the absence of the attachment is known.

To attain continuous variation of image scale, the telescope has to include a zoom lens as the objective, an erector system, and an eyepiece. Such a telescope is sometimes called "pancratic". A specific characteristic of pancratic systems is the ratio of extreme magnifying powers

$$M = \Gamma_{max}/\Gamma_{min} \qquad (14.28)$$

A zoom lens must involve at least two separate components in order to vary its equivalent focal length by changing the space between the components. Indeed from (3.27) with reference to Fig. 14.23a (the components are seen to move from setting I to setting II) we have

$$\phi_I = \phi_1 + \phi_2 - \phi_1\phi_2 d_I$$

$$(14.29)$$

$$\phi_{II} = \phi_1 + \phi_2 - \phi_1\phi_2 d_{II}$$

where the quantities ϕ_I, ϕ_{II}, d_I, and d_{II} are specified. Simultaneous solution of this set of equations gives the expressions to determine ϕ_2 from

$$\phi_2^2(d_I - d_{II}) + \phi_2(\phi_I\, d_{II} - \phi_{II}\, d_I) + \phi_{II} - \phi_I = 0$$

and then ϕ_1 as

$$\phi_1 = (\phi_1 - \phi_2)/(1 - \phi_2\, d_1) \qquad (14.30)$$

If the focal lengths of the lens components are of the same magnitude but opposite sign, as in the upper sketch of Fig. 14.23, i. e., $f_1' = -f_2'$, then

$$f_1' = \sqrt{f_1'\, d_I} \quad \text{or} \quad f_1' = \sqrt{f_{II}'\, d_{II}}$$

A zoom lens design constituted by two positive components, as in Fig. 14.23b, is also feasible. For example, if the focal lengths of the elements are the same, i. e. $f_1' = f_2'$, then Eq. (14.30) gives

$$\phi_1^2\, d_I - 2\phi_1 + \phi_I = 0 \qquad (14.31)$$

If, for example, the specification tells us $f_1' = 100$ mm and $d_I = 80$ mm, then solving (14.31) yields $f_{1,1}' = f_{2,1}' = 55.28$ mm or $f_{1,2}' = f_{2,2}' = 144.72$ mm.

The other least value of f_{II}' at the least d_{II} is found from the expression

$$f_{II}' = f_1'^2/(2f_1' - d_{II})$$

which for the same initial data as above gives $f_{II}' = 32.66$ mm at $d_{II} = 17$ mm and $f_1' = 55.28$ mm.

When one of the components of a zoom lens attachment is moved, this causes the entire lens to be moved to compensate for image defocusing as the first movement changes the distance between the last component and

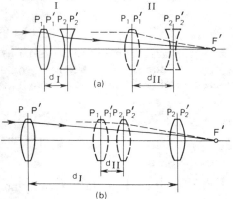

Fig. 14.23. Zoom lenses

the focal plane defined as

$$a_2' = f'(1 - d/f_1')$$

where the relevant nomenclature is suggested by Fig. 14.24.

In agreement with Eqs. (14.2) and (14.28) we obtain for the magnifying power ratio $M = f_{max}'/f_{min}'$.

It is worth to be noted that the zoom system diagrammed at (a) in Fig. 14.23 is preferable to the system at (b) as its relative magnification range M is from 2 to 6, whereas for the positive-lens system it never exceeds 2.

As a rule, pancratic telescopes make use of pancratic erector system. This system is either a two-component mechanically compensated (the second component moves in synchronism with the first to avoid defocusing) zoom system, or a three-element optically compensated zoom system. The latter construction will be discussed in Chapter 15, thus far we note only that in such systems two alternate lenses are linked and moved together with respect to the lenses between them.

Figure 14.24 shows the optics of a two-lens erector zoom system. Here, point A may be the second focal point of the objective lens of the telescope, while point A' is to coincide with the first focal point of the eyepiece. The lengths of this zoom system must be constant, i. e.,

$$L = -a_1 + d + a_2' = \text{constant} \tag{14.32}$$

Thus, the lenses of focal lengths f_1' and f_2' constituting the system may be at a various distance d from each other but such that the length of the system remains the same. In layout, the specified parameters are the length of the system L, the greatest β_{max} and smallest β_{min} magnifications, and the focal lengths f_1' and f_2'. The air space d satisfying a given linear magnification β can be determined as

$$d = 0.5L - 0.5\sqrt{L^2 - 4[L\,(f_1' + f_2') + f_1'f_2'\,(1 - \beta)^2/\beta]}$$

As a rule d is computed for two extreme values of β to check that in motion the lenses do not interfere.

The distance from the first principal plane of the first component to the

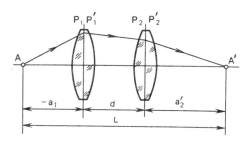

Fig. 14.24. A two-lens pancratic erector system

object plane is

$$a_1 = f_1' [f_2' (1 - \beta) + \beta\, d]/\beta[f_1' + f_2' - d]$$

With a_1 and d at hand, the distance a_2 is calculated by Eq. (14.32).

This type of zoom erector systems may be designed for power ratios, this time defined as β_{max}/β_{min}, as high as 20, but ordinarily this ratio is between 4 and 8.

Being the mechanically compensated zoom system, this two-element design is at a serious disadvantage due to a curvilinear rule of motion by which one of the components is to be shifted to avoid defocusing. This specifically implies a delicate profile of the slot in the cam through which the pin of the component's mount moves. A technologically more attractive design will be the one where the components move in agreement with a linear rule.

The requirement for a linear movement of components is satisfied in a four-element zoom erector indicated in Fig. 14.25. Two positive components of the system are linked to move along the optical axis with respect to a fixed negative component between them.

The first component of the system, designed to shorten the system's overall length, images the object point A in a point A_2 which is then reimaged by the zoom subsystem of the three subsequent components into an image point A'. The second and the fourth components are linked to move together with respect to the third element thus producing some varifocal effect. The image shift associated with a motion z is characterized by the shift of the image plane δz. With an appropriate choice of powers and spaces this defocusing can be made rather small, say, for a power ratio of 4, δz may be kept within fractions of a millimetre.

In another optically compensated zoom system, two negative elements are linked and move with respect to a positive component fixed in between.

The equation of length for such systems has the form

$$L = -a_1 + d_1 + d_2 + d_3 + a_4'$$

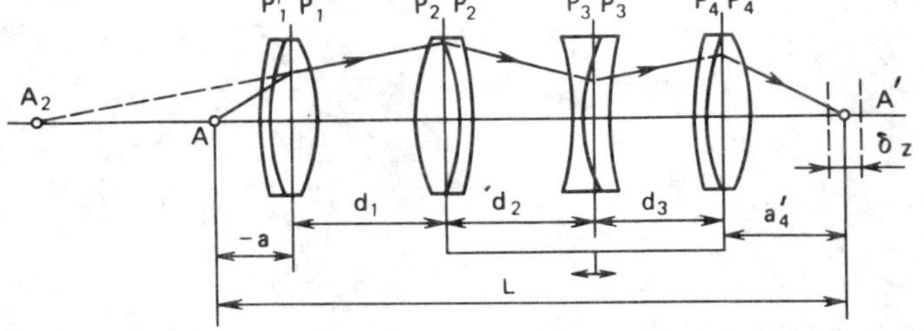

Fig. 14.25. A four-lens, optically compensated, pancratic erector system

and the equation for the active subsystem is

$$L_a = -a_2 + d_2 + d_3 + a_4'$$

where a_2 is calculated by the conjugate distance equation for the known a_1 and f_1' because $a_2 = a_1' - d_1$.

14.11 Stereoscopic Telescopes

The ability of the eye for three-dimensional perception of space, called stereoscopic vision, gave rise to a special class of stereoscopic telescopes designed as binocular systems.

Owing to a finite distance between the nodal points of the right and left eyes (interocular distance b) the images of a distant point occur at the different distances from the centres of retinae. This fact provides the base for three-dimensional perception of space in humans.

Imagine now that the eyes of a human are replaced by two objective lenses spaced at a distance B (base) as indicated in Fig. 14.26. In the image plane these objectives focus the images of a distant point at different distances y_l' and y_r' from their respective optical axes.

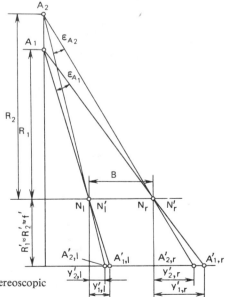

Fig. 14.26. Construction to illustrate the stereoscopic parallax

17*

For two object points A_1 and A_2 the difference of these distances, called the linear parallax, is, respectively,

$$p_1 = y'_{1,\,r} - y'_{1,\,l}$$
$$p_2 = y'_{2,\,r} - y'_{2,\,l}$$

Similarly, the angular parallax for A_1 is the angle ε_{A_1} between the axes of sight $A_1A'_{1,\,l}$ and $A_1A'_{1,\,r}$, and so is the angular parallax ε_{A_2} for point A_2.

For sufficiently large distances R_1 and R_2, we may let approximately $R'_1 \approx R'_2 \approx f'$. Then $R \approx f'B/p$ and if $p/f' \approx \tan \varepsilon$, the distance $R = B/\tan \varepsilon$. Therefore, the distance to a point can be determined by measuring the linear or angular parallax for this point. This finding is used to advantage in range finders.

In binocular stereoscopes each eye of the observer receives ray bundles from an object through identical individual optical systems. This construction can be found in binoculars, stereoscopic telescopes, and binocular range finders.

A schematic representation of a branch of a stereoscopic telescope can be found in Fig. 14.15g. The distance between the axes of sight, B, can be altered in such instruments and the natural stereoscopic effect occurs when each branch of the instrument produces an erect image.

Figure 14.27 shows a schematic diagram of a stereoscopic telescope. The erect image is secured by the penta prisms 2 and roof prisms 4. A wide base B of the instrument improves the perception of depth in object space. The ratio of the instrument base B to the interocular distance b

$$P_0 = B/b$$

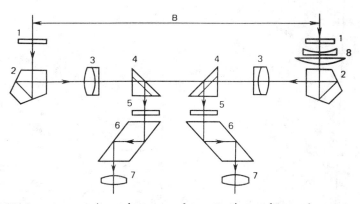

Fig. 14.27. A stereoscopic telescope: *1* protecting glass, *2* penta prism, *3* objective, *4* 90° roof prism, *5* reticle, *6* rhomboid prism, *7* eyepiece, *8* compensator system

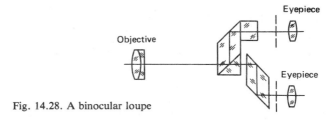

Fig. 14.28. A binocular loupe

characterizes the gain in stereoscopic effect produced by the instrument as compared with the perception by the unaided eye. The product of P_0 by the magnifying power of the instrument

$$P = \Gamma_t P_0$$

indicates how many times the actual distance in the viewed terrain exceeds its image apparent in the instrument.

We recall from the earlier chapters that the three-dimensional perception is valid within the radius of stereoscopic vision. For stereoscopic instruments, this radius can be derived as

$$R_s = B/\Delta\varepsilon_{min}$$

Because the limit angle of stereoscopic perception is $\eta = \Delta\varepsilon'_{min} = \Delta\varepsilon_{min}\Gamma_t$,

$$R_s = B\Gamma_t/\eta$$

By way of example, binoculars with $B = 125$ mm, $\Gamma_t = 8\times$, enable, for $\eta = 4.9 \times 10^{-5}$ rad (10 seconds of arc), $R_s = 20.4$ km; recall that for the eye R_{max} is about 1.3 km.

Stereoscopic instruments are always of a binocular structure, but the contrary is not true — binoculars may not be stereoscopic. For example, a binocular instrument shown in Fig. 14.28 fails to produce a three-dimensional image. The distance separation of object points is then judged from additional factors of spatial perception such as extension of objects, apparent perspective contraction of size, directions of the shadows and so on.

14.12 Electro-Optical Image Conversion

Electro-optical transducers are invoked when telescopic systems produce retinal illumination close to a threshold value of 5×10^{-9} lx for the eye, or when objects under observation radiate beyond the visible range, for example, in the ultraviolet or infrared.

The electro-optical transducer is used to transform the radiation entering the instrument into the visible range and/or to improve the radiance of the image formed by the basic optical system on the photocathode of the transducer. For visual observation, the transducer converts the visual image first into electron signals to reproduce it finally on a CRT screen.

Figure 14.29 shows a schematic diagram of a telescope employing an image converter tube. The objective lens forms an image of the object on the photocathode of the photomultiplier (PMT) transducer, and the viewer sees the reconstructed image on a screen through an eyepiece playing the role of a magnifying glass. Since the electronic signals (video signals) can be easily handled into any orientation for the image, these systems avoid the necessity of incorporating optical erector systems. These systems also lack the conjugation of rays ahead of the objective lens in object space and behind the eyepiece in image space. Accordingly, there is no mutual compensating of the aberrations of the objective and eyepiece. It is required therefore that these optical components of the system be corrected for aberration in such a way as to completely utilize a low resolving power of the transducer (compared to that of the objective and the eyepiece).

To characterize a telescope with an electro-optical image converter Soviet designers refer to the following parameters: the magnifying power Γ_t, real angular field 2ω, entrance pupil diameter D, system length L, electro-optical (linear) magnification of the transducer $\beta_{eo} = -D_s/D_c$, CRT screen diameter D_s, PMT photocathode diameter D_c, resolving power of the screen N_s, spectral sensitivity of the photocathode ($S_{\lambda',c}$, s_λ, or S_{λ_m}) or integral sensitivity S, photonic gain (efficiency) K defined as the ratio of the light flux Φ_s emitted by the CRT screen into the hemisphere to the flux $\Phi_{e,c}$ incident on the photocathode, luminance of the screen L_s which is a function of the irradiance $E_{e,c}$ of the photocathode and screen brightness factor η_L, and, finally, distance between the photocathode and the screen, i. e., transducer length L_{eo}.

The magnifying power of a telescope with an electro-optical transducer is therefore

$$\Gamma_t = -(f_o'/f_e')\beta_{eo} \qquad (14.33)$$

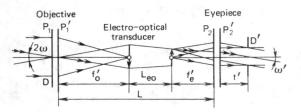

Fig. 14.29. Principle of a telescope with an electro-optical transducer

and the overall length of the system is

$$L = f_o' + L_{eo} + f_e' \tag{14.34}$$

Combined solution of (14.33) and (14.34) gives the focal lengths of the objective and the eyepiece

$$f_o' = \frac{L - L_{eo}}{\Gamma_t - \beta_{eo}} \Gamma_t$$

$$f_e' = \frac{L - L_{eo}}{\beta_{eo} - \Gamma_t} \beta_{eo}$$

For a given true field 2ω and PMT photocathode diameter D_c, we get

$$f_o' = -D_c/2\tan \omega$$

$$f_e' = -f_o'\beta_{eo}/\Gamma_t$$

The apparent field $2\omega'$ in image space, derived by the expression $\tan \omega' = \Gamma_t \tan \omega$, must be matched with the focal length of the eyepiece and the screen diameter D_s, viz.,

$$\tan \omega' = D_s/2f_e'$$

The entrance pupil aperture is adjusted so that for a given radiance of the image L_e the image on the photocathode receives a sufficient irradiance $E_{e,c}'$.

The diameters of the objective and eyepiece can be determined with Eqs. (14.18) and (14.20) for an ordinary telescope.

The image quality on the transducer's screen is inferior to that of the image on the photocathode. It depends on the size of the spread spot δ_0 of the electrons on the screen. The diameter of this spot is found to be [10]

$$\delta_0 = 1.2L_{eo}(U_{max}/E_c)$$

with electrostatic focusing, and

$$\delta_0 = 2L_{eo}(U_{max}/U_a)$$

where L_{eo} is the distance between the photocathode and CRT screen, U_{max} the highest initial energy of electrons, E_c the field intensity at the photocathode, and U_a the anode voltage.

The CRT screens of electro-optical transducers are known for their comparatively low resolving power in the range 20 to 40 mm^{-1}, so that $\delta_0 \approx 0.025$ to 0.05 mm.

The resolution (in mm^{-1}) at the photocathode $N = \beta_{eo}/\delta_0$ must be consistent with the resolving power of the objective so that $N_o \geq N$.

In laying out or selecting objectives and eyepieces for telescopes with image converter systems, the chromatic aberrations must be controlled in the range of sensitivity of the photocathode and the screen. The relative sensitivity plots for the photocathode and screen of an image converter system are exemplified in Fig. 14.30.

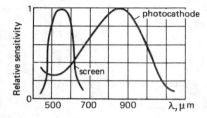

Fig. 14.30. Spectral sensitivity of a photocathode and a CRT screen

15

PHOTOGRAPHIC OBJECTIVES

The photographic objective is an optical system which forms a real image of objects on a sensitive layer of film, PMT photocathode, or TV transmission tube.

Unlike some other optical systems, photographic objectives are corrected for all types of aberrations. The objective lens is a key element of any camera and its properties are of decisive importance for the quality of image.

15.1 Principal Characteristics

The principal characteristics of any objective lens are its focal length f', aperture ratio D/f' (or f-number f'/D), and angular field coverage 2ω. Other important characteristics include resolving power, modulation transfer function (MTF), the distribution of illumination over the image field, the spectral transmissivity, and some other characteristics of the system and materials.

The focal length of objective lens determines the size of image, length of the system, and its power. Distant objects of height y have a smaller height $y' = -f' \tan \omega$ in the image plane. For nearby objects, the scale of image is determined by the transverse magnification $\beta = -f/z$, or with the same medium in front of and behind the lens, $\beta = f'/z$.

For the same object distance z, the image size $y' = y\beta$ will be larger the longer the focal length. Therefore, large-scale imagery calls for long-focus objectives.

Objectives are classified by their value of the ratio of focal length to frame diagonal. For normal camera lenses this ratio is for the most part in the range 0.9 to 1.5. Objectives for which this ratio is less than 0.9 are termed short-focus, and those having this ratio in excess of 1.5 are termed long-focus lenses.

The focal length of modern camera lenses ranges from a few millimetres (for example, the $f/2.5$ OKS-7-1 objective for 16-mm film has $f' = 7$ mm and $2\omega' = 87.5°$) to one metre as in the $f/10$ MTO-1000 objective which has $f' = 1000$ mm, and $2\omega = 2.5°$.

Zoom lenses, as will be recalled, are systems with continuously variable focal length. These lenses are capable of continuous expansion and contraction, within a certain range, of the image on the film plane. For example, the $f/2.8$ Jantar-5 has $f' = $ 40-80 mm and $2\omega' = $ 57°-30° and the $f/1.8$ Vario-Goir-2 has $f' = $ 6.6-66 mm and $2\omega' = $ 58°-6°.

As we have learned in earlier chapters, the illumination of any image produced by an objective is controlled by the aperture ratio (inverse relative aperture) D/f' of the objective. From Eq. (7.66) it will be recalled that the illumination of the image of an axial object point having a luminance L $(n = n')$ is given as

or
$$E' = \frac{\tau\pi L}{4}\left(\frac{D}{f'}\right)^2\left(\frac{\beta_p}{\beta_p - \beta}\right)^2$$

$$E' = \frac{\tau\pi L}{4F^2}\left(\frac{\beta_p}{\beta_p - \beta}\right)^2$$

where $F = f'/D$ is the f-number, or aperture ratio.

For the object at infinity, $\beta = 0$ and

$$E' = \frac{\tau\pi L}{4}\left(\frac{D}{f'}\right)^2 = \frac{\tau\pi L}{4F^2}$$

Observing the ratio $\tau\pi/4$ is a constant, we see from this equation that the transmitting (light gathering) power of a lens varies as the square of its aperture ratio.

The relative aperture of an objective lens defined with account of the transmission factor τ is termed the effective f-number

$$F_{eff} = f'/D\sqrt{\tau} = F/\sqrt{\tau}$$

For optical engineering purposes the values of effective relative apertures are made into a standard series of $f/0.7$, $f/1$, $f/1.4$, $f/2$, $f/2.8$, $f/4$, $f/5.6$, $f/8$, $f/11$, $f/16$, $f/22$, $f/32$, and $f/64$, where a change from one relative aperture to the next doubles or halves the illumination of the image. With reference to the above expressions for image illuminance, this implies that the relative aperture has to change $\sqrt{2} = 1.4$ times.

For extra-axial point of luminance L, the illuminance of the image $E'_{\omega'}$ depends on the amount of vignetting in the optical system, estimated by the vignetting ratio (unvignetted to total aperture ratio) k_ω, and the apparent field $2\omega'$ (see Eq. (7.72))

$$E'_{\omega'} = k_\omega E' \cos^4\omega'$$

The disadvantage of the fourth-power variation is especially sensed in wide-angle systems. This power can be reduced to equalize the illumination

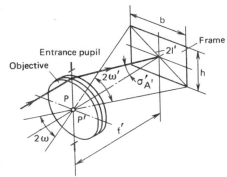

Fig. 15.1. Relation of the angular field with the film gate in a photographic camera

over the image field with the help of the aberrational vignetting technique developed by Rusinov [18].

Recalling that the term 'speed' is used sometimes as a synonym of aperture ratio (inverse relative aperture), the photographic objectives may be loosely classified as extra high-speed (f-number < 1.4), high speed ($1.4 \leqslant F < 2.8$), speed ($2.8 \leqslant F < 5.6$), and low speed ($5.6 \leqslant F$) objectives. The f-number of objectives is varied by changing the aperture of an iris diaphragm which performs as an aperture stop.

The angular field of objective in image space (apparent field) $2\omega'$ defines the size of film format in the film plane. The film format is predominantly of rectangular form with a height h, width b, and the diagonal $2l = \sqrt{h^2 + b^2}$ (Fig. 15.1). A film gate of this size placed ahead of the film plays the role of the field stop.

For a known focal length, the apparent field is $2\omega' = 2 \arctan (l/f')$. The angular field 2ω in object space (true field) is related to the apparent field $2\omega'$ and the angular magnification between the pupils (ratio of entrance to exit pupil angular subtenses) by the formula (see the relevant nomenclature in Fig. 15.2)

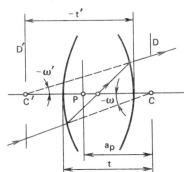

Fig. 15.2. Relationship between the angular field and the angular magnification

$$\tan \omega = \frac{\tan \omega'}{\gamma_p}$$

where $\gamma_p = 1 + a_p/f'$

If the objective is immersed in a uniform medium and the ratio of entrance to exit pupil apertures is unity, the true field and the apparent field are the same ($2\omega = 2\omega'$).

The angular fields in the horizontal and vertical planes can be found with reference to Fig. 15.1 as

$$\tan \omega_g' = b/2f'$$
$$\tan \omega_v' = h/2f'$$

so that $\tan \omega' = \sqrt{\tan^2\omega_v' + \tan^2\omega_g'}$

According to the angular field, the photographic objectives may be categorized into narrow angle ($2\omega' < 40°$), normal ($40° \leqslant 2\omega' < 60°$), wide angle ($60° \leqslant 2\omega' < 100°$), and extra-wide angle ($100° \leqslant 2\omega'$) units.

It is quite natural that at a given extent of field coverage different focal lengths will produce different formats. Some of more popular formats are summarized in Table 15.1.

Expressions relating the principal optical characteristics of objectives suggest that an increase in aperture ratio entails a decrease in angular field, while an increase in the focal length imposes limitations on the aperture ratio and the angular field (to secure a satisfactory imagery of the system).

Volosov [2] investigated the relationship between the optical characteristics of best photographic objectives and has established that for large groups of object lenses there exists a certain invariant within each group for a set of parameters,

D = aperture diameter   f ' = focal length
ω = ½ angle of view

$$C_m = \frac{D \tan \omega}{f'} \sqrt{\frac{f'}{100}} = \frac{\tan \omega}{F} \sqrt{\frac{f'}{100}}$$

This invariant depends on the photographic resolution, falloff of illumination over the image field, elaboration of the optical system, and some other parameters. Quite appropriately, it may be called the quality factor of the objective. Modern anastigmats have C_m in the range 0.22 to 0.24. The design of a lens will be a straightforward matter if C_m is less than 0.20.

Table 15.1. Applications of Film Formats

Format, mm	Application
3.55 × 4.9	8-mm film shooting
4 × 5.36	Super-8 film shooting
7.42 × 10.05	16-mm film shooting
8 × 11	in compact cameras
10 × 14	
14 × 21	
16 × 22	35-mm ordinary film shooting
18 × 24	in semi-format cameras
24 × 36	in mini-format cameras
23 × 52.2	70-mm wide-gauge shooting
6 × 6	'reporter' photography
9 × 12	
13 × 18	technical photography
18 × 18	technical and aerial photography
18 × 24	technical photography
30 × 30	aerial photography
30 × 40	in printing arts
50 × 50	aerophotography
50 × 60	
70 × 80	in printing arts

15.2 Resolving Power and Modulation Transfer Function

Resolving power is a widely used criterion of image evaluation. Numerically it indicates how many object lines or points the image forming system can image without overlapping within a 1-mm stretch. In resolution tests it is usually measured by examining the image of a pattern of alternating bright and dark lines or bars. A target consisting of several sets of bar patterns of graded spacing is used and the finest pattern in which the bars can be distinguished is taken as the limiting resolution of the system under test.

The outlined approach to testing the limiting resolution is seen to be fairly universal and may be used to characterize the performance of objective lenses (N_o), photographic materials, image tubes, etc. (N_{ph}), or the entire objective-film system (N_s), each time the spatial frequency N being measured in mm^{-1}. Sometimes two types of bar targets, or resolution charts, are in use, namely, high- and low-contrast charts.

The number of lines per millimeter in the visual image, N_o, for the perfect lens as tested with a high-contrast resolution chart is defined as

$$N_o = 1/\delta_0'$$

where $\delta_0' = 1.22\lambda F$ is the visual limiting resolution in mm, based on the Rayleigh criterion $1.22\lambda/D$. For $\lambda = \lambda_e = 0.5461 \ \mu m$,

$$N_o = 1500D/f' = 1500/F \tag{15.1}$$

The values of N_o for real objective lenses differ from the value given by (15.1) because of various amount of residual aberrations, unequal image contrast, different test patterns, and so on. For the Jupiter type of anastigmatic lenses in current use, a better approximation can be obtained with the expression

$$N_o \approx 560/F$$

The resolving power of photographic materials, N_{ph}, depends on the target contrast k_t, exposure, development, and some other conditions. If we denote the resolving power of material tested at $k_t = 1$ (absolute contrast) by $N_{ph}^{(1)}$, then for low-contrast measurements ($k_t < 1$), the resolving power can be estimated in terms of spatial frequency [11] as

$$N_{ph} = N_{ph}^{(1)} \sqrt{k_t}$$

The figures of resolving power for some Soviet-made films are summarized in Table A3 [11].

The resolving power of any particular lens-film combination, N_s, is normally connected with the appropriate N_o and N_{ph} values by the approximate formula

$$1/N_{lf} = 1/N_o + 1/N_{ph}$$

or in terms of the δ_0' and photosensitive layer $\delta_{ph} = 1/N_{ph}$

$$\delta' = \delta_0' + \delta_{ph}$$

Practical lenses exhibit somewhat lower values of resolving power because of aberration and light scattering phenomena. For example, the $f/2.8$ Jupiter-12 lens having $f' = 35$ mm and $2\omega = 62°$ is capable (by Eq. (15.1)) of a resolving power of $N_o = 200 \ mm^{-1}$ and when used in conjunction with a film of $N_{ph} = 135 \ mm^{-1}$ is expected to have at the centre of the field

$$N_{lf} = N_o N_{ph}/(N_o + N_{ph}) = 80 \ mm^{-1}$$

whereas the testing reveals $N_{lf} = 60 \text{ mm}^{-1}$.

Table 15.2 correlates the photographic resolving power for high speed lenses with image quality ratings [2].

It is recognized that the objective lenses having uniform resolution over the entire field exhibit better performance, but a characteristic setback of the lenses is that the resolving power falls with the distance from the center of the lens (see Table A4).

The presence of residual aberrations in real optical systems disturbs the purely diffractional distribution in the image illumination of a luminous object point (this ideal pattern called the Airy disc is shown at (b) in Fig. 15.3) so that the intensity in the central spot diminishes and that in the non-central rings rises to produce a point spread pattern such as that of Fig. 15.3a. The background beyond the central spot lowers the contrast and deteriorates the quality of imagery. The final situation may, however, retain a satisfactory resolving power. Hence, resolving power alone cannot be used as a figure of merit in image quality evaluation.

A more substantial representation of lens performance can be obtained with the modulation transfer function (MTF) to be explained below. As a rule, in evaluating the limiting resolution the object is represented as a set of luminous points or bright and dark bars. If we express the contrast in the image (i. e. the ratio of bright minus dark to bright plus dark bar brightnesses) as a 'modulation' we can plot the modulation as a function of the number of lines per millimeter. The smallest amount of modulation which the system can detect will then give the limiting resolution of the system under evaluation.

Any performance evaluation based on bar resolution patterns is in fact based on a square wave brightness distribution whose image illumination

Table 15.2. Lens-Film Resolving Power vs. Image Quality

Image quality	N_{lf} (mm^{-1}) over the field		
	at the center	all over	at the edge
excellent	$\geqslant 40$	> 35	> 35
good	$\geqslant 50$	> 35	30 to 25
fair	50 to 45	35 to 30	25 to 22
satisfactory	45 to 40 or	30 to 25	20
poor		25 to 20	or < 20

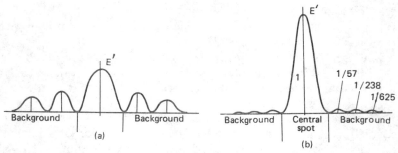

Fig. 15.3. Radiance distribution (point spread function) in the image of a point object (a) in real cameras, (b) in the Airy disc

distribution is distorted by the characteristics of the optical system under study. However, if the object pattern is in the form of sine (or cosine), wave, the distribution in the image is also described by a sine (cosine) wave, regardless of the shape of the spread function (the line spread function is convolved with the object brightness distribution function to produce the image function). This fact has led to the widespread use of the optical modulation transfer function to characterize the performance of a lens system. Physically, the modulation transfer function is the ratio of the modulation in the image to that in the object as a function of the spatial frequency (cycles per unit of length) of the sine wave pattern, $\text{MTF}(N) = M_i/M_o$. A plot of MTF versus frequency N is thus an almost universally applicable measure of the performance of an image forming system. It can be applied not only to lenses, but to films and image forming tubes.

If we assume the brightness (luminance, radiance) of a surface to be imaged as consisting of alternating light and dark bands the brightness of which varies according to a cosine (or sine) function, as indicated by the upper part of Fig. 15.4, the distribution of brightness can be expressed mathematically as

$$L(x) = L_{av} + L_a \cos 2\pi N x \tag{15.2}$$

where L_{av} is the average brightness, L_a the amplitude of brightness variation along the spatial coordinate x perpendicular to the bands, and N the spatial frequency characterizing the size of the bars and being the inverse period of brightness variation.

In a real object surface the sine components of contributing gratings of brightness bars may differ from one another in the amplitude proportional to the distribution of brightness, in phase, i. e. orientation in the plane, and in spatial frequency. The sine function, invariable on passing through

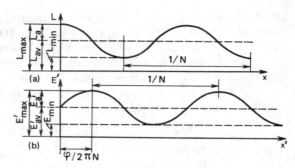

Fig. 15.4. Sine variation of brightness in an object pattern (*a*), and image illuminance distribution (*b*)

the lens system, receives an amplitude and phase dependent on the properties of the optical system. The image distribution of illumination is obtained by summing up the sine components of illumination in the image plane. The distribution of illumination in the image of an elementary object (point or line) at x' has the form

$$E'(x') = E_{av} + E_a T_N \cos(2\pi N x' - \varphi)$$

where T_N is the *modulation transfer factor*, and φ is a phase shift.

The quality of imagery in this case can be assessed from the amplitude and phase (shift over the field) that a certain spatial frequency receives on the passage through the lens under test. A figure of merit used to estimate the performance of lens systems is the complex *optical transfer function* (OTF). This function is specifically tailored for use with continuous tone imagery and is described with the help of the Fourier transformation. The Fourier transform of the image energy distribution, $\tilde{E}'(N_x, N_y)$, is the product of the optical transfer function $\tilde{A}(N_x, N_y)$ and the Fourier transform of the object brightness distribution $\tilde{L}(N_x, N_y)$, namely,

$$\tilde{E}'(N_x, N_y) = \tilde{A}(N_x, N_y)\tilde{L}(N_x, N_y)$$

For a unidimensional object (point or line), the optical transfer function $\tilde{A}(N)$ can be found by taking the Fourier transform of the point (line) spread function (image irradiance distribution) as follows [8]

$$A(N) = \int_{-\infty}^{\infty} A(x') \exp(-i2\pi N x')dx'$$

$$= \int_{-\infty}^{\infty} A(x')\cos(2\pi N x')dx' - i \int_{-\infty}^{\infty} A(x')\sin(2\pi N x')dx'$$

$$= T_c(N) - iT_s(N)$$

$$= |A(N)|e^{i\varphi(N)} = T(N)e^{i\varphi(N)}$$

where $T(N) = \sqrt{T_c^2(N) + T_s^2(N)}$ and $\varphi(N) = \arctan T_s(N)/T_c(N)$ are the *modulation transfer function* and the *phase transfer function*, or phase shift, introduced if the line spread function $A(x')$ is asymmetrical. The OTF reduces to the MTF whenever the PTF is zero.

The modulation transfer function $T(N)$ describes how the modulation transfer factor T_N depends on spatial frequency.

The modulation (contrast) of object surface luminance distributed according to the sine law (15.2) is

$$M_o = (L_{max} - L_{min})/(L_{max} + L_{min}) \qquad (15.3)$$

where, as can be seen from Fig. 15.4a, $L_{max} = L_{av} + L_a$ and $L_{min} = L_{av} - L_a$. Incorporating these expressions for L_{max} and L_{min} in (15.3) we get $M_o = L_a/L_{av}$. Physically this coefficient indicates the variation of the amplitude value of brightness about its average level.

The modulation of image irradiance can be seen from Fig. 15.4b

$$M_i = (E_{max} - E_{min})/(E_{max} + E_{min})$$

where $E_{max} = E_{av} + E_a$ and $E_{min} = E_{av} - E_a$.

The imagery in a lens is a linear process, therefore we may let $E_{av} = aL_{av}$ and $E_a = aT_N L_a$ so that $M_i = L_a T_N/L_{av}$, or in view of $M_o = L_a T_N/L_{av}$,

$$T_N = M_i/M_o$$

One particular advantage of the modulation transfer function is that it can be cascaded by simply multiplying the MTF's of two or more components to obtain the MTF of the combination, say lens and a detector.

The MTF plots for four Soviet made objectives are illustrated in Fig. 15.5.

To derive the limiting resolution of a lens-film combination the concept of the *threshold modulation curve* (TM) is most widely used by designers of photographic cameras. The TM curve for a photographic emulsion represents the sinusoidal modulation which must be present in a three-bar target at its fundamental spatial frequency if the target is to be resolvable in the developed photograph. The high contrast limiting resolution for a lens-

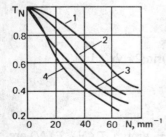

Fig. 15.5. Modulation transfer functions for some Soviet lenses: *1* Jupiter-12, *2* Vega-7, *3* Hindustar-50, *4* Vega-3

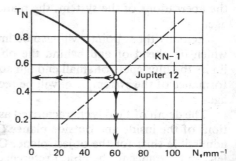

Fig. 15.6. Graphical evaluation of the limiting resolution for a lens-film combination

film combination is determined by plotting the lens MTF, such as that shown in Fig. 15.5, and the film TM curves on the same graph and noting the spatial frequency at which they intersect. By way of example, Fig. 15.6 demonstrates the evaluation of the limiting resolution for a Jupiter-12 lens used in conjunction with the KN-1 film (see Table A3).

15.3 Depth of Field

The concepts of depth of focus and depth of field relate to the axial distances within which the object and image planes can be shifted without distorting a tolerable quality of the image.

The tolerable shift of the object plane defines the depth of field. Referring to Fig. 15.7, let Q be the object plane conjugated with some reference plane Q' (e. g. film or reticle) in image space, and p be the distance from the entrance pupil D of the optical system to the object plane. If we neglect

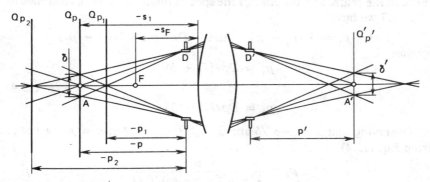

Fig. 15.7. Depth of field and depth of focus

the aberrations of the system, the plane Q' will be the plane of point image.

Actually those points will also be imaged as blurs of small enough size which lie ahead of and behind the object reference plane. Within some limits the blur spots of small enough size will not adversely affect the performance of the system, i. e. will be recognized as sharp enough point images.

The depth of field is characterized as the difference $p_1 - p_2$ of the positions of the inside and outside planes Q_1 and Q_2 confining the amount of admissible shift for the object plane. Ordinarily the characteristics of the lens and distance to the object reference plane are specified so that p_1 and p_2 must be computed to determine the depth of field for this system.

Let δ' be the largest admissible blur spot size (diameter) and δ be the corresponding size of the object spot (see Fig. 15..7). The size δ' is determined by the conditions of observation, i. e. by the angular resolution of the eye ψ_{eye} and the observation distance p_{eye}, and also the scale to which the photograph was made. In the general case, if β_e is the linear magnification of the enlarger providing the given scale, then

$$\delta' = -p_{eye}\psi_{eye}/\beta_e$$

If the film is viewed directly, then $\beta_e = -1$.

In practical estimations of depth of field, it is common to take δ' equal to 0.03 mm for 35-mm film shooting, 0.015 mm for narrow-gauge film shooting, and 0.05 mm for photography.

More accurate approximations for δ' can be obtained from the blur spot derived with account of the aberrations and the distribution of energy in the spot.

To determine the distances p_1 and p_2 we examine the similar triangles formed by the rays touching the edge of the entrance pupil and passing through the centre and the edge of the spot of diameter δ. With reference to Fig. 15.7 we have

$$D/-p_1 = \delta/(p_1 - p) \quad \text{and} \quad D/-p_2 = \delta/(p - p_2)$$

whence

$$p_1 = Dp/(D + \delta)$$

$$(15.4)$$

$$p_2 = Dp/(D - \delta)$$

Observing that $\beta = -\delta'/\delta$ and $\beta = f'/(s_1 - s_F)$ at $n = n'$, we have from Eq. (15.4)

$$p_1 = \frac{p}{1 - \delta'(s_1 - s_F)/f'D}$$

$$p_2 = \frac{p}{1 + \delta'(s_1 - s_F)/f'D}$$

or

$$p_1 = \frac{p}{1 - \delta'(s_1 - s_F)/FD^2}$$

$$p_2 = \frac{p}{1 + \delta'(s_1 - s_F)/FD^2}$$

or

$$p_1 = \frac{p}{1 - \delta'F(s_1 - s_F)/f'^2}$$

$$p_2 = \frac{p}{1 + \delta'F(s_1 - s_F)/f'^2} \qquad (15.5)$$

where $F = f'/D$ is the aperture ratio, or f-number, of the system.

Example 15.1. Determine the depth of field for the photography of objects at a distance $p = -2$ m by an $f/4$ camera with $f' = 50$ mm, $s_F = -46$ mm, $t = 40$, and $\delta' = 0.05$ mm.

Solution. Observing that $s_1 = p + t = -2000 + 40 = -1960$ mm and $s_1 - s_F = -1960 + 46 = -1914$ mm we get $p_1 = -1735$ mm and $p_2 = -2365$ mm. The depth of field is therefore equal to 630 mm.

The mass produced photographic and cine cameras are permanent-focus designs. We determine, by way of example, the depth of field for a camera lens whose film plane coincides with the second focal plane of the lens. In this case the object plane is at infinity $(p = -\infty)$. The outside plane at the object side is also at infinity, while the inside object plane is at a distance $p_{1\infty} = -f'^2/\delta'F$ (see Eq. (15.5)). With this type of focusing a portion of the depth of field is out of use as both p and p_2 are at infinity. In order to gain the maximum depth of field we assume that only $\bar{p}_2 = -\infty$ and the object reference plane is at a distance $\bar{p} = -p_{1\infty}$, called the *hyperfocal distance*. Now the inside object plane is at the distance $\bar{p}_1 = -f'^2/2\delta'F$. We see that the inside object plane appears half way closer than in the former case. This distance, $\bar{p}_1$, may therefore be referred to as the 'front edge of infinity'.

Turning now to the depth of focus we define it as the amount by which the plane of the film may be shifted longitudinally with respect to the reference plane of sharp point imagery (Fig. 15.8). The spread of a point image due to defocus is characterized by the blur spot diameter δ' which in-

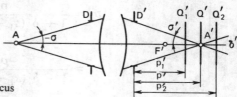

Fig. 15.8. Depth of focus

creases away from the reference plane Q'. With reference to Fig. 15.8, the depth of focus is $p_2' - p_1' = \Delta p_{2,1}' = 2(p' - p_1')$ where $p' - p_1' = \delta'/2 \tan \sigma'$, i. e. $\Delta p_{2,1}' = \delta'/\tan \sigma'$.

Since $\tan \sigma' = D'/2p'$, $D' = D\beta_p$, and $p' \approx f'$, then for $\beta_p = 1$

$$\Delta p_{2,1}' = 2\delta' F$$

or

$$\Delta p_{1,2}' = \pm\delta' F$$

15.4 Light Exposure in Photography

The light exposure of a film is defined as the product of the illuminance at the emulsion E' and exposure time t (see also Eq. (7.18))

$$H = E't \qquad (15.6)$$

In photographic cameras, a light exposure is controlled by setting an f-number and a speed of the shutter (exposure time). The opacity (optical density) D of developed emulsion is proportional to the light exposure. In turn, the light exposure is related to the emulsion speed S of the film (in GOST units) as

$$S = 10/H_D$$

where H_D is the light exposure providing an optical density $D = D_0 + 0.85$ exceeding the one of the unexposed film D_0 by 0.85.

For many types of black and white emulsions, D_0 is in the region 0.05 to 0.15, and for the foto-250 film (see Table A3) used at artificial illumination, $D_0 = 0.2$.

The opacity of exposed areas of a film is not directly proportional to light exposure in all the region of exposures. A typical D vs. log H curve for a photographic emulsion is plotted in Fig. 15.9. Designers of photographic equipment often use the slope of this curve to characterize a particular emulsion; $\tan \alpha = \gamma$ is referred to as the *contrast coefficient of a photosensitive layer*,

$$\gamma = (D_2 - D_1)/(\log H_2 - \log H_1)$$

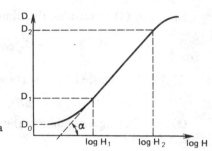

Fig. 15.9. Optical density plot for a photographic emulsion

It is quite obvious that the portion of the curve below the linear segment related to *underexposure*, whereas that above the linear segment relates to *overexposure*.

The contrast coefficient for ordinary negative roll films is within 0.8 to 1.1, for cine negative films it is within 0.65 to 0.85, for sound films it is 2.4, for positive films it equals 2.4, for the Mikrat-200 film it is 3, and for phototechnical copy film it is in the range from 1.8 to 3.6. For colour negative film this coefficient lies within 0.65 to 0.85, reversible colour film has it in the range 1.4-1.9, and positive colour film has it in the range 2.1 to 3.3.

To correctly convey the energy distribution on the object surface an object area with the least luminance, L_{min}, must correspond to the area of lowest illuminance, hence with the least opacity, in the developed film, and this opacity must exceed the optical density of the unilluminated fog.

The image illuminance E'_ω of extra-axial points is known to be lower than that of axial points E'_0 (see Eq. (7.72)), therefore in assigning a time of exposure this fact should not be overlooked. Combining Eqs. (15.5), (15.6) and the expressions for axial point illumination, we obtain the expression for exposure time in photography

$$t = \frac{40F^2}{k_\omega \tau \pi \, L_{min} \, \cos^4 \omega' S} \left(\frac{\beta_p - \beta}{\beta_p} \right)^2$$

Example 15.2. Determine the time of exposure in photographing a printed material of reflectivity $\rho = 0.6$ illuminated from a distance of 450 mm by an electric lamp of luminous intensity $I = 60$ cd at an angle of incidence $\varepsilon = 30°$. The film used is foto-65 (i. e. $S = 65$ units of the GOST classification). The objective data are as follows: $2\omega' = 47°$, $F = 2$, $\beta = -0.15$, $\beta_p = 1$, and no vignetting $k_\omega = 1$. The transmission factor of the objective is $\tau = 0.9$.

Solution. (1) Determine the illuminance at the text surface

$$E = \frac{I}{r^2} \cos \varepsilon = 257 \text{ lx}$$

(2) The luminance of the surface having a reflectivity ρ is

$$L = \rho E / \pi = 49 \text{ cd/m}^2$$

(3) Substituting these data in the expression for exposure time yields the shutter speed

$$t = \frac{40 \times 4(1 + 0.15)^2}{65 \times 1 \times 0.9 \times 3.14 \times 49 \times 0.71} \approx \frac{1}{30} \text{ s}$$

15.5 The Principal Types of Photographic Lenses

Optical system designers classify photographic objectives according to the number and form of the lenses, sign of their focal length, arrangement of powers and spacings in a specific design, i.e. by the optical formulation. The optical data used to classify photographic lenses include the focal length f', relative aperture f'/D (or aperture ratio D/f' if in a Soviet specification), angular field coverage 2ω, or film format (see Section 15.1).

By their specific application, photographic lenses may be classed into those for photography, aerial photography, motion picture applications, television camera lenses, infrared objectives, and radiographic lenses.

A classification may also be based on the lens design geometry. Accordingly, normal lenses are those having the effective focal length longer than the back focal length and shorter than the distance from the front surface to the image plane. If the focal length of a photographic objective is equal or exceeds the distance from the front surface to the image plane, this design is called a telephoto lens. Conversely, if the effective focal length of the lens is equal or less than the back focal length, then this design is a reverse telephoto lens.

According to the degree of correction of aberrations involved in a particular design, the lens can be an achromatic, apochromatic, aplanatic, or anastigmatic design form. Commercial photographic objectives are mainly anastigmats with an achromatic or even apochromatic degree of correction of the chromatic aberrations. Accordingly these objectives are three, or more, lens formulations.

In this section, we outline the optical characteristics of certain types of photographic objectives. Some Soviet-made photographic lenses, which will be referred to for illustration of design, are depicted in Fig. 15.10 and summarized in Table A4. The largest group is constituted by objectives qualified as 'universal'. These units are noted for the moderate optical characteristics, i.e., their relative apertures never exceed $f/2.8$ and the angular field $2\omega' < 60°$. This group includes the Triplet, Vega and Industar types of objective.

The Triplet type is a three-lens objective widely used in simple commercial cameras (Fig. 15.10a). Most objectives of this group have a relative aperture of $f/4$. Lanthanum glasses have led to faster formulations of relative aperture $f/2.8$. We direct the readers who wish to know more about the design of Triplet forms to the books by Klimkov [7] (in Russian) and by Slyusarev [26].

The more popular anastigmatic lens is the Hindustar format which is given a higher degree of correction than to the Triplet lens and therefore notable for a better performance. The Hindustar-61 objective (Fig. 15.10b) made use of lanthanum glasses and demonstrates a rather high limiting resolution.

The Vega objectives are, by their performance characteristics, midway between the universal group and the high-speed objectives. They provide a

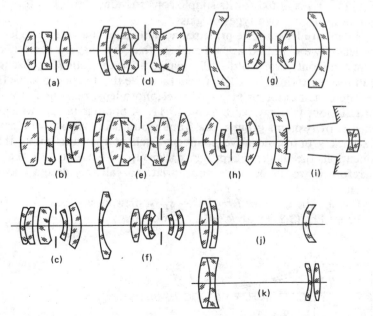

Fig. 15.10. Photographic objectives: (a) Triplet T-43, (b) Hindustar-61, (c) Vega-1, (d) Jupiter-8, (e) Helios-44, (f) Mir-1, (g) MR-2, (h) Orion-15, (i) MTO-500, (j) Tair-3, (k) Telemar-22

fairly good imagery at the relative aperture $f/2.8$ and an angular field coverage up to 50°. The Vega-1 (Fig. 15.10c) and Vega-3 forms are five-element objectives.

A large group is constituted by high-speed photographic objectives having relative apertures of $f/2$ and covering angular fields of 40° to 50°. The Jupiter and Helios types of objective belong to this group. The Jupiter-8 type of lens is shown in Fig. 15.10d, and the Helios-44 lens in Fig. 15.10e. These are six-lens designs; the former is compounded of five types of glass while the latter of three types of glass but the last design is a good field flattener. This group of objective lenses also includes formulations of somewhat increased focal length equal to 85 mm; these are the Jupiter-9 and Helios-40 lenses.

The wide-angle objective group consists of such design forms as Mir, Orion, Jupiter-12, and MR-2. The Mir-1 (Fig. 15.10f), Mir-10 and Jupiter-12 designs may be related to high-speed wide-angle objectives, the Mir formulations having somewhat higher resolution at the edge of the field. The Mir-10 lens is compounded of comparatively few types of glasses. The MR-2 objective designed by M.M. Rusinov has the wider angular coverage in this group (Fig. 15.10g) and the Orion-15 type (Fig. 15.10h) is noted for its simple construction — this is a four-lens formulation of only two types of glass.

Telephoto lenses. The principal advantage of the telephoto lenses over the lenses we qualified as 'normal' is that they have a shortened length from the front vertex to the film plane, L. In the following discussion we shall use the ratio $k_t = L/f'$ to characterize this telephoto lens effect.

The construction principle of a telephoto lens, reckoned as a thin lens formulation, is suggested by Fig. 15.11. A popular telephoto lens design consists of two lens groups, the front group being a positive member and the back group a negative member. Each of the indefinitely thin components in the diagram represents a group of lenses of the real system, therefore it would be more appropriate to call this design a two-group system.

Usually the distance $L = \Sigma d + s'_{F'}$ somewhat exceeds $L_0 = d + a'_2$.

From Eq. (3.27) with $k_t = L/f'$ we have

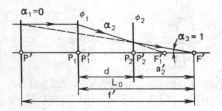

Fig. 15.11. A telephoto lens system

$$\phi_1 = \phi + (1 - k_t)/d \tag{15.8}$$

$$\phi_2 = (k_t - 1)/d(k_t - \phi d) \tag{15.9}$$

$$a_2' = (k_t - \phi d)/\phi \tag{15.10}$$

or, rewriting in terms of the focal length f' instead of powers ϕ,

$$f_1' = \frac{f'd}{d + f'(1 - k_t)} \tag{15.11}$$

$$f_2' = \frac{d(f'k_t - d)}{f'(k_t - 1)} \tag{15.12}$$

$$a_2' = k_t f' - d \tag{15.13}$$

Let us determine the largest f_2' by differentiating it with respect to d in (15.12) and equating the derivative to zero. Then

$$\bar{d} = f' k_t/2 \tag{15.14}$$

and upon substitution into (15.8)-(15.10)

$$\bar{f_1'} = \frac{f'k_t}{2 - k_t} \tag{15.15}$$

$$\bar{f_2'} = \frac{f'k_t^2}{4(k_t - 1)} \tag{15.16}$$

$$\bar{a_2'} = f'k_t/2 \tag{15.17}$$

The formulae (15.14)-(15.17) suggest the optimal parameters of a telephoto lens derived on the condition that the power of the back group is a minimum by absolute value.

Figure 15.12 illustrates the paths of rays, and the positions of entrance pupil D, exit pupil D', and aperture stop in a two-member telephoto lens. The clear aperture of the front group is usually equal to the clear aperture of the objective, i.e. $D_1 = D = f'/F$. The lower ray of an oblique bundle of rays then traverses the front component at a height $D_1/2$, and the principal ray defines the position t of the entrance pupil, which with account of

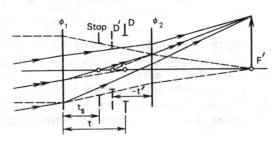

Fig. 15.12. Positions of the pupils and aperture stop in a telephoto lens

the vignetting ratio k_ω (relative clear aperture) is as follows

$$t = D(k_\omega - 1)/2\tan \omega_1$$

The position of the aperture stop is given by Eq. (3.7) as

$$t_s = tf_1'/(t + f_1')$$

The position of the exit pupil t' with respect to the second group can also be derived by the equation of conjugate distances

$$t' = (d - t_s)f_2'/(d - t_s + f_2')$$

The clear aperture of the second group can be traced by the upper oblique ray whose height at the first component can be determined by tracing the ray with equations for angles (4.14) and heights (4.15).

Telephoto lenses find their use predominantly where large focal lengths are essential. As a rule this type of objective lens has $k_t \approx 0.8$, angular fields of $2\omega \leqslant 30°$, and relative apertures up to $f/2.8$.

To step down the ratio k_t, the back group must be dispersive (of negative power). Sometimes the back group is selected to have a positive power, but this choice makes k_t very nearly unity (Jupiter-11 and Jupiter-16, for example).

Let us now look at three-group formulations. We use in succession the formulae for angles (3.21) and heights (3.22) to derive the expressions for the power of two systems (see Section 3.9) at $\alpha_1 = 0$ and $\alpha_4 = 1$ and arrive at the principal equations of three-group system constituted by indefinitely thin components

$$\phi = \phi_1(1 - \phi_3 d_2) + (1 - \phi_1 d_1)(\phi_2 + \phi_3 - \phi_2 \phi_3 d_2)$$

and

$$a_3' = f'[(1 - \phi_1 d_1)(1 - \phi_2 d_2) - \phi_1 d_1]$$

For $f' = 1$, k_t is numerically equal to $L = d_1 + d_2 + a_3'$, and

$$\phi_1 = \frac{1 - \phi_2 - \phi_3 + \phi_2 \phi_3 d_2}{1 - \phi_2 d_1 - \phi_3 d_1 + \phi_2 \phi_3 d_1 d_2}$$

$$\phi_2 = \frac{k_t(1 - \phi_3 d_1 - \phi_3 d_2) + \phi_3(d_1 + d_2)^2 - 1}{k_t d_1(1 - \phi_3 d_2) + d_1^2(\phi_3 d_2 - 1) + \phi_3 d_1 d_2^2}$$

The three-group formulation involves a greater number of parameters, $\phi_1, \phi_2, \phi_3, d_1$, and d_2, therefore some of them, say ϕ_3, d_1 and d_2, are normally specified in advance.

The diagrammatic sections at (j) and (k) in Fig. 15.10 are for the Tair-3 and Telemar-22 objectives made as the split front and split back lens formulations. Their k_t ratios are respectively equal to 0.96 and 0.8.

Some reflecting lenses may also be related to the telephoto lens systems. The L/f' ratio for these forms is considerably lower than for lens designs. For instance, the MTO-500 reflecting objective (Fig. 15.10i) has k_t as low as 0.32.

The reflecting objectives are superior to lens designs in that they are free from chromatic aberrations. The Cassegrain objective system (Fig. 15.13) is used in a great variety of applications because of its compactness and the fact that the second reflection places the image behind the primary mirror where it is readily accessible.

The telephoto lens ratio for a two-mirror reflecting system is defined as

$$k_t = s'_{F'}/f'$$

The focal length of the reflecting objective system is given by

$$f' = \frac{r_1 r_2}{2(r_1 - r_2 - 2d)}$$

and the back focal length by

$$s'_{F'} = \frac{r_2(r_1 - 2d)}{2(r_1 - r_2 - 2d)}$$

The constructional data of a Cassegrain system depend on the specified values of f', $s'_{F'}$, and d. For $f' = 1$, $h_1 = 1$, $\alpha_1 = 0$ and $\alpha_3 = 1$, we have from the formula for incidence heights $h_2 = h_1 - \alpha_2 d$, or in view of $h_2 = s'_{F'}$, $\alpha_2 = (1 - s'_{F'})/d$.

The formula for the radius (4.16) yields, for $n_1 = n_3 = 1$ and $n_2 = -1$, $r_1 = 2/\alpha_2$ and $r_2 = 2s'_{F'}/(1 + \alpha_2)$.

The determination of the relative aperture for the Cassegrain system is based on the annular form of the entrance pupil. The area of the entrance pupil (see Fig. 15.13) equals $\pi(h^2_{1u} - h^2_{1l})$ which gives for the equivalent

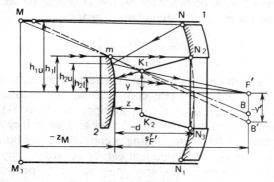

Fig. 15.13. A Cassegrain reflecting system with baffles

circular entrance pupil

$$D_{eq} = 2\sqrt{h_{1u}^2 - h_{11}^2}$$

Now in radiative transfer computations one may use the relative aperture in the form f'/D_{eq}. The angular limiting resolution can be determined from the diameter of the axial bundle $D = 2h_{1u}$.

To prevent stray radiation from flooding the image area, Cassegrain systems have buffles installed both inside and outside the reflecting system. Fig. 15.13 indicates the type of buffles frequently used to overcome the stray radiation problem. The exterior hood NMM_1N_1 is a cylindrical extension of the main exterior tube of the instrument. This hood is often used in addition to the internal cone buffle $N_2K_1K_2N_3$.

The dimensions of the buffles are determined by tracing the critical ray MmK_1B. The rays crossing the axis at smaller inclinations than this ray will not be passed by the secondary mirror and internal baffle. The rays having greater inclinations will not be passed by the exterior cylindrical baffle.

In the coordinate system with the origin at the vertex of the secondary mirror, point K_1 will have the coordinates z and y (Fig. 15.13) which can be derived with the following approximate expressions

$$z = \frac{(h_{2u} - h_{21})s_{F'}}{(h_{11} - h_{21})s_{F'} + h_{2u}d}\, d$$

$$y = \frac{s_{F'} - z}{s_{F'}}\, h_{2u}$$

The ray MB is seen to pierce the image plane at a distance

$$F'B = (y - h_{11})s_{F'}/z + h_{11}$$

If the size of the useful image y' does not exceed $|F'B|$, then the exterior cylindrical baffle is not necessary. However, if $|y'| > |F'B|$ then a cylindrical exterior baffle of diameter $2h_{1u} = D$ should be installed, its extension from the vertex of the secondary mirror being

$$z_M = \frac{h_{1u} - h_{11}}{y' - h_{11}}\, s_{F'}$$

To achieve a high quality imagery Cassegrain systems make use of aspheric mirrors or additional lens components. The catadioptric system (Fig. 15.14) may then consist of a front compensating lens (I) operating in parallel rays, a mirror system (II) of a primary and secondary mirror, and a rear lens compensator (III) intercepting a converging ray bundle.

The front compensating component can be constituted by a few lenses (up to three) with spherical surfaces, one Schmidt corrector lens with aspheric surface, or an achromatic Maksutov meniscus convex toward the

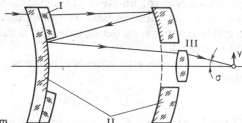

Fig. 15.14. Catadioptric objective system

object. It is important that these front compensating systems installed to correct the spherical aberrations of the two-mirror system should not introduce chromatic aberrations.

The rear correcting element tends to correct the coma and field curvature. Most often this is a singlet or doublet. Additional degrees of freedom for correction of aberrations can be achieved by making both mirrors as second surface reflectors, but the resultant construction is more sophisticated. This problem will be examined at length in Section 21.14.

Zoom lenses. Zoom objectives are variable power systems which allow continuous variation of focal length, in a certain range, as a result of which a continuous scaling of the image is produced. Variation of image scale is achieved by moving certain groups of lenses within the objective along the axis. The image (film) plane must remain fixed during the scaling manipulations. To achieve a fixed relationship for the film plane, at least two groups of lenses should move simultaneously in mutually compensating directions.

In some systems, called 'variobjectives' in the Soviet literature, a variation of power is achieved by moving all groups of lenses in the zoom system. To contrast, the systems called 'transfokators' consist of a zoom lens attachment, which is a variable power afocal device, and a fixed image forming lens.

Zoom systems in which the defocusing of the image (arisen in varying the magnification) is eliminated by the nonlinear compensating motion of one of the other two elements of the system are referred to as *mechanically compensated*. Since the shift of the compensating element is nonlinear, it is usually effected by a cam arrangement, hence the modifier 'mechanical' in the name of the method. The zoom systems in which the image plane defocus is compensated by a linear shift of certain components are called 'optically compensated' systems.

A key characteristic of zoom systems is the ratio m showing the 'magnifying effect' of a particular system:

$$m = \beta_{max}/\beta_{min} \quad \text{or} \quad m = f'_{max}/f'_{min}$$

Objectives for amateur photography have m in the range 2 to 4, and amateur cine cameras have m predominantly within 4 to 6. The current trend in zoom system development is to attain magnifying effects as high as 20 or even 40.

Afocal zoom attachments are characterized by the angular magnification ratio

$$m = \gamma_{max}/\gamma_{min}$$

If the focal length of the fixed prime lens is f'_p then for the overall system $f'_{max} = f'_p\gamma_{max}$ and $f'_{min} = f'_p\gamma_{min}$.

A simple varifocal system may be imagined as consisting of two thin components of focal lengths f'_1 and f'_2 spaced at a distance d. For this system Eq. (3.27) yields

$$f' = f'_1 f'_2/(f'_1 + f'_2 - d)$$
$$a'_{II} = f'(f'_1 - d)/f'_1$$

The distance a_{II} defines the separation of the second lens from the image plane. A continuous variation of the interlens spacing d results in a continuous variation of the effective focal length of the system.

Two-component zoom systems can provide a magnifying effect up to 20 with an appropriate selection of the component powers, therefore this type of system formed a base for most zoom designs.

Zoom systems with a nonlinear movement of a system member imply a sophisticated lens mount, therefore linear motions are mechanically more attractive. One way to linear motion is by linking two components to move together with respect to a fixed lens between them. As a rule, the active subsystem is followed by a fixed lens, thus increasing the number of zoom system components to four. In a real zoom system, each component may be compounded of a few lenses each of which plays its own role in forming internal images and in the production of a real image of a desired scale in the film plane.

In order to avoid defocusing of the image plane when two linked lenses (or groups) move about in a zoom system, the powers and spaces of the system components are so chosen that the image remains in exact focus or close to this point. Such systems are therefore called 'optically compensated'. Two forms of such systems are more popular than the others. Fig. 15.15a shows a design with two positive components moving with respect to a negative lens between them, and Fig. 15.15b shows a system with two negative components linked to move with respect to a positive lens inbetween. For example, the five-lens zoom system Rubin-1 shown in Fig. 15.16 consists of three fixed groups, labelled I, III, and V, and two linked positive groups, II and IV, which can linearly move with respect to the first groups; optical characteristics for this objective are listed in Table A4.

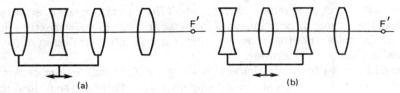

Fig. 15.15. Four-component zoom lens designs

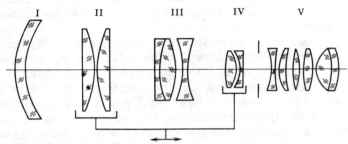

Fig. 15.16. The Rubin-1 zoom lens

Reverse telephoto lenses. By reversing the basic power arrangement of the telephoto lens, one can achieve a back focal length which is longer than the effective focal length. This arrangement is convenient to introduce, in the space between the objective and film, a mirror to bend a portion of the bundle into a sighting device of the camera. Fig. 15.17 suggests that such an objective can be constructed by selecting a negative front group and a positive rear group.

The reverse telephoto lenses for photographic and cine applications are designed such that the aperture stop and the exit pupil almost coincide in the thin lens design and both appear in the principal plane of the second group, as indicated in Fig. 15.17a. In the reverse telephoto lenses for colour television, the aperture stop must be located near the first focal point

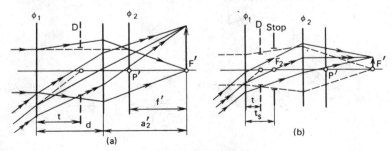

Fig. 15.17. Schematic diagrams of reverse telephoto lenses

of the second group, as indicated in Fig. 15.17b. This arrangement ensures a telecentric course of principal rays in image space. This power setting is optimal for colour photography, too, but it is difficult to realize in objectives with large aperture ratios.

In addition to the basic characteristics, f', f'/D, and 2ω, the performance of reverse telephoto lenses may be assessed in terms of the following parameters: (1) back focal length ratio $k_s = s'_{F'}/f'$, (2) telephoto lens ratio $k_t = (\Sigma d + s'_{F'})/f'$ which is more than unity and often attains values as high as 3 or 7, (3) lens diameter ratio $k_D = D_1/f'$, and (4) dimensional ratio $k_d = k_s/k_t$.

In real reverse telephoto lenses, k_d is about 0.13 to 0.48. Larger k_d values mean compact constructions, but the higher this ratio at a given k_s the harder the system lends itself to a good control of aberrations.

Wide-angle lens systems. It is quite obvious that in order to image an angle of 180° or more on a finite sized flat film, a large amount of distortion is unavoidable. Quite appropriately, lens systems serving this purpose are referred to as 'distorting objectives' in the Soviet literature on the subject. When a lens has a considerable amount of negative distortion, the angular field coverage may be made in excess of 180°. Such lens forms find their use in meteorologic and space applications.

The size of image is now determined with the formula $y' = -f' \sin \omega$ rather than with $y' = -f' \tan \omega$. For $-\omega = 90°$ the sine formula gives $y' = f'$, that is, the image diagonal will be twice the focal length.

The principal design form of the wide angle lens is depicted in Fig. 15.18a. The early (1930) design of this system was devised by Gill in the form shown in Fig. 15.18b. The lens had a relative aperture of $f/22$ and

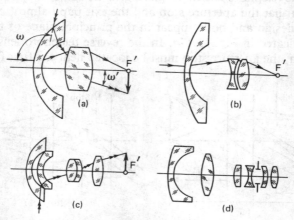

Fig. 15.18. Optics of extra-wide-angle "distorting" objectives

covered a total angular field of $180°$.

Distorting objectives are designed in the same power formulations as reverse telephoto lenses. The front group consists of one or two elements and is a strongly distorting member (Fig. 15.18c). The second group is designed so as to correct the aberrations to achieve a sharply defined image.

The effect of the cosine fourth power law is a stumbling block in the way to extremely wide-angle lens development. However, the provision of negative distortion at the edge of the image field takes care, as it were, of ray bundle concentration in this portion of the field so that the optical density at the periphery is practically not inferior to that in the centre of the field.

Concentric objectives. Concentric lenses, as the name implies, have all their spherical surfaces centred on a common origin. These systems form the image at a concave spherical surface. The entrance and exit pupils of the system are also centred at the system origin. A principal ray passes through the system unrefracted, i.e. acquires the properties of a ray passing along the optical axis. The axial and oblique bundles are also identical (Fig. 15.19), therefore the control of aberrations in this system reduces to keeping the spherochromatic aberration as low as possible. A concentric objective does not suffer either from come, or astigmatism, or distortion.

These systems are notable for their wide field coverage ($2\omega \approx 130°$) and comparatively large relative aperture (about $f/2$). Sutton devised such a spherical objective as far back as 1859.

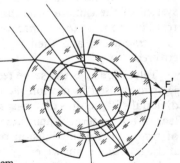

Fig. 15.19. A concentric lens system

OPTICS OF TELEVISION SYSTEMS

16.1 Camera and Picture Tubes

In very general terms, a telecommunication system (Fig. 16.1) consists of a transmitting terminal, *1*, a communication link, *2*, and a receiving terminal, *3*.

At the transmitting end, the basic element of the terminal is a television camera. It converts the visual information on a scene being televized, thrown by the camera objective lens on the photocathode of the camera tube, into electric signals (video signals) proportional to the light intensity of the respective object points. These signals are amplified and controlled in the respective circuits of the transmitting terminal to be transmitted via a communication link to the receiver which detects the signals and produces an image on the screen of a cathode-ray tube.

The basic principle of telecasting is the one of sequential point-by-point sensing by a narrow electron beam of the image elements cast on the camera tube to convert their light intensities into electric signals. This sequential interrogation of the elements of the image by an electron beam controlled to move in a certain pattern is termed *scanning*. The deflection system of the camera tube causes the electron beam of the tube to sweep from left to right (*horizontal* or *line scan*) and from top to bottom (*vertical* or *frame scan*) and then rapidly retrace as the entire field or frame is covered.

Television uses the same technique for the transmission of moving objects as is used in motion pictures. Physically, it transmits a series of individual still pictures differing in the phase of the motion and caught by the lens of the camera. When these still pictures are reproduced the persistence of vision creates an illusion of a continuous motion. As the number of stills, or *frames*, shown every second is increased, the frequency spectrum, or bandwidth, needed for television signals has to be increased, which is undesirable. In order to minimize the bandwidth needed, the *number of frames per second* should be kept to a reasonable minimum at which, however, the viewer would not notice the intermittent nature of the still frames. Experiments show that this minimum is 10 to 15 frames per second. Unfortunately, at this frequency another unpleasant factor, known as *flicker*, comes in when the whole of the picture produces the sensation of

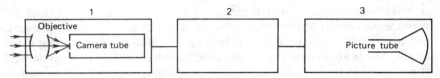

Fig. 16.1. The principal constituents of a television system

a pulsating motion. To avoid this flicker effect the frame frequency has to be increased three to three and a half times.

From the foregoing it might be concluded that the frame frequency for broadcasting television should be 50 frames per second, which is a convenient choice as it is the mains frequency with which the scanning sawtooth generator can be locked. Unfortunately, a 625-line picture (by the USSR scanning standard, there are 625 lines per frame) transmitted at this frame-scane frequency would require a bandwidth of over 12 MHz. This is too huge a bandwidth to transmit without marked distortion and process (detect and amplify) without difficulty in a receiver. One way to overcome this limitation is by means of *interlaced scanning*. This technique makes it possible to halve the frame-scan frequency (and to narrow the bandwidth required in proportion) while retaining the picture quality (resolution, contrast, etc.) and to exceed the critical flicker frequency by a comfortable margin.

With interlaced scanning there are two fields per frame, and each field scan is arranged to explore half the lines that make up a frame. The first field scan explores, say, odd lines, and the second field scan even lines. Frames, each made up of two fields, are thus scanned at a rate of 25 per second. Although each field explores only half the total lines in the image (312.5 lines if in the Soviet broadcasting TV), the persistence of vision produces an illusion that the viewer sees a full frame containing 625 lines. Today, interlaced scanning is widely used in the Soviet Union and abroad. Some camera tubes, image dissectors for example, operate without interlacing at a rate of 25 frames per second.

The television standards for the number of lines per frame are different in different countries. The United Kingdom has 405 lines per frame, the United States and Japan have 525, the USSR and GDR have 625, and France has 819 lines. In actual broadcasting this number of lines consists of z_a active lines which actually take part in the production of image patterns and those (about 7.5 to 8% of the standard number) lost in retrace and flyback. For the Soviet standard of 625 lines per frame, $z_a = 577$ lines.

At the receiving end, the signals oncoming through the transmission link are amplified and conveyed with the electron beam to the phosphorescent screen of the picture tube. The line and frame sweep generators of the receiver keep the beam of the picture tube in synchronism and in phase with the movement of the beam in the camera tube at the transmitting end, the intensity of the beam being varied in proportion to the incoming video signal. The beam sweeps the picture tube from left to right (line scane) and from top to bottom (frame scane) and in so doing strikes the phosphorescent screen applied to the vacuum side of the picture tube and causes it to phosphoresce with the intensity proportional to that of the beam. As a result, a television picture is formed.

Camera tubes. The camera tube is a high-vacuum electronic device. Its photocathode is placed in the image plane of the camera objective lens. The tubes which depend in their operation on the outer photoemissive effect are called *image orthicons* and *iconoscopes*, while those which depend on the inner photoelectric, or photoconductive, effect are called *vidicons*.

In the material that follows we examine the design and operation of the camera tube based on a vidicon with reference to Fig. 16.2. In outline, the vidicon is an evacuated glass bulb *1* the *faceplate* of which is covered (from the vacuum side) by a translucent layer of metal, *2*, and a film of photoresistor, *3*, called the *photoconductive target*. The metallic layer plays the role of a current collecting, or signal, plate. It is connected electrically to the load resistor R_L from which the video signal is picked up. The photoresistor film is 1 or 2 μm thick, its specific dark resistance is about $10^8 \ \Omega \ cm^{-2}$.

The electron beam is formed by the gun, *6*, and on its way to the photoconductive target it undergoes the control of the anodes, *4* and *5*. The deflection and focusing of the beam is effected by the magnetic coils mounted on the envelope of the tube.

Unless the photocathode is uniformly illuminated its transverse resistance (conductance) is the same throughout the entire field. Incident light causes the photocathode to emit electrons in quantities proportional to the light intensities of the picture elements. The photocathode is made

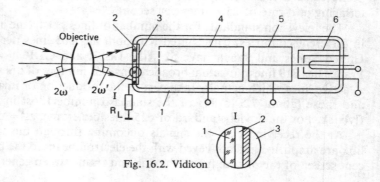

Fig. 16.2. Vidicon

semitransparent so that at least part of the incident light can pass onto the vacuum side (on the right in the diagram) and cause photoemission inside the tube. On breaking away from the photocathode electrons leave on its surface positive potentials varying in magnitude in accordance with the light intensities of the respective scene elements. In this way a charge, or conductance, image is produced on the vacuum side of the photocathode. The deflection coils cause the electron beam, produced by the hot (thermionic) cathode, to sweep rapidly from left to right (along the lines) and also, but more slowly, from top to bottom and retrace with a much slower sawtooth pattern, thereby scanning the charge image on the surface of the photocathode. As the beam scans the charge image an alternating current collected by the signal plate flows through the load resistor, R_L. This is the video signal current the instantaneous value of which corresponds to the potential at each point of the charge image and, hence, to the light intensity of each scene element.

In general, vidicons are compact high-sensitivity camera tubes. Since these tubes depend on the inner photoelectric effect for their operation, they are more inertial, in terms of photoelectric lag, than image orthicons.

To ensure a high quality of imagery, camera tubes should exhibit high light-transfer characteristics, correctly reproduce gradations of shade over a wide range of object luminance, and have a high resolution and a high signal-to-noise ratio (SNR). In what follows we look at the principal optical characteristics of camera tubes.

(1) The spectral sensitivity of the photocathode is normally represented as a plot of the spectral sensitivity against wavelength (Fig. 16.3). The plot normally shows the relative spectral sensitivity, s_λ, normalized to the maximum sensitivity attained at a wavelength λ_m. The wavelengths λ_1 and λ_2 encompass the range of interest for the specific application. For the most part, the spectral sensitivity characteristics of camera tubes correspond to that of the photocathodes.

(2) The light-transfer characteristic (Fig. 16.4) is a plot of the photoemissive current, i_{ph}, in microamperes as a function of photocathode illumination E'_{ph} in luxes. In comparing the light-transfer characteristics of various devices, the designer should focus on the minimal and maximal il-

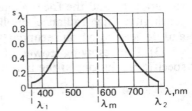

Fig. 16.3. Spectral sensitivity characteristic of a vidicon

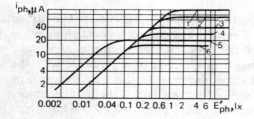

Fig. 16.4. Light-transfer characteristics of Soviet-made image orthicons: *1* LI-216, *2* LI-215, *3* LI-201, *4* LI-213, *5* LI-218, *6* LI-17

luminations of the working range, estimate the span of the working range, i.e. the range $E'_{max}-E'_{min}$ in which the tube will operate properly. The slope of the characteristic determines the rate of change of the video signal for the same increment of the image illuminance. This rate is essential in deciding whether or not one can achieve in the television picture the same gradations of luminance as in the televized scene.

It is apparent that the minimum illumination for the photocathode is limited by the associated noise. To provide a necessary illuminance at the photocathode, E'_{ph}, the illumination of the scene being televized, E_s, may be estimated with the formulae (7.68) and (7.71) assuming that the reflection factor of the scene is ρ, and the optical system of aperture ratio D/f' has lateral magnification β and transmission factor τ, viz.,

$$E_s = \frac{4E'_{ph}(1 - \beta)^2}{\tau\rho(D/f')^2} \tag{16.1}$$

Camera tubes are also characterized by the sensitivity which is inverse of the minimum illumination required to transmit a standard TV chart at a given definition (see Table A5), a given magnitude of video signal and a signal-to-noise ratio. This will be an illumination of 10 to 30 lx for image iconoscopes, 1 to 5 lx for image orthicons, and 5 to 10 lx for vidicons.

By way of illustration, if an $f/2$ objective lens of $\tau = 0.8$ images a 1.7-m high object at $\beta = -0.014$ on a photocathode of $h = 24$ mm height having an average diffuse reflectivity of $\rho = 0.6$, then the illumination at the object must be (at least) 1000 lx for an image iconoscope, 200 lx for an image orthicon, and 350 lx for a vidicon.

Radiance amplifiers can be inserted between the photocathode of the camera tube and the focal plane of the objective lens to improve the tube sensitivity. Such radiance amplifiers may be electronic image converters, and devices based on secondary electron conduction [10].

(3) The resolving power of camera tubes can be estimated with the specific resolution, in the number of lines per millimeter, defined as

$$N_t = z_a/h$$

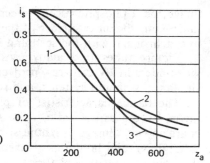

Fig. 16.5. Amplitude response (aperture) characteristics of camera tubes

where z_a is the number of active lines per frame and h is the height of the photocathode.

The resolution of camera tubes is decided for the most part by the quality of focusing of the scanning beam, and by the focusing of the beams involved in the transfer of the electron image. A more accurate definition of tube resolution can be obtained from the dependence $z_a = f(i_s)$ indicated in Fig. 16.5. This plot, called the amplitude response (aperture) characteristic, is in fact the modulation transfer function of a specific tube. It indicates that higher spatial frequencies are transmitted by video signals of lower intensities, i.e. at less pronounced modulations.

(4) The height, h, and width, w, of the photocathode (in mm) define the angular field of the objective lens as

$$\tan \omega' = l'/f' \quad \text{or} \quad \tan \omega' = l'/a'$$

where $l' = \sqrt{h^2 + w^2}/2$ is the semi-diagonal of the photocathode.

(5) Signal-to-noise ratio (SNR). The output current of a camera tube may be considered as consisting of two components, a signal current, i_s, proportional to the light flux incident on the photocathode, and a noise fluctuation current, i_n, whose magnitude varies at random. Experimental evidence indicates that for a good quality imagery the signal-to-noise ratio SNR $= i_s/i_n$ must be at least 20.

Through a proper choice of target material, it is possible to build vidicons sensitive to X-rays, ultraviolet radiation with $\lambda_{min} = 200$ nm (amorphous selenium), and infrared light with $\lambda_{max} = 2000$ nm (PbS).

The camera tubes with line scanning (noninterlaced), known as image dissectors, find their use mainly as instantaneous action tubes with short duration of the light pulse (LI-601 type).

Some optical characteristics for a number of Soviet-made camera tubes are listed in Table A5.

Picture tubes. A television picture tube, or kinescope, is a cathode-ray tube in which electrical signals are translated into a visible picture, or image, on a phosphorescent screen. The electron beam of the tube scans line by line the inner surface of the tube coated by a thin film of phosphor in synchronism with the scanning in the camera tube.

Picture tubes find many uses as CRT tubes of TV sets, displays, viewfinders in TV cameras, projection tubes, flying spot TV systems, facsimile transmission systems, and screen photography.

The basic characteristics of picture tubes are the size of the screen (given as the aspect ratio $= w/h$), emission colour, spectral emission characteristic, image brightness, picture contrast, phosphor persistence (afterglow), and luminous efficiency. The last characteristic is defined as the light output of a picture tube (flux) for the power of the impinging beam of electrons. It is measured in lm/W, or, when it is taken as the ratio of the screen luminous intensity to the beam power, in cd/W.

The spectral emission characteristic, or simply the colour, of a picture tube depends on the phosphor composition as indicated in Table 16.1.

The luminance of an image on the screen depends on the high-tension voltage at the main anode of the tube. The average minimum luminance is taken to be 40 cd/m². The highest luminance for non-projection picture tubes is 200 cd/m². In projection tubes, the average screen luminance can be as high as 1000 cd/m².

The contrast of a picture on the screen, normally determined by the dark-highlight range of luminance, depends not only on the luminance of the highlights (L_{max}) and dark areas (L_{min}), but also on the luminance (L_s) due to light scattering inside the tube. Indeed, the electron beam striking the phosphor causes it to emit light in two directions: outside, towards the viewer, and inside the bulb. Partly reflected from the internal coating of

Table 16.1 Emission Characteristics of Screen Phosphors

Phosphor composition	Emission colour	Persistence, s	λ_{max}, nm	cd/W	Destination
Mixture of ZnS and CdS activated with silver	white	2×10^{-3}	455-470	6	Picture tubes
Willemite (Zn_2SiO_4Mn)	green	10^{-3}		2	
ZnS activated with silver	blue	2×10^{-3}			Colour picture tubes
Zinc phosphate activated with manganese	red	10^{-2}		0.8	
Zinc oxide activated with Zn	greenish	2.5×10^{-6}	510	0.9	Picture tubes for flying spot systems
Gelenite	UV to blue	10^{-7}	450		

the bulb, the light rays are scattered and reach the screen again where they brighten up the darker portions of the image, thereby reducing the contrast of the picture.

Another cause of reduced picture contrast is *halation*. The effect of halation which affects mostly fine and medium detail looks like the intensity distribution in Fig. 15.3. When the electron beam striking the screen at a point excites the phosphor, some of rays going out of this point travel forward, towards the viewer, and some are reflected at the interface between the glass and the air and go back to illuminate the phosphor around the point of beam incidence. As a result, the viewer can see a bright luminous spot surrounded by a less bright ring, the halo. To account for the light scattering inside the envelope and halation effects, the picture contrast is defined as the ratio $(L_{max} + L_s)/(L_{min} + L_s)$. In modern picture tubes, the highest picture contrast reported is between 30 and 40.

The phosphor persistence of a screen is measured as the time interval during which the luminance decays to 1% of its peak value after the beam has moved on. Soviet designers divide phosphor persistence time (in seconds) into five categories: very short (less than 10^{-5}), short (10^{-5} to 10^{-4}), moderate (10^{-2} to 10^{-1}), long (10^{-1} to 16 s), very long (over 16 s).

The optical characteristics of some Soviet-made picture tubes are summarized in Table A6.

16.2 Objectives of TV Cameras

The optical systems of a photographic camera, microscope, or telescope can be used for forming the image of a televized object on the target of the camera tube. The schematic illustration of this process may be suggested by Fig. 16.6. Of course, for each particular lens-tube combination the optical characteristics of the components should be carefully matched. Objective lenses used in telecasting are for the most part similar to photographic objectives and are, as usual, characterized by the focal length f', relative aperture f'/D and angular field coverage 2ω.

The aspect which makes the television objective differ from other types of lens concerns the account of the effect of the glass plates involved (transparent optical flat of the camera tube, and a variety of filters) traversed by convergent bundles ahead of the photocathode.

The focal length of an objective defines its transverse (linear) magnification, β. The television channel from the photocathode of a camera tube to the screen of a picture tube may also be characterized by a linear magnification, often expressed in terms of the image scale indicated

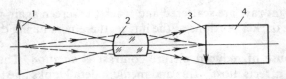

Fig. 16.6. Illustrating the formation of an image at the photocathode of a camera tube: *1* object, *2* lens, *3* photocathode, *4* camera tube

as $1:m_t$, namely,

$$\beta_{TV} = 1/m_t = h_{pt}/h_{ct}$$

where h_{pt} is the height of the picture tube, and h_{ct} is the height of camera tube photocathode.

Accordingly, the overall scale of a TV image $(1:m)$ will be

$$1/m = -\beta(1/m_t)$$

When the photocathode receives an image of distant objects, the scale of the TV image may be estimated as

$$1/m = -\frac{f'}{a}(1/m_t)$$

where a is the distance from the objective (more accurately, from the first principal point) to the objects, and f' is the focal length of the lens.

The aperture ratio of the lens $(D/f' = 1/F$, where F is the f-number or relative aperture) should be selected with account of the spectral sensitivity of the photocathode, luminance of objects being televized, L, and admissible illumination of the photocathode E'_{ph}, as

$$\frac{D}{f'} = 2n\sqrt{E'_{ph}/\tau\pi L}$$

where n is the index of refraction in object space.

The luminance of objects (in cd/m^2) is determined as

$$L = \frac{\rho I \cos \varepsilon}{\pi r^2}$$

where I is the luminous intensity of a light source illuminating the object (in cd), r the distance from the source to the object (in m), ρ the diffuse reflectance of the object, and ε the angle between the optical axis of the lens and the direction of light from the source to the object.

Whenever a weakly illuminated images on the photocathode are envisaged, the operator should take care of their illumination somewhat exceeding the minimum allowable level. The illumination on the photocathode, E'_{ph}, as a function of that at the scene being televized, E_s, may be estimated with the formula (16.1).

The linear field size of the image formed by the lens must be equal to, or somewhat exceed, the size of the photocathode.

By way of illustration we list a few Soviet-made zoom systems which find wide use in studio and field TV cameras. Note that the letter T in the lens designation implies a TV formulation.

	F	f'	2ω
Vario-Goir-1T	f/4	40-400	54°-6°
Alkor-6	f/4	80-400	28°-3°
	f/8	160-800	
Meteor-7	f/1.9	25-100	36°-9°
Vario-Goir-2T	f/2.4	15-150	54°-6°

Optical characteristics of some Soviet-made picture tubes are summarized in Table A6.

16.3 Resolution and MTF of Television Systems

The definition of a television picture has to do with the maximum number of fine details discernible in the image. It is an important aspect of picture quality. With poor definition, the picture lacks crispness or sharpness — long and medium shots of faces, say, in the scene being televized become unrecognizable.

Above all, the definition of the television picture depends on the resolving power, or resolution, of the picture-imaging medium, that is, on its ability to discern and reproduce fine details present in the televized scene. Reproduction of fine details and edge gradations on a television screen is above all dependent on the number of picture elements (or lines) chosen for the scanning standard. For television systems, resolution is estimated by the number of white and black lines, z_n, fitted in the frame height, h.

In particular, the definition of a television picture transmitted by a camera tube depends on the sharpness of the image produced by the objective lens on the photocathode and on the number of lines in the scanning standard. To be compatible with a particular camera tube, an objective lens should be designed so that the point spread size δ' due to the residual aberrations does not exceed the height of a scanning line, i.e., $\delta' \leqslant h_{ph}/z_n$.

For objectives used in conjunction with camera tubes, the size of blur spots, δ', may be estimated as

$$\delta' = \frac{2}{\pi N'} \sqrt{2.5(1 - T)}$$

where N' is the spatial frequency of the image being televized, and T the modulation (contrast) transfer factor (see Section 15.2). To illustrate, at $N' = 13$ mm^{-1} and $T = 0.6$ to 0.8, the blur spot diameter is $\delta' = 0.05$ to 0.035 mm.

In telecasting the effect of contrast increases, therefore the limiting resolution characteristics, such as N_{sf} for a lens-film combination or N_o for visual resolution derived in Section 15.2, can no longer be used for direct estimations of the desired resolving power of a television system, but may be applied only as approximate criteria. A more objective criterion is the modulation transfer function (MTF) which, as will be recalled from Section 15.2, is the dependence of the modulation transfer factor on spatial frequency. The MTF of a television system, $T(N)$, may then be derived as the product of the MTFs of the objective lens, $T_o(N)$, camera tube, $T_{ct}(N)$, and the electronics (amplifiers, picture tube, etc.) of the TV channel, $T_e(N)$, namely, $T(N) = T_o(N)T_{ct}(N)T_e(N)$.

By way of example, Fig. 16.7 shows the MTF plots, called transition MTF by television engineers, for the $f/4.5$ Tair-44-T lens of $f' = 300$ mm (derived for the D, G', C and h spectrum lines) and the EM1-9677 vidicon. As is seen, the MTFs of the lens for the red and blue spectrum bands (see Section 7.1 for the nomenclature of the lines) are inferior to that of the vidicon plotted for white light.

In black-and-white television, each element of an image is conveyed to the viewer by means of only one coordinate, luminance. In colour television, the image is formed by combining three primary colours: red (R), green (G), and blue (B), accurately proportioned in luminance, i.e. each colour has its own weighting (trichromatic) coefficient in the luminance sum, to match the spectral sensitivity of the eye. More specifically, if we

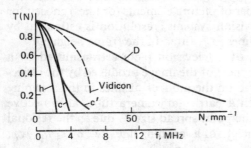

Fig. 16.7. Modulation transfer functions of the Tair-44-T lens (measured at the spectral lines indicated) and the EM1-9677 vidicon

denote the luminances of the primary colours by L_R, L_G, and L_B, then they should be combined in the following proportions

$$L = 0.299L_R + 0.587L_G + 0.114L_B \qquad (16.2)$$

In the existent systems of colour television, the white light coming from the scene being televized to the television camera is divided into the primary components and these are further handled in the specific branches of the system. Accordingly, illumination engineers should take into account the spectral characteristics of the camera tubes and picture tubes. The levels of light flux in each colour are controlled by light filters so as to produce the visual sensation in agreement with Eq. (16.2).

Because the light from the televized scene is decomposed into the primary colours, the objectives of the respective camera tubes operating in a monochromatic light should be corrected for the chromatic aberrations corresponding to the three wavelengths of the peak sensitivity of the camera tubes in these colours. Monochromatic image (blue, green, and red) are then handled by camera tubes and subsequent electronics to be combined again in the picture tube of a colour TV receiver.

Among the requirements imposed on the optical system for colour imagery there is a need for a large back focal length to fit between the system and its image plane the dichroic mirrors splitting the incoming light flux into the primary colours and directing the component primaries into individual vidicons. A different approach to the optical system arrangement of a television camera is realized in the Soviet-made KT-103 TV camera schematized in Fig. 16.8.

The prime objective, *1*, of the KT-103 camera (ordinary lens or zoom lens) forms the optical image of the scene being televized at a field lens, *3*. In between these lenses there is a beam-splitting prism, *2*, which divides the oncoming light beam into two parts, one being directed into the luminance channel (upwards in the diagram), the other part being passed through to the colour channels. The field lens, *3*, is followed by a dichroic mirror, *5*, reflecting the blue portion of the image onto the mirror, *6*, which conveys it through a relay lens, *7*, and a blue filter, *8*, to a vidicon, *9*, labelled *B*. The mirror *5* passes the green and red portions of the image, that is, for these beams this mirror is a plane-parallel plate. The red portion of the image is reflected by the semitransparent mirror, *11*, (upwards in the diagram) and is conveyed by the relay objective, *7*, mirror, *6*, and a red filter, *12*, to the vidicon, *9*, labelled *R*.

The green rays of the image are passed by the dichroic mirror, *11*, and the relay lens, *7*, forms the optical image at the target of the vidicon, *9*, labelled *G*. This flux is controlled by the light filter, *10*, ahead of the green

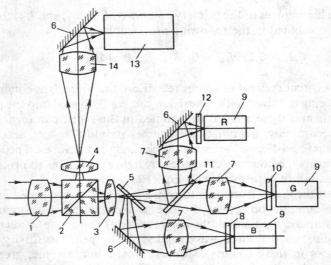

Fig. 16.8. Optical diagram of a colour TV camera

camera tube. It is worth noting that all the relay lenses, 7, are identical and operate at the magnification $\beta \approx -0.4$.

The portion of the primary beam reflected into the luminance channel is passed through a field lens, 4, to a relay lens, 14, operating at a magnification of -1, and on reflection from a mirror, 6, forms an optical image at the target of the image orthicon, 13.

16.4 Flying Spot Projection Systems

The optical system of a flying spot TV setup is designed for telecasting transparent slides (transparencies and motion picture frames) and non-transparent flat pictures. The optical principles of this arrangement are suggested by Fig. 16.9. In brief, it projects the bright spot of a picture tube through a transparency onto a photomultiplier (PMT).

To be more specific, the cathode-ray tube, 1, having a phosphor persistence time of 1×10^{-7} s forms on its screen a sharply focused bright spot of uniform luminance. On scanning, this spot traverses over the entire screen in horizontal lines (horizontal scan) which form a pattern called a raster when the entire frame is over. The objective lens, 2, projects the screen of the picture tube into the plane of the transparency, 3, placed ahead of the condensor, 4, which projects the exit pupil of the objective on-to the photocathode of the photomultiplier. As a result, the photocathode

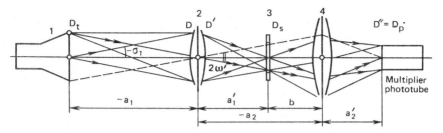

Fig. 16.9. Optical performance of a flying-spot projection system

plane receives the integral image of the picture tube. In projection of transparencies, the PMT current is proportional to the optical density of an elementary area of the transparency, *3*, illuminated by the image of the light spot from the CRT at this time instant.

For the observer's eye having a considerable persistence on this time scale, the CRT screen appears to be continuously illuminated, whereas at a specific time the optical system picks up only the luminance of an elementary area equal to the section of the electron beam of the tube.

Observing that the diameters (diagonals) of the picture tube, D_t, transparent slide, D_s, the entrance pupil of the lens, D, and photocathode, D_p, are known, we have, from the formula of transverse magnification, for the objective lens $\beta_1 = -D_s/D_t$, and for the condenser $\beta_2 = -D_p/D\beta_p$, where β_p is the magnification between the pupils of the objective lens (ratio of pupil diameters). To illustrate, if the diagonal of a 90 × 120 mm CRT screen is 150 mm and that of a 18 × 24 mm film format is 30 mm, then $\beta_1 = -0.2$.

The focal length of the objective lens is selected such that the decrease of illuminance at the edges of the image by the cosine fourth law be consistent with an angular field $2\omega'$ not exceeding 40°, which allows for a focal length of $f_1' = 200$. The back focal length (from the objective to the slide plane) is given as $a_1' = (1 - \beta_1)f_1'$, and the distance from the objective to the picture tube as $a_1 = a_1'/\beta_1$.

When it is sought to use a condenser of smaller size, it should be placed closer to the transparency plane, *3*, that is, b should be selected as short as possible, say $b = 10$ to 20 mm. Because $-a_2 = a_1' + b$ is known, so is the transverse magnification β_2, and the focal length of the condenser is then

$$f_2' = -(b + a_1')\beta_2/(1 - \beta_2)$$

The illumination analysis of the flying spot system boils down to the evaluation of a minimal relative aperture F at which the PMT current will exceed the noise level.

Given the diameter of the CRT beam section, δ, at the plane of the screen, and the luminance of the spot, one can derive the luminous intensity of the scanning spot

$$I = \pi\delta^2 L/4 \tag{16.3}$$

From this spot as a light source, the objective lens at a distance a_1 from the screen will receive a light flux Φ within the solid angle subtended by the entrance pupil of the objective, so that

$$\Phi = \frac{\pi D^2}{4a_1^2} I \tag{16.4}$$

On passing through the optical system and the area of the transparency being projected, this flux will be modulated (attenuated) in accordance with the transmittancies of the system, τ, and the slide, τ_s, so that the flux arriving at the photocathode of the photomultiplier is

$$\Phi' = \tau\tau_s\Phi \tag{16.5}$$

In view of $D/2a_1 = \tan\sigma_1$ we find from Eq. (3.11)

$$\tan\sigma_1 = \frac{\beta_1}{2(1 - \beta_1)F}$$

From this expression and Eqs. (16.3)-(16.5) we get

$$F = \frac{\beta_1\pi\delta}{4(1 - \beta_1)} \sqrt{\frac{\tau\tau_s L}{\Phi'}}$$

If the minimum integral responsivity of the light detector is specified as $S_{min} = i_{min}/\Phi_{min}$, then we may obtain the relative aperture F of the objective lens as a function of the PMT current i_{min}, viz.,

$$F = \frac{\beta_1\pi\delta}{4(1 - \beta_1)} \sqrt{\frac{\tau\tau_s L S_{min}}{i_{min}}}$$

A more compact arrangement for this type of projection system can be

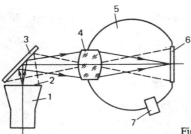

Fig. 16.10 A compact television episcope

achieved by inserting first-surface mirrors between the CRT screen and objective lens and between the objective lens and the light detector. The quality of image will, however, be somewhat deteriorated owing to the scattering of light at the mirrors.

Figure 16.10 shows a setup for projection of images of opaque black-and-white objects. The screen, *2*, of a flying-spot picture tube, *1*, is projected by a first surface mirror, *3*, and an objective lens, *4*, onto the surface of the object, *6*, placed inside a photometric sphere, *5*. The light flux integrated by the sphere is sensed by the light detector, *7* (PMT). More often than not, the size of the object is 9 × 12 cm, and the linear magnification of the lens is $\beta = -1$.

Coloured pictures can be projected with a similar arrangement schematized in Fig. 16.11. In this case, the flux integrated by the photometric sphere is separately sensed by three photomultipliers screened by special buffles from the direct reflections from the picture and fitted by colour filters for extracting the primary colours. The amplifiers of the PMTs assist in weighting the luminances of the three signals in agreement with Eq. (16.2).

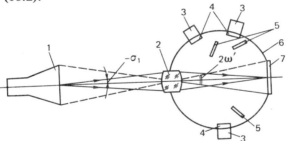

Fig. 16.11. Optical performance of a colour TV episcope: *1* flying-spot CRT, *2* lens, *3* light detectors (PMT), *4* colour filters, *5* baffles, *6* photometric sphere, *7* colour picture

17

PROJECTION SYSTEMS

17.1 Fundamentals

Projection systems, as the name implies, deal with projecting on a viewing screen transparent slides, motion picture films, negatives, drawings, graphical materials, printed matter, and small objects. Accordingly, the projection systems include epidiascopes, motion picture projectors, photographic enlargers, projection instruments for photogrammetry, reading setups for microfilms, projection systems for microscopes and microelectronics processors.

The optical system of a projection instrument can be conveniently divided into *illumination* and *projection parts*.

The illumination systems of general purpose were outlined in Chapter 12. Below we shall see how these systems can be fitted into the overall arrangement of projection instruments.

The projection part of the instrument converts diverging bundles of oncoming rays having concentric wavefronts into converging bundles of rays. This action is usually performed by a *projection lens*. Both projection and illumination parts of the instrument should be matched in order to achieve the desired illumination on a screen and the desired distribution of this illumination at a given scale of the image.

Depending on the nature of objects being projected, i.e. whether they are opaque or transparent, the projection systems used are episcopes or diascopes. An *episcopic projection* is formed by the beam of light reflected from the opaque object. A *diascopic projection* is formed by the beam of light passing through the transparent object. The projection principle employed in motion picture projectors relates to the latter type.

Optical systems capable of both types of projections are known as *epidiascopes*.

The principal characteristics of the optical systems of projection systems include the scale of image or linear magnification, illumination of image, and the size of the projected frame, and, occasionally, the size of the viewing screen. These characteristics are defined as functions of (1) projection distance, (2) focal length of lens, (3) relative aperture of lens, (4) luminance of the source, (5) transmittance of the entire system, and (6) construction of the illumination system.

The *projection distance* is the distance from the projection lens to the viewing screen. It can be fixed, as in stationary motion picture projectors, or variable as in enlargers.

The function of the projection lens imposes a number of requirements. To be more specific, it must ensure a constant contrast and resolution over the transparency (film gate) and, respectively, a satisfactory contrast and resolution in episcopic projection; a low amount of vignetting which, when present, disturbs the distribution of illuminance on the screen; and stringent control of distortion, which is especially important when the projected image is to be used for measurement as is the case with photogrammetry.

It is worth noting that for a projection lens the control of aberrations is designed for the object being situated at a finite distance from the lens, more specifically in the range of image scales from 1 over 25 to about 25 over 1. For large object (or image) throws, the aberrations are corrected assuming the object (or image) is at infinity.

In practical design of projection lenses the illumination on the screen is specified for the situation when there is no slide in the film gate (for slide projection systems), or when there is a diffuse reflecting surface on the mounting table of the episcope.

The luminance, L, of a viewing screen depends on its illumination E and reflectance, ρ. For commercial motion-picture projection, Soviet designers take the luminance of screen equal to about $100 \, \text{cd/m}^2$ (allowing for the losses in the shutter); for slide projectors, $L \approx 50 \, \text{cd/m}^2$; for episcopes, L can be taken up to $20 \, \text{cd/m}^2$; and for measuring instruments exploiting projection principles, $L = 15$ to $25 \, \text{cd/m}^2$.

For an ideally white screen, the diffuse reflectance $\rho = 1$; for a screen covered with a layer of magnesium carbonate, $\rho = 0.89$; for a screen of technical zinc oxide or with baryta coating, $\rho = 0.8$; and for a matt plastic screen, $\rho = 0.72$.

At $\rho = 0.8$, the recommended amount of illumination at the central portion of screens, as derived with Eq. (7.71), is as follows: for commercial motion picture, $E = 400 \, \text{lx}$; for slide projectors, $E \approx 200 \, \text{lx}$; for episcopes, $E \leqslant 80 \, \text{lx}$; and for measuring projection instruments, $E = 60$ to $100 \, \text{lx}$.

The illumination of a screen, having the illuminance E_0' at the centre, falls with the distance from the centre as

$$E' = k_\omega E_0' \cos^4 \omega'$$

where k_ω is the vignetting ratio (relative unvignetted aperture), and ω' the angle between the principal ray and the optical axis of the projection lens in image space.

To avoid uneven illumination of the screen associated with a nonuniform luminance of the source or due to the spherical aberration of the illumination system, the slide frame is preceded by a ground glass. Another solution is to matt the last surface of the condenser. These methods will obviously increase the light losses in the system. An alternate method is to image the source of light (filament or arc) in the entrance pupil of the projection lens.

17.2 Episcopes

We examine the construction of an episcope with reference to Fig. 17.1. The beams of light produced by the lamps, 2, are reflected from the surface of an opaque object, 1, and redirected by the mirror, 3, into the projection lens, 4, which projects the image of the object on a screen. The mirror is essential to produce an erect image. This must be a first-surface reflector to avoid doubling.

If the surface of the object is a diffuse (Lambertian) reflector (see Eq. (7.71)) then for its luminance we have

$$L = \rho E / \pi \qquad (17.1)$$

where E is the illumination on the object, and ρ its reflectance.

For an arrangement with m symmetrically spaced identical sources of light, each of a luminous intensity I spaced a distance l from the centre of the object at an angle ε between the ray to the centre of the object and the normal at this point (see Fig. 17.1), the illumination of the object is

$$E = \sum^{m} E_i = \sum^{m} \frac{I \cos \varepsilon}{l^2} = \frac{mI \cos \varepsilon}{l^2} \qquad (17.2)$$

The illumination of the viewing screen can be estimated as

$$E' = \tau \pi L \sin^2 \sigma'_A. \qquad (17.3)$$

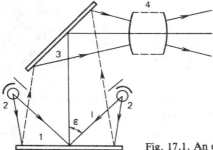

Fig. 17.1. An episcope

where τ is the overall transmission of the system consisting of a mirror and a lens ($\tau = \rho_M \tau_1$), and $\sigma_{A'}'$ the angular aperture of the projection lens.

If the projection throw, approximately equal to the distance p' from the exit pupil of the projection lens to the screen, exceeds the exit pupil diameter D' of the lens many times, and the lateral magnification between the pupils $\beta_p = 1$, then

$$\sin \sigma_{A'}' \approx D'/2p' = D/2p' \tag{17.4}$$

because $D' = D$, the diameter of the entrance pupil.

Also,

$$p' \approx s' \approx f' (1 - \beta) \tag{17.5}$$

where s' is the distance from the last surface of the projection lens to the screen, f' the focal length of the lens, and β the lateral magnification of the lens in projecting an object on the screen.

Equations (17.1), (17.3)-(17.5) solved in conjunction yield for the illuminance of the screen

$$E' = \frac{\tau \rho E}{4} \left(\frac{D}{f'}\right)^2 \frac{1}{(1 - \beta)^2} \tag{17.6}$$

The focal length, f', of the projection lens is found by Eq. (17.5) and the relative aperture f'/D by Eq. (17.6). Since the product $\tau \rho E$ is comparatively small, illumination engineers tend, first of all, to keep the magnification β as low as possible and, second, use an objective with a large relative aperture.

The angular field 2ω can be estimated from

$$\tan \omega = -\beta \frac{y}{p'} \approx -\beta y/(1 - \beta)f' \tag{17.7}$$

where y is the semi-diagonal of the object (βy is then the semi-diagonal of the screen).

The focal length, relative aperture, and angular field thus derived lead to an appropriate projection lens for an episcope. It commonly occurs that projection lenses of episcopes have relative apertures $f/1.5$ to $f/2.5$ and angular field $2\omega \leqslant 45°$.

Equation (17.2) can then be used to choose the suitable lamp from a catalogue. A condenser may be brought into the episcope arrangement to better utilize the light flux of the illumination lamp.

To conclude it is worth mentioning that the diffuse reflectance of drawings, photographs, and printed materials is in the range of 0.6 to 0.8.

17.3 The Projection Lantern

The most widely used types of projection apparatus are those for the projection of transparencies, such as slides and motion picture film images. Because this is the projection in transmitted light, it provides a higher illumination on the screen than any episcope does.

Two arrangements for projection are in popular use. In one, the source of light is imaged in the aperture of the projection lens. In the other, the source of light is imaged in the plane of the transparency and, consequently, is projected by the lens into the screen where it appears overlapped on the image in the projected slide.

The image of a filament of the light source on the screen can be avoided by using a radiant emitter of uniform luminance. In stationary motion picture projectors, this is an arc. In addition, if the film gate is of a large size, the illumination system must be of large magnification type, which means a large size of the projector proper.

The second arrangement for projection cannot be used where the slide (or film) is to reside in the film gate for a long time because of a danger of overheating. This arrangement is recommended for use in cine projection where film formats are small and the film rate is sufficiently high.

In Fig. 17.2a, the source of light, 1, is imaged by the condenser, 2, into the entrance pupil of the projection lens, 4. In Fig. 17.2b, the source is imaged in the plane of the transparency, 3, to be projected.

Given that the sine condition (see Chapters 7 and 9) is satisfied, the linear magnification of the illumination system for the first arrangement is

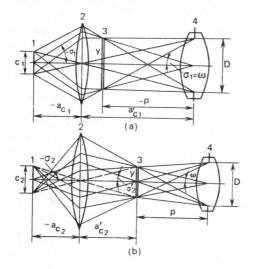

Fig. 17.2. Schematics of a projection condenser system. The condenser forms an image of the source (a) in the aperture of the projection lens, and (b) in the plane of the film gate

(the subscript "c" refers to the condenser)

$$\beta_{c_1} = \frac{\sin \sigma_1}{\sin \sigma_1'} = -\frac{D}{c_1} = a_{c1}'/a_{c1} \qquad (17.8)$$

where D is the diameter of the entrance pupil of the projection lens, c_1 the useful size of the light source, σ_1 the entrance angular aperture of the illumination system equal to the semi-angle subtended by the condenser at the source $2\sigma_s$, and σ_1' the exit angular aperture of the illumination system equal to, or exceeding, the semi-angular field of the projection lens, i.e. ω.

For $\sin \sigma_1' \approx \tan \omega$ we have from (17.7)

$$\beta_{c1} = -\frac{1 - \beta f'}{\beta y} \sin \sigma_1$$

The angular subtense of the condenser in the first arrangement, when the source of light is imaged in the entrance aperture of the projection lens, can be determined as

$$\sin \sigma_{s1} = \sin \sigma_1 = -\frac{\beta_{c1} \beta y}{(1 - \beta)f'} \qquad (17.9)$$

We assume here that the linear magnification, β, of the projection lens is specified, its focal length f' is determined by Eq. (17.5) and y is the semi-diagonal of the film gate.

The entrance pupil diameter, D, of the projection lens (defining β_{c1}) is evaluated by an illumination analysis (see Section 17.4).

With reference to Fig. 17.2b, assuming the sine condition is satisfied, we have for the linear magnification of the illumination system (condenser) in the second arrangement

$$\beta_{c2} = \frac{\sin \sigma_2}{\sin \sigma_2'} = -\frac{2y}{c_2} = a_{c2}'/a_{c2} \qquad (17.10)$$

where y is the semi-diagonal of the slide, c_2 the size of the source of light, σ_2 the aperture angle of the condenser in object space equal to the semi-angular subtense, i.e. σ_s, and σ_2' the aperture angle of the condenser in image space equal to, or exceeding, the aperture angle of the projection lens in object space.

Using Eq. (17.4) derived on the condition that the linear magnification between the pupils of the lens $\beta_p = 1$, i.e. $D' = D$, and Eq. (17.5) we get

$$\beta_{c2} = -2 \frac{1 - \beta}{\beta} \frac{f'}{D} \sin \sigma_2$$

Hence, the angular subtense of the condenser for the second arrangement (the source is imaged in the plane of the slide) is determined with the formula

$$\sin \sigma_{s2} = \sin \sigma_2 = - \frac{\beta_{c2}\beta}{2(1 - \beta)} \frac{D}{f'} \qquad (17.11)$$

If $|\beta| \gg 1$, as is the case in stationary motion picture projectors,

$$\sin \sigma_{s2} = \sin \sigma_2 = \frac{\beta_{c2}}{2} \frac{D}{f'}$$

By incorporating the linear magnifications (17.8) and (17.10) into (17.9) and (17.11), respectively, we obtain

$$\sin \sigma_{s1} = \sin \sigma_1 = \frac{\beta}{1 - \beta} \frac{y}{c_1} \frac{D}{f'}$$

and

$$\sin \sigma_{s2} = \sin \sigma_2 = \frac{\beta}{1 - \beta} \frac{y}{c_2} \frac{D}{f'}$$

Thus, the angular subtense of the condenser at the source of the illumination system is evaluated with the same dependence for either arrangement.

Let us now determine the diameter of the entrance pupil of the projection lens.

Equations (17.3)-(17.5) suggest that the necessary illumination on the screen is

$$E' = \frac{\pi\tau L}{4(1 - \beta)^2} \left(\frac{D}{f'}\right)^2 \qquad (17.12)$$

where L is the luminance of the source of light, τ is the transmission of the illumination and projection subsystems in cascade, β the linear magnification of the projection lens, and f'/D the relative aperture of the lens.

Having computed the focal length f', angular field 2ω, and relative aperture f'/D, the designer is in a position to choose a lens for the specific projector. Objectives of slide projectors and enlargers have a relative aperture in the range $f/4.5$ to $f/9$, and the angular field can in some cases be as high as 122° (for example in a multi-chamber photogrammetric projector). Motion picture projectors have relative apertures in the range $f/1.2$ to $f/2$ and angular fields up to 16°.

The clear aperture of the illumination system can be decreased by placing a collecting lens at the film gate (see Section 14.5).

17.4 Size and Illumination Analysis for an Enlarger

Let us design a photographic enlarger with the following performance data: the linear magnification ranges from $-\beta_{min}$ to $-\beta_{max}$, the diagonal of the largest film format is $2y$, the screen illuminance is E', and the maximum projection throw is p'_{max}.

(1) Equation (17.5) gives the focal length of the projection lens as

$$f' \approx p'/(1 - \beta) = p'_{max}/(1 - \beta_{max}) \qquad (17.13)$$

(2) The angular field 2ω of the lens will be determined by Eq. (17.7)

$$\tan \omega = \beta_{max} y/(1 - \beta_{max})f' \qquad (17.14)$$

Any other linear magnification smaller than $|\beta_{max}|$ will decrease the angular field actually used, increase the absolute value of the distance $a \approx p$, and decrease $a' \approx p'$ because $\beta = a'/a \approx p'/p$.

(3) We assume that the transmission τ of the illumination system and the projection system in conjunction is $\tau = \rho_M \tau_1$. Then for the given illumination of the screen, E', and luminance of a lamp L, Eq. (17.12) yields for the aperture ratio

$$\frac{D}{f'} \geqslant 2(1 - \beta_{max})\sqrt{E'/\tau \pi L} \qquad (17.15)$$

(4) With the found values of f', 2ω, and f'/D we already can select a suitable lens in a catalogue. Some mismatch in the characteristics of the choice and the computed lens is sometimes allowable; for example, the angular field 2ω can be chosen somewhat higher than the computed value. Once the choice has been made, the transmission τ_1 for this lens is corrected.

It will be recalled here that the light exposure, an essential characteristic of the enlarger, depends, among other things, on the illumination of the screen and, hence, on the relative aperture of the projection lens.

(5) The illumination system, schematized in Fig. 17.3, is to image the source of light in the plane of the entrance aperture of the lens. The reflector is an ellipsoid with the filament of the lamp at the first focus, F_1, and the image of the filament at the second focus, F_2, coinciding with the centre of the entrance pupil of the lens.

The aperture of the reflector, D_c, must correspond to the angular field 2ω of the lens. The distance g from the entrance pupil to the edge of the

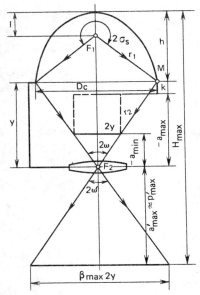

Fig. 17.3. Construction to design an enlarger with an ellipsoid reflector

reflector is a constant given as

$$g = -a_{max} + k = -\frac{1 - \beta_{min}}{\beta_{min}} f' + k \qquad (17.16)$$

where k is the distance between the film in its extreme position (for $-\beta_{min}$) and the edge of the reflector chosen from constructional considerations and for convenience of operation.

With reference to Fig. 17.3, the reflector aperture is

$$D_c = 2g \tan \omega \qquad (17.17)$$

The distance between the foci F_1 and F_2 of the ellipsoid is

$$F_1 F_2 = 2\sqrt{a^2 - b^2} \qquad (17.18)$$

where a and b are respectively the lengths of the semimajor and semiminor axes of the ellipse.

The distance from the vertex of the ellipse to F_1 is

$$l = a - \sqrt{a^2 - b^2} \qquad (17.19)$$

Recall also that the sum of the distances from the foci to any point on the ellipse is constant,

$$r_1 + r_2 = 2a \qquad (17.20)$$

For the point M on the diameter D_c, defined by Eq. (17.17), we have from the triangle F_1MF_2

$$r_1 \sin \sigma_s = r_2 \sin \omega \qquad (17.21)$$

From Eq. (17.18) we get

$$-r_1 \cos \sigma_s + r_2 \cos \omega = 2\sqrt{a^2 - b^2} \qquad (17.22)$$

By virtue of (17.21) we obtain

$$r_1 \cos \sigma_s = \pm\sqrt{r_1^2 - r_2^2 \sin^2 \omega}$$

and in view of (17.22)

$$\mp \sqrt{r_1^2 - r_2^2 \sin^2 \omega} = 2\sqrt{a^2 - b^2} - r_2 \cos \omega$$

With Eqs. (17.19) and (17.20) we obtain for the semimajor axis

$$a = \frac{l(l + r_2 \cos \omega)}{2l - r_2(1 - \cos \omega)} \qquad (17.23)$$

where $r_2 = D_c/2\sin \omega$.

From (17.19) it follows that the semiminor axis is

$$b = \sqrt{2al - l^2} \qquad (17.24)$$

The distance l is selected such that the reflector height is kept as low as possible provided at the same time a convenient room for the lamp with a focusing socket.

The apical subtense of the reflector, $2\sigma_s$, is determined from the expression

$$\sin (180° - \sigma_s) = D_c/2r_1$$

where $r_1 = 2a - D_c/2\sin \omega$. Hence,

$$\sin (180° - \sigma_s) = \frac{D_c \sin \omega}{4a \sin \omega - D_c} \qquad (17.25)$$

Referring to Fig. 17.3, the reflector height is

$$h = \frac{D_c}{2} \cot \sigma_s + l$$

or

$$h = 2a - l - g \qquad (17.26)$$

(6) The overall maximum height of the enlarger is then

$$H_{max} = p'_{max} + g + h \qquad (17.27)$$

(7) The linear magnification of the reflector

$$\beta_r = -D/c$$

where D is the diameter of the entrance pupil of the lens, and c the size of the filament.

On the other hand,

$$\beta_r = -\frac{F_1 F_2 + l}{l} = -\frac{2\sqrt{a^2 - b^2} + l}{l}$$

Hence,

$$c = \frac{Dl}{2\sqrt{a^2 - b^2} + l} \qquad (17.28)$$

(8) An appropriate lamp is to be looked up in a catalogue by the required value of luminance, L, and the filament size c.

Example 17.1. Design an enlarger by the given magnification, β, from -1.5 to -10; film format, 6×6 cm; $E' \geqslant 100$ lx; and $p'_{max} = 825$ mm.

Solution. We assign $\tau = 0.75$ and $L = 2.5 \times 10^5$ cd/m^2 and find by Eqs. (17.13)-(17.15) that $f' = 75$ mm, $2\omega = 54°20'$, and $D/f' = 1{:}3.6$.

From the catalogue of Soviet-made objectives the Soviet designer may see that the best fit to this constructional data is the $f/3.5$ Hindustar-58 lens having $f' = 75$ mm, and $2\omega = 60°$.

The luminance 2.5×10^5 cd/m^2 can be obtained from a 100-W incandescent lamp having a matt envelope and a filament size of 4-5 mm.

The constructional dimensions of the ellipsoidal reflector and the height of the enlarger, derived by Eqs. (17.16)-(17.28) with the efficient angular field $2\omega = 54°20'$, $k = 35$ mm, and $l = 40$ mm, are as follows: $g = 160$ mm, $D_c = 165$ mm, $a = 133.2$ mm, $b = 95.1$ mm, $2\sigma_s \approx 211°$, $h = 66.5$ mm, $H_{max} \approx 1050$ mm, and $c = 3.8$ mm.

OPTICAL PHOTOELECTRIC SYSTEMS

18.1 Characteristics of Optical Detectors

We shall refer to an optical system as photoelectric if it uses a photoelectric detector for recording and measuring of incident light. The family of such systems should also include those instruments which exploit thermal and opto-acoustical integral detectors.

A detector of radiant energy (radiometer) converts the energy of a light flux incident on the instrument into an electric energy. In order to assess and compare the performance of photoelectric detectors they are described in terms of certain characteristics and parameters measured at some set conditions. The parameters involved are integral sensitivity, sensitivity threshold, size and shape of the photosensitive layer, etc. The characteristics normally describe properties of a particular detector in terms of several quantities and may be represented analytically or as tables and plots. These include the spectral response, frequency response, output voltage of a radiometer, etc.

Below we outline the principal parameters and characteristics of detectors which are essential in matching the performance of an optical system and detector for the work with a particular source of light.

A characteristic called the *spectral responsivity* is the ratio of the detector output signal (response) to a monochromatic flux, $d\Phi_e$. Depending on the electric circuit in which the detector is connected, the response can be a variation of current, voltage, or some other electric parameters. It may accordingly be referred to as a current responsivity, voltage responsivity, and the like. We shall denote the output of a detector by i. Then the spectral responsivity of this detector is

$$S(\lambda) = i/d\Phi_e$$

The dimensions of this quantity may be A W^{-1} or V W^{-1}.

The function $S(\lambda)$ for most detectors is a single-peak curve as indicated in Fig. 18.1a. For simplicity of photometric computations, it is normalized to the maximum value, S_{max}, so that the relative spectral responsivity becomes

$$s(\lambda) = S(\lambda)/S_{max} \qquad (18.1)$$

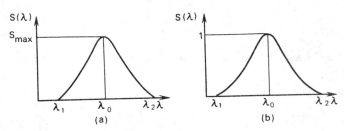

Fig. 18.1. Spectral responsivity of a light detector

The dependence of $s(\lambda)$ on wavelength is called the spectral characteristic of the detector and plotted in the respective performance specifications as shown in Fig. 18.1b. This characteristic indicates the range in which a particular detector is operable and, given $S_{\max}$, the ab-solute spectral responsivity may be readily derived as $S(\lambda) = s(\lambda)S_{\max}$.

For a detector whose responsivity is independent of wavelength, the spectral characteristic will be a straight line parallel to the axis of wavelengths.

The *integral sensitivity* of a detector is the ratio of the detector response to the entire radiant flux,

$$S = i/\Phi_e \tag{18.2}$$

If the source of light has a continuous spectral distribution of intensity, then the sensitivity will be given as

$$S = \frac{i}{\Phi_e} = \frac{S_{\max} \int\limits_{\lambda_1}^{\lambda_2} s(\lambda)\varphi_e(\lambda)\,d\lambda}{\int\limits_0^{\infty} \varphi_e(\lambda)\,d\lambda} \tag{18.3}$$

The units of (integral) sensitivity are $A\ W^{-1}$ and $V\ W^{-1}$.

This expression indicates that the sensitivity of a detector depends on the spectral content of the source of light. Therefore, in calibration a standard source of light must be used. One of such sources is an incandescent lamp with a tungsten coil operating at a colour temperature of 2854 K. Occasionally the sensitivity of detectors is measured with a lamp having a tungsten filament operating at a colour temperature of 2360 K. It will be recalled that the colour temperature of a source of radiation is the actual temperature of the black-body which has the same effective spectral distribution (in the visible range).

In the visible range, a standard source is characterized by its luminous flux (see Section 7.3), and the sensitivity of a photometric detector is derived as the ratio of its response to this luminous flux, viz.,

$$S = \frac{i}{\Phi} = S_{max} \int_{\lambda_1}^{\lambda_2} s(\lambda)\varphi_e(\lambda)\,d\lambda / 680 \int_{0.38}^{0.77} V(\lambda)\varphi_e(\lambda)\,d\lambda \qquad (18.4)$$

the respective units being A lm^{-1} or V lm^{-1}.

In Eqs. (18.3) and (18.4), in place of the spectral density of radiant flux, $\varphi_e(\lambda)$, one may use the spectral density of radiant exitance, $m_e(\lambda)$.

When the spectral characteristic $s(\lambda)$ of a detector and its luminous sensitivity (18.4) are known, the radiant spectral responsivity of the detector may be found as

$$S(\lambda) = s(\lambda)S_{max} = s(\lambda)S 680 \int_{0.38}^{0.77} V(\lambda)\varphi_e(\lambda)\,d\lambda / \int_{\lambda_1}^{\lambda_2} s(\lambda)\varphi_e(\lambda)\,d\lambda \qquad (18.5)$$

the respective units being A W^{-1} or V W^{-1}. The integral in this expression can be evaluated by one of the graphical methods.

A parameter which will be called the *sensitivity threshold* is defined as the minimum radiant flux that produces an output signal in the detector such that its ratio to the noise of the detector is just equal to the given signal-to-noise ratio. This threshold flux will be denoted as Φ_t or $\Phi_{e,t}$, the respective dimensions being lm or W. The noise level in a detector depends on the area of the sensitive layer, A_d, and the operating bandwidth, Δf; therefore, $\Phi_{e,t} \sim \sqrt{A_d \Delta f}$. Accordingly, the dimension of this threshold will be W cm^{-1} $Hz^{-1/2}$ or lm cm^{-1} $Hz^{-1/2}$.

It is worth emphasizing that like the integral sensitivity, the sensitivity threshold of a detector depends on the spectral distribution of the source of light. Therefore, the sensitivity threshold of any particular detector is determined with a standardized source of light (at 2854 K or 2360 K). For detectors operating in the far infrared, the standard source operates at a colour temperature of 500 K.

The size and shape of the sensitive surface of a detector influence the sensitivity threshold and must be involved in any evaluation of the characteristics of an optical photoelectric system. If an irradiance E_e is produced on the light sensitive surface of area A_d, then the radiant flux incident on the detector is

$$\Phi_e = E_e A_d$$

and the response of the detector, by (18.2), is

$$i = S\Phi_e = SE_e A_d$$

Hence, the irradiance on the detector surface that must be produced by the optical system to obtain a given response R is

$$E_e = i/SA_d$$

18.2 Evaluating the Entrance Pupil Diameter

We shall examine the optical photoelectric system with reference to its diagram in Fig. 18.2. The radiant energy from the source, *1*, passes through a number of optical media and the optical system and arrives at the detector, *2*. Some optical systems have light filters, *3*, to change the spectral composition of the light energy arriving at the detector.

The characteristics of the optical system should be designed such that the response of the detector to the flux from a particular source of light is at least equal to some minimum level i_{min} related to the sensitivity threshold of this detector, i.e.,

$$i_{min} = ki_t$$

where $k \geqslant 1$, and i_t is the level corresponding to the sensitivity threshold.

In the examination that follows we assume that the source of light and a detector are specified or selected, that is, the radiance and area of the source and the minimum detector response and its integral sensitivity are known. We also assume that the transmission factors of all media traversed by light in our system are specified.

If the source of light is on the optical axis and has a radiance L_e identical in all directions, the entrance pupil of the optical system will intercept a radiant flux

$$\Phi_e = \tau_a \pi L_e \sin^2 \sigma_A A_e$$

where τ_a is the transmission of the atmosphere between the source and the entrance pupil of the optical system, σ_A the aperture angle of the optical

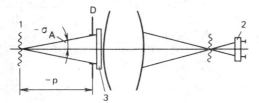

Fig. 18.2. Schematics of an optical photoelectric system

system in object space (see Fig. 18.2), and A_e is the area of the source (emitter).

If τ_f is the transmission of the filter used, τ_s the transmission of the optical system, then in the absence of vignetting in the system the radiant flux leaving the system will be

$$\Phi_e' = \tau_f \tau_s \Phi_e$$
$$= \tau_a \tau_f \tau_s \pi L_e \sin^2 \sigma_A A_e \qquad (18.6)$$

Now we suppose that all the flux Φ_e' arrives at the sensitive surface of the detector with a sensitivity S, and determine the threshold response of the detector

$$i_{min} = S\Phi_e'$$
$$= \tau_a \tau_f \tau_s \pi L_e \sin^2 \sigma_A A_e S \qquad (18.7)$$

Thus, the aperture angle of the system in object space at which the detector ensures the threshold response for a given source of light will be given as

$$\sin \sigma_A = \sqrt{\frac{i_{min}}{\tau_a \tau_f \tau_s \pi L_e A_e S}} \qquad (18.8)$$

Now, the aperture of the entrance pupil of the system is

$$D = 2p \tan \sigma_A \qquad (18.9)$$

If only a fraction of the flux leaving the optical system arrives at the detector, $\Phi_e'' = \Phi_e' k$, the respective correction with $k < 1$ should be made in Eq. (18.7).

Equation (18.8) is derived on the assumption that the transmission factors of optical media and the responsivity of the detector are determined for a given spectral distribution of radiance. If it is not the case, the evaluation of the entrance pupil of an optical system should take account of the spectral characteristics of the source of light, the optical media involved, and the detector.

18.3 The Effect of Spectral Characteristics

It commonly occurs that the sources of light for photoelectric systems are heated bodies emitting continuous spectra. Most sources of thermal radiation radiate energy in a manner which can be readily described in terms of a black-body emitting through a filter, making it possible to use

21*

the black-body radiation laws as a starting point for many radiometric calculations.

Planck's law describes the spectral radiant exitance of a perfect black-body as a function of its temperature T (in degrees Kelvin) and the wavelength of the emitted radiation (in micrometres)

$$m_e^* (\lambda) = \frac{c_1}{\lambda^5 (e^{c_2/\lambda T} - 1)} \qquad (18.10)$$

where $m_e^* (\lambda)$ is the radiation emitted into a hemisphere in power per unit area per wavelength interval, i.e., in $W\,cm^{-2}\mu m^{-1}$; c_1 is a constant $= 3.74 \times 10^4\ W\ cm^{-2}\ \mu m^4$ when the area is in square centimetres and the wavelength in micrometers; and c_2 is a constant $= 14.380 \times 10^4\ \mu m\ K$.

If we differentiate the Planck equation (18.10) with respect to wavelength and set the result equal to zero, we can determine the wavelength at which the spectral exitance $m_e^* (\lambda)$ is a maximum. This is *Wien's displacement law* which gives the wavelength (in μm) for maximum $m_e^* (\lambda)$ as

$$\lambda_m = 2896\ T^{-1} \qquad (18.11)$$

Substituting this value of λ_m in Planck's equation (18.10) we obtain the amount of $m_e^* (\lambda)$ at this wavelength

$$m_e^* (\lambda_m) = 1.301\ (T/1000)^5 \qquad (18.12)$$

If we integrate Eq. (18.10), we can obtain the total radiant exitance at all wavelengths

$$M_e^* = \int_0^\infty m_e^* (\lambda)\,d\lambda$$

The resulting equation is known as the *Stefan — Boltzmann law*

$$M_e^* = \sigma T^4 \qquad (18.13)$$

where $\sigma = 5.672 \times 10^{-12}\ W\ cm^{-2}\ K^{-4}$.

Planck's equation is awkward to use and for this reason a number of tables, charts and graphs are available which allow the user to simply look up the values of $m_e^* (\lambda)$ for the appropriate temperature and wavelength. The computations with Planck's law can be simplified by introducing the new normalized variables as follows

$$x = \lambda/\lambda_m \qquad (18.14)$$

and

$$y = m_e^* (\lambda)/m_e^* (\lambda_m) \qquad (18.15)$$

The normalization constants λ_m and $m_e^* (\lambda_m)$ are determined by Eq. (18.11) and Eq. (18.12) respectively.

In the new variables Planck's equation becomes

$$y = 142.32 x^{-5} (e^{4.9651/x} - 1)^{-1} \qquad (18.16)$$

and the respective plot is sketched in Fig. 18.3. Now to determine the spectral density of radiant exitance for a particular λ one is first to compute λ_m and $m_e^* (\lambda_m)$ by (18.11) and (18.12) in order to find the appropriate x for the chosen λ. The value of y for this x is looked up in tables or extracted from a plot in the normalized coordinates. The desired value of $m_e^* (\lambda)$ can be determined by Eq. (18.15) as

$$m_e^* (\lambda) = y m_e^* (\lambda_m)$$

Because $m_e^* (\lambda_m)$ is a function of temperature, the procedure outlined above enables one to construct the black-body spectral distribution curve for any temperature.

Most real thermal radiators are not perfect black-bodies. They radiate less energy than a black-body at the same temperature. Accordingly, many of them are called grey-bodies. A grey-body is one which emits radiation in exactly the same spectral distribution as a black-body at the same temperature, but with reduced intensity. A straightforward way to compare a grey-body radiation with that of the respective black-body is by means of the ratio of its function of radiant exitance to that of a perfect black-body at the same temperature. This ratio, known as *emissivity*, is thus a measure of the radiation (and absorption) efficiency of a body.

For many materials the emissivity is a function of wavelength. In regions of the spectrum where this occurs emissivity becomes *spectral* emissivity $\varepsilon(\lambda)$ defined as

$$\varepsilon(\lambda) = m_e(\lambda)/m_e^* (\lambda) \qquad (18.17)$$

and treated just as any other spectral function.

It should be noted that most materials show a variation of emissivity with temperature as well as wavelength. These materials are known as

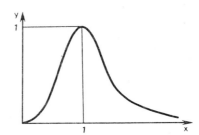

Fig. 18.3. Spectral distribution of black-body radiation

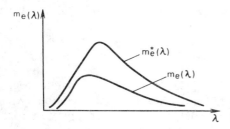

Fig. 18.4. Spectral densities of radian exitance for a black-body and a real body

"selective" emitters in the Soviet literature. Fig. 18.4 sketches the plots of the spectral density of radiant exitance of a black-body and a "selective" emitter at the same temperature. It will be seen that these curves peak at different wavelengths. Specifically, metals have shorter maximum wavelengths than a black-body.

The situation is simpler with grey-bodies whose emissivity

$$\varepsilon(T) = m_e(\lambda)/m_e^*(\lambda) \qquad (18.18)$$

is independent of wavelength and varies only with temperature. Therefore, the respective integral quantities may be used to define the ratio (18.18) and the emissivity will then be qualified as total. When dealing with grey-bodies, it is necessary to insert the emissivity factor ε into the black-body equations. Planck's equation (18.10), the Stefan — Boltzmann law (18.13), and the Wien displacement law (18.11) should be modified by multiplying the right-hand side by the appropriate value of ε.

For many real emitters the emissivity is determined experimentally and may be found in the form of tables and plots [10, 12].

To recapitulate our findings, once the spectral density of radiant exitance of a black-body has been calculated, the values of $\varepsilon(\lambda)$ or $\varepsilon(T)$ can be looked up to determine the spectral density of radiant exitance for a "selective" emitter as

$$m_e(\lambda) = \varepsilon(\lambda)m_e^*(\lambda) \qquad (18.19)$$

or for a grey-body as

$$m_e(\lambda) = \varepsilon(T)m_e^*(\lambda)$$

When an optical photoelectric system is designed for a particular thermal radiator, its spectral characteristics are taken into account by specifying its temperature, spectral emissivity, and dimensions of the emitting surface. The principal layout (Fig. 18.2) remains obviously the same as in the design with integral characteristics.

We wish to illustrate the procedure assuming that the specification list contains the characteristics of the emitter (area A_e, temperature T, and spectral emissivity $\varepsilon(\lambda)$), the characteristics of the detector (minimal output signal i_{min}, relative spectral response $s(\lambda)$, and maximum spectral response S_{max}), the spectral transmittance of the atmosphere $\tau_a(\lambda)$, the spectral transmittance of the filter $\tau_f(\lambda)$, and the spectral transmittance of the optical system $\tau_s(\lambda)$. We note in passing that the spectral transmittancies of optical media are ordinarily given in graphical form.

Let the radiation source lie on the optical axis and exhibit the same radiance in all directions. In the absence of vignetting, the monochromatic radiant flux emerging from the optical system will be

$$d\Phi_e' = \tau_a(\lambda)\tau_f(\lambda)\tau_s(\lambda)dL_e \pi \sin^2 \sigma_A A_e$$

where $dL_e = l_e(\lambda)d\lambda$.

Because the radiance of the source is identical in all directions we may write by virtue of (18.19)

$$dL_e = dM_e/\pi = m_e(\lambda)d\lambda/\pi = \varepsilon(\lambda)m_e^*(\lambda)d\lambda/\pi$$

Consequently,

$$d\Phi_e' = \tau_a(\lambda)\tau_f(\lambda)\tau_s(\lambda)\varepsilon(\lambda)m_e^*(\lambda)d\lambda \sin^2 \sigma_A A_e$$

If the whole of the flux is intercepted by the light sensitive surface of the detector then the incremental output (response) of the detector is

$$di = S(\lambda)d\Phi_e'$$
$$= \sin^2 \sigma_A A_e S_{max}\tau_a(\lambda)\tau_f(\lambda)\tau_s(\lambda)s(\lambda)\varepsilon(\lambda)m_e^*(\lambda)d\lambda \qquad (18.20)$$

The total response of the detector to the radiant flux distributed in some way between λ_1 and λ_2 will be found as (we assume that this is the minimum useful signal)

$$i_{min} = \sin^2 \sigma_A A_e S_{max} \int_{\lambda_1}^{\lambda_2} \tau_a(\lambda)\tau_f(\lambda)\tau_s(\lambda)s(\lambda)\varepsilon(\lambda)m_e^*(\lambda)d\lambda$$

$$= \sin^2 \sigma_A A_e S_{max}I$$

Now the sine of the aperture angle in object space is

$$\sin \sigma_A = \sqrt{i_{min}/S_{msx}A_e I} \qquad (18.21)$$

and the diameter of the entrance pupil is $D = 2p \tan \sigma_A$.

The integration limits for the integral I depend on the spectral ranges of the detector and transmittances of the optical media. The integral I can be evaluated graphically. A relevant technique will be outlined below.

Consider the ratio

$$I \bigg/ \int_0^\infty m_e^*(\lambda)d\lambda = k \qquad (18.22)$$

where the integral in the denominator is seen to be the radiant exitance of a black-body determined by the Stefan — Boltzmann law (18.13) as

$$\int_0^\infty m_e^*(\lambda)\,d\lambda = \sigma T^4 = 5.672 \left(\frac{T}{1000}\right)^4 \qquad (18.23)$$

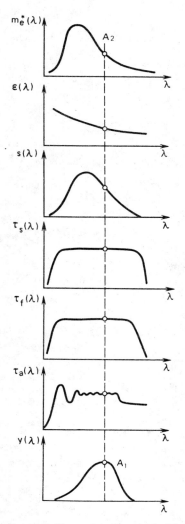

Fig. 18.5. Illustrating the graphical evaluation of the integral I

In order to evaluate the value of k we plot all the functions in the integrand of I. Because the temperature of the emitter is given, the spectral density of radiant exitance of the black-body can be derived from the respective curve plotted in the normalized coordinates. The spectral curves of the other dimensionless functions are normally known from the performance specification and data collected for the particular system design. Plotted to a unified λ scale these functions are indicated in Fig. 18.5.

Having set the appropriate integration limits we partition the range thus defined into a number of incremental intervals and for each partition evaluate the value of the product

$$y(\lambda) = \tau_a(\lambda)\tau_f(\lambda)\tau_s(\lambda)s(\lambda)\varepsilon(\lambda)m_e^*(\lambda)$$

Plotting a curve through the determined values of y we obtain the area A_1 (bottom plot) proportional to the integral I. Notice that the area A_2 under the top plot is proportional to the integral in the denominator of Eq. (18.22). Assuming the plots of $y(\lambda)$ and $m_e^*(\lambda)$ are identically scaled, the ratio of the integrals may be replaced by the ratio of the respective areas, $k = A_1/A_2$. By virtue of (18.23) the desired integral can be derived as

$$I = 5.672\,(T/1000)^4 k$$

In our further examinations of various optical arrangements of photoelectrical systems we shall refer to Eq. (18.8), keeping in mind, though, that where the respective spectral characteristics will have to be taken into account, this will be replaced by Eq. (18.21).

18.4 Arrangements with the Detector in the Image Plane

A commonly met arrangement of optical photoelectric systems is the one with the light sensitive surface of the detector being placed in the image plane or near this plane. Some general recommendations for the design of this type of systems are as follows.

Owing to purely technological deficiencies some areas of the detector may have a higher responsivity than the others. Therefore, to secure a stable operation of the entire system it is desirable that the image of the source of light occupy as large of the detector surface as possible. For an efficient utilization of the flux intercepted by the entrance pupil of the optical system, this must have no vignetting and the image of the source must be consistent with the surface of the detector. Some controversies among the stated conditions would be handled in an optimal manner if the detector surface is geometrically similar to the form of the source.

Let us have a closer look at various photoelectric systems having their detectors situated in the plane of the image.

A single-lens system with a source at a finite distance (Fig. 18.6). Let us assume that the radiation source, *1*, has a radiance L_e and an area A_e, and the detector, *2*, , has an integral sensitivity S and area A_d. We also assume that the minimum response of the detector is i_{min} and the transmission of the optical system is τ_s. Now the aperture angle in object space can be found by Eq. (18.8) as

$$\sin \sigma_A = \sqrt{i_{min}/\tau_s \, \pi L_e A_e \, S} \qquad (18.24)$$

For a short distance between the source and the optical system we may let approximately $\tau_a = 1$. The system has no filter and is free from vignetting. The formula (18.24) is valid so long as the image of the emitter fits into the light sensitive surface of the detector.

The image-detector size matching is achieved by an appropriate choice of magnification. If the source of light is a $b \times c$ rectangle and the detector surface is of round shape, the magnification of the optical system should be

$$\beta = -d_d/\sqrt{b^2 + c^2} \qquad (18.25)$$

where d_d is the diameter of the detector. The minus sign implies that the image is inverted.

If the distance L from the source to the detector is specified, Eqs. (3.12)-(3.14) yield the respective values of f', a, and a'.

In the course of the primary layout, we may assume the optical system being a thin lens, i.e., let $\Delta_{PP'} = 0$.

It commonly occurs that the entrance pupil of the optical system coincides with the front surface. Then for a thin lens, $a = p$, and from (18.9) the entrance pupil diameter is $D = 2a \tan \sigma_A$.

The formulation of the optical system depends on the angle $2\sigma_A$. If $2\sigma_A$

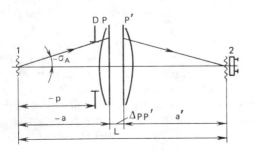

Fig. 18.6. An optical photo-electric system with a source at a finite distance

$\leqslant 30°$ is sufficient, a single-lens system will do; for $2\sigma_A \leqslant 60°$, a two-lens system will be a better choice; and for $2\sigma_A \leqslant 90°$, a three-lens formulation will be optimal. Having adopted a design form, the system transmission, τ_s, and the lengths a and a' must be refined.

A single-lens system with a source at infinity (Fig. 18.7). The detector, *2*, is situated now at the second focal plane of the system. If the maximum angle subtended by the source, *1*, at the first principal point is 2ω, then its image at the second focal plane will be of size $d'_e = 2f'$ tan ω. This image must fit into the detector surface, i.e., $d'_e \leqslant d_d$. Consequently, the optimal focal length of the system should be $f' = d_d/2$ tan ω.

If experiments show that the image of the source is considerably smaller than the sensitive surface, the detector should be shifted away from the focal plane.

When the source is known and the choice of detector is made, Eqs. (18.8) and (18.9) give the aperture angle in object space and the entrance pupil aperture, respectively.

Recognizing that $|p| \gg D$ for an infinitely distant source, we may assume sin $\sigma_A = $ tan σ_A, so that

$$D = 2p \sin \sigma_A$$
$$= 2p\sqrt{i_{\min}/\tau_a \tau_f \tau_s \pi L_e A_e S} \tag{18.26}$$

We recall that the size of an infinitely distant object is characterized by its angular subtense 2ω. If our source is a round shape of angular subtense 2ω radians, the area of this source is $A_e = \pi p^2 \omega^2$. Substituting this value of A_e in Eq. (18.26) yields the diameter of the entrance pupil as

$$D = \frac{2}{\pi\omega} \sqrt{\frac{i_{\min}}{\tau_a \tau_f \tau_s L_e S}} \tag{18.27}$$

Celestial-body recording systems. The progress in space research in recent decades has evolved a variety of photoelectric systems for the recording of the radiation coming from the stars. Astronomy and astrophysics use their own photometric units based on the notion of star magnitude, or stellar brightness. This specific feature of the field has to be taken into account in the design of appropriate systems.

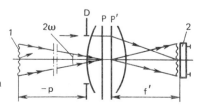

Fig. 18.7. An optical photoelectric system with a source at infinity

From the earth's surface, any star is a perfect point source which can be characterized by the irradiance it produces at the earth's surface or at the upper boundary of the earth's atmosphere. Stellar brightness is a measure of the illumination the star produces on a plane set at right angles to its light. The scale of stellar magnitudes is defined by the equation

$$m = -2.5 \log E - 13.89 \qquad (18.28)$$

where E is the illumination (lx) produced by the star light at the upper boundary of the earth's atmosphere.

According to Eq. (18.28) a first-magnitude star produces an illumination of 1.11×10^{-6} lx, and a second-magnitude star an illumination of 1.11×10^{-8} lx. This formula may also be used to describe the radiation of emitters of finite size, say the moon, sun, and a variety of distant terrestrial sources. By way of example, at full moon, its light produces an illumination of about 0.2 lx at the earth's surface, which corresponds to $m = -12.55$.

Photoelectric system design calls for the photometric quantities characterizing stellar brightnesses to be converted to the radiant quantities. We recall from Section 7.3 that the ratio of the respective luminous quantities (stellar brightness or E in lx) and radiant quantities (E_e in W m^{-2}) is known as the luminous efficacy (lm/W)

$$K = \frac{\Phi}{\Phi_e} = \frac{E}{E_e} \qquad (18.29)$$

In astrophysics it is assumed that stars radiate as black-bodies of various temperature. They are divided into spectral classes denoted by capital letters. To illustrate, a class A star of surface temperature 10 000 K has a luminous efficacy of 61.35 lm/W, and the maximum luminous efficacy 84.18 lm/W has a class G star of surface temperature 6000 K, i.e. the same temperature as the sun's.

The principal arrangement of a stellar radiometer is shown in Fig. 18.8. For a star of stellar brightness m, Eq. (18.28) gives the illumination E at the boundary of the earth's atmosphere. The class of the star leads us by virtue of (18.29) to the irradiance at the boundary of the earth's atmosphere $E_e = E/K$.

Suppose now that our optical photoelectric system is at the earth's surface. Given that the transmission of the atmosphere is τ_a, the radiant flux incident from the star, I, to the system entrance pupil of diameter D is

$$\Phi_e = \tau_a E_e \pi D^2/4$$

The image of the star will be produced at the second focal plane of the optical system and, provided it is free from aberrations, this image of a dis-

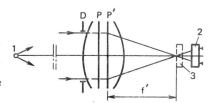

Fig. 18.8. An arrangement for detecting the radiation of stars

tant point will be a diffracted blur (Airy disc). Therefore, to exploit as large as possible area of the detector this is located some distance behind the focal plane. The diffracted spot of light should not exceed the detector surface. Occasionally this shift is caused by a need for inserting an analyser, *3*, at the second focal plane.

If τ_s stands for the transmission of the optical system with the analyser, then the radiant flux arriving at the detector is

$$\Phi_e' = \Phi_e \tau_s = \tau_a \tau_s E_e \pi D^2 / 4 \qquad (18.30)$$

Suppose that the integral sensitivity of the detector is S, then the response (output signal) of the detector

$$i_{min} = \Phi_e' S = \tau_a \tau_s E_e \frac{\pi D^2}{4} S$$

Hence to extract the signal of a faint star of a given stellar brightness m from the noise of the chosen detector, the respective optical system must have the entrance pupil of aperture

$$D = 2\sqrt{i_{min} / \tau_a \tau_s \pi E_e S} \qquad (18.31)$$

The focal length of this optical system does not affect the size of the stellar image, therefore its choice in this case is controlled by the relative aperture.

A two-lens optical system. A thin-lens optical representation of this system is indicated in Fig. 18.9. In this arrangement, the source of light, *1*,

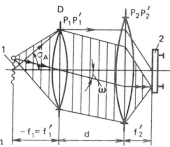

Fig. 18.9. The optics of a two-lens photoelectric system

is placed at the first focal plane of the front lens while the detector, 2, is placed at the second focal plane of the rear lens. The magnification of the system defined in this arrangement by the ratio

$$\beta = -f_2' / f_1' \qquad (18.32)$$

is adjusted so that the image of the source lies just within the sensitive surface of the detector. Hence, if the source and detector are chosen, the magnification of the two-lens system can be determined, say, by Eq. (18.25). Then the focal length of one component is subject to that of the other in view of Eq. (18.32).

The aperture angle in object space can be obtained as

$$\sin \sigma_A = \sqrt{i_{\min} / \tau_s \pi L_e A_e S}$$

where τ_s is the transmission factor of the two-lens system.

The diameter of the front lens is

$$D_1 = 2f_1' \tan \sigma_A$$

The diameter of the second lens is found subject to the absence of vignetting for the point of the source that is the farthest from the optical axis. If the source of light is a $b \times c$ rectangle the maximum inclination of the parallel ray bundle emerging from the front lens can be found from

$$\tan \omega = \sqrt{b^2 + c^2} / 2f_1'$$

Once the spacing d between the lenses is chosen, the necessary diameter of the second lens is

$$D_2 = D_1 + 2d \tan \omega \qquad (18.33)$$

For appreciable spacings between the lenses, the diameter of the rear component derived by Eq. (18.33) might grow prohibitively large. In such cases some amount of vignetting is unavoidable.

In the presence of vignetting, the widely separated lens system (Fig. 18.10) can be designed as follows. The source of light, 1, and the front lens may be regarded as a searchlight of radiant intensity

$$I_{es} = \tau_{s1} I_e (D_1' / d)^2$$

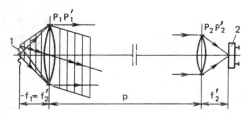

Fig. 18.10. A two-lens system with widely spaced elements

where τ_{s1} is the transmission of the front lens, I_e the radiant intensity of the source of light, d the diameter of the source, and D_1' the diameter of the entrance pupil of the front lens.

If the spacing p between the lenses exceeds the critical distance of the beam formation, then the radiance in the entrance pupil of the second lens will be

$$E_e = \tau_a I_{es}/p^2$$

where τ_a is the transmission of the atmosphere at the distance p.

If we denote the entrance pupil diameter of the second lens by D_2 the radiant flux entering the second lens is

$$\Phi_e = E_e \pi D_2^2/4$$

and the radiant flux arriving at the detector, 2, is

$$\Phi_e' = \tau_{s2}\Phi_e$$

where τ_{s2} is the transmission of the second lens of the optical system.

The output signal (response) of the detector in the second focal plane of the rear lens is

$$i = S\Phi_e'$$

18.5 Arrangements with the Source Image Overlapping the Detector

In our preceding evaluations of photoelectric systems, all the flux intercepted by the entrance pupil of the optical system arrived at the light sensitive surface of the detector. This situation holds when the system is free from vignetting and the image exactly fits the sensitive area of the detector. The last condition is met by a suitable choice of the magnification or focal length of the optical system.

Some practical designs, however, prevent the engineer from adjusting the desired magnification or focal length of the optical system. In the circumstances, the flux incident on the entrance pupil of the optical system fails to entirely fit the sensitive surface of the detector, thus invalidating the expressions derived above for the entrance pupil of the optical system.

When the image of the source of light exceeds the size of the light sensitive area of the detector, it is advisable to design the system starting from image space as follows.

The irradiance produced by the source on the light sensitive area of the detector can be determined by Eq. (7.63) as

$$E_e' = \tau_a \tau_f \tau_s \pi L_e \sin^2 \sigma_A'.$$

Since the image of the source overlaps the detector, the flux incident on the detector will be given as

$$\Phi'_e = E'_e A_d$$
$$= \tau_a \tau_f \tau_s \pi L_e \sin^2 \sigma'_A \cdot A_d \qquad (18.34)$$

where A_d is the light sensitive area of the detector.

Recalling that the output signal of the detector is

$$i_{min} = \Phi'_e S \qquad (18.35)$$

we obtain by virtue of (18.34) and (18.35) the expression for the aperture angle in image space

$$\sin \sigma'_A = \sqrt{i_{min}/\tau_a \tau_f \tau_s \pi L_e A_d S} \qquad (18.36)$$

If the source of light is at infinity, then $\sin \sigma'_A = D/2f'$ and in view of (18.36) we obtain for the aperture ratio of the optical system

$$\frac{D}{f'} = 2\sqrt{\frac{i_{min}}{\tau_a \tau_f \tau_s \pi L_e A_d S}} \qquad (18.37)$$

This and penultimate formulas are applicable also for the case of the source image exactly fitting the sensitive area of the detector. It is quite obvious that the area of the detector then should give way to the area of the emitter image A'_e, namely,

$$\sin \sigma'_A = \sqrt{i_{min}/\tau_a \tau_f \tau_s \pi L_e A'_e S}$$

will be the right equation for the source at a finite distance, and

$$\frac{D}{f'} = 2\sqrt{\frac{i_{min}}{\tau_a \tau_a \tau_f \tau_s \pi L_e A'_e S}} \qquad (18.38)$$

may be used for the source at infinity.

18.6 Arrangements with the Detector at the Exit Pupil

However uniform the detector surface may be fabricated, some applications find it inhomogeneous in response over the surface. In such cases the placement of detector at the image plane is undesirable, as a motion of a small image over the detector would produce unstable response. This setback can be alleviated by placing the detector at the exit pupil of the optical system. In the absence of vignetting, the plane of the exit pupil will have a uniform irradiance, thus ensuring an even irradiance of the detector for any position of the light source.

The simplest arrangement of a photoelectric system, having its detector placed at the exit pupil, must be a two-lens form. The thin-lens diagram of this system is indicated in Fig. 18.11. The front lens projects the source of light, *1*, into the plane of the field stop. The angular size of the source, corresponding to the field of the optical system in object space, or the angular field where the light source may be moving is denoted by 2ω. The rear component, a field lens, images the exit pupil of the front lens into the exit pupil of the optical system where the light sensitive surface of detector, *2*, is situated.

We wish to illustrate the design of such a system and suppose that the characteristics of the source of light and the detector are specified.

If the source of light is at a finite distance from the system, the aperture angle in object space can be determined by Eq. (18.8) as

$$\sin \sigma_A = \sqrt{i_{min}/\tau_a \, \tau_f \, \tau_s \, \pi L_e \, A_e \, S}$$

where τ_s is the transmission of the two-lens system assigned in advance to be refined later.

The first stage of the procedure is a thin-lens design. The mount of the front lens plays the role of the entrance pupil of the optical system. Given the throw from the system to the source (see Fig. 18.11), the diameter of the system's entrance pupil will be $D = 2a_1 \tan \sigma_A$. For an infinitely distant source, the entrance pupil diameter may be determined by Eq. (18.27) or Eq. (18.31).

Once the magnification of the front lens, β_1, is selected, we can proceed to the distance a_1' from the front lens to the field stop and to the focal length of the front lens as follows

$$a_1' = a_1 \beta_1$$

$$\frac{1}{f_1'} = \frac{1}{a_1'} - \frac{1}{a_1}$$

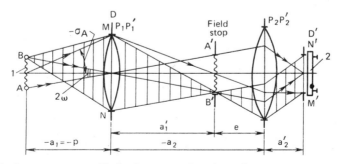

Fig. 18.11. An arrangement with the detector at the exit pupil

The field stop of diameter $D_{fs} = 2a_1'$ tan ω is placed at the source image produced by the front lens. This plane is an appropriate place to set an analyser.

If the source of light is at infinity, the field stop diameter is $D_{fs} = 2f_1'$ tan ω.

The magnification of the second lens should be such that the exit pupil of the system just fits the light sensitive surface of the detector. For a circular sensitive area of diameter d_d, the magnification of the rear lens can be derived from the expression

$$\beta_2 = -d_d/D \qquad (18.39)$$

where D is the diameter of the entrance pupil of the optical system equal to the diameter of the front lens, assumed to be thin at this stage.

The spacing e between the field stop and the rear lens is subject to the designer's choice. This choice should provide a sufficient space for an analyser at the image plane, if necessary. If this is not the case, we may let $e = 0$, then the rim of the second lens will play the role of the field stop.

Having determined the magnification of the rear lens and the spacing e, we are in a position to write the expressions to compute the distance from the rear lens to the exit pupil, a_2', and the focal length of the rear lens, viz.,

$$-a_2 = a_1' + e$$
$$a_2' = a_2\beta_2$$
$$1/f_2' = 1/a_2' - 1/a_2$$

The diameter of the rear lens is calculated subject to the freedom from vignetting at the edge of the field. This condition will be satisfied if the optical system propagates the ray MB' through the upper edge of the entrance pupil and the lower rim of the field stop. With reference to Fig. 18.11,

$$D_2 = D_{fs} + e \frac{D + D_{fs}}{a_1'}$$

It is apparent that for $e = 0$ $D_2 = D_{fs}$.

18.7 Some Basic Arrangements

Photoelectric instruments are agglomerates of optical, electronic, and electromechanical facilities put together to convert the energy of an incident light flux into an electric signal. After processing, this signal may be used for recording or control or to produce a sensitive impression on the human observer. The optical system is of prime importance here as it is the first to handle the arriving information. Accordingly, the optical part of a photoelectric system must ensure the necessary intensity of the flux arriving at the detector, the desired size and quality of an optical image, and spectral filtering of a useful signal from extraneous noise.

Some of the problems solvable by modern photoelectric optical systems are listed below.

(1) The analysis of an object inserted in the beam travelling from the source of light having a known distribution of intensity to the detector whose spectral characteristics are also known. The objective of this analysis may be the evaluation of a transmission factor, a spectral characteristic of the object, an absorption factor in reflection, etc. so as to record various parameters of the object under testing and control them if the object is a process.

(2) The investigation of a light-emitting object to measure the power and spectral distribution of this radiation with the purpose, say, of recognition and detection of objects.

(3) The measurement of the characteristics and parameters of a detector.

(4) The evaluation of the coordinates of an object or its alignment.

(5) The measurement of parameters and characteristics of optical systems for the purpose of recording, control, observation, etc.

The recent additions to this list are the problems of data transmission and computing.

In the rest of this section we touch upon some basic arrangements of photoelectric instruments developed to solve aforementioned problems.

Figure 18.12a sketches a setup developed to analyse an object, 4, inserted in a collimated beam of light travelling between the lenses 2 and 6. A light filter, 3, comes in when the spectral content of the source, 1, is evaluated. To extract a narrow interval of wavelengths, the arrangement provides a room to insert an interference filter in the collimated beam. The detector, 7, may be placed either in the image plane of the source (or near it) or in the exit pupil of the optical system. The intensity of the beam is varied by the compensator, 5-5′, made, for example, as two sliding wedges producing a plane-parallel plane of variable thickness. In fact, this is a

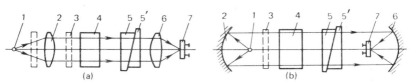

Fig. 18.12. Optical setups for measurements in transmitted light: (a) two-lens arrangement, (b) two-reflector arrangement

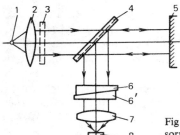

Fig. 18.13. Construction to measure the absorptance in reflection

neutral filter of variable density. This arrangement can be used for measurement work with zero reading.

In some fabrications the lens elements may have selective transmissions and result in erroneous readings. It should not be overlooked that most types of optical glass used for lens manufacture are transparent in the range from 0.35 μm to 3 μm. Therefore, for some applications, the use of reflecting systems, 2 and 6, in Fig. 18.12b, might be a more appropriate choice.

Figure 18.13 shows a zero-reading arrangement designed to measure the reflectance and absorptance in reflection at the surface of an object, 5. The collimated beam of light formed by the lens 2 and the filter 3 arrives at the object under test, 5, after passage through the beamsplitter, 4. The light reflected by the object returns to the beamsplitter and upon reflection at its surface goes through the compensator, 6-6', and the lens, 7, to be detected by the detector, 8. This setup can be used for measurements of reflectance, too.

Figure 18.14 schematizes the optics of an instrument developed for tracking an infinitely distant object. The objective, 1, forms an image of the object at the second focal plane where an analyser, 2, is placed. The condenser, 3, collects the flux arriving from the object onto the light sensitive surface of the detector, 4, which can be situated, say, in the exit pupil of the optical system. As the position of the object with respect to the optical axis varies, the radiant flux through the analyser also varies, and the detector produces an alternating signal.

Figure 18.15 shows an arrangement with differential switching of detectors. The light from the source, 1, reflected from the mirrors 2 and 2' is

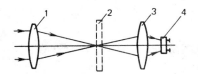

Fig. 18.14. The optics of an instrument for tracking a radiating object

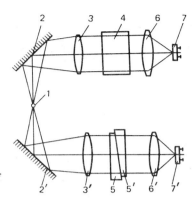

Fig. 18.15. A setup with a differential switching of detectors

directed in the lens *3* of the measurement branch, where the object being tested, *4*, is placed, and in the objective *3′* of the reference branch. To alter the flux in the reference branch, it has a compensator, *5-5′*, of variable transmission or an iris diaphragm. The lenses *6* and *6′* focus the light onto the detectors *7* and *7′*. The circuits of the detectors are connected in opposition to exclude non-zero response unless the fluxs incident on the detectors are indentical.

The reading of the quantity being measured is taken from the vernier of the wedge compensator. An obvious advantage of this arrangement is independence of measurements of variation in the source intensity. It requires, however, detectors with identical and stable characteristics.

OPTICAL SYSTEMS FOR LASERS

19.1 Properties of Laser Light

Laser radiation is known for its high monochromaticity, directivity, and power. It is coherent in time and space and polarized. These properties are characteristic of all lasers regardless of a specific laser type and its technical data.

The monochromaticity of a laser beam is characterized by its linewidth, $\Delta\lambda$, measured as the full width at half-maximum power (FWHM). For gas lasers, the FWHM is 10^{-3} to 10^{-4} nm, for solid-state lasers it is 10^{-1} to 10^{-2} nm, and for semiconductor lasers 1-10 nm, but for gas lasers the linewidth can be narrowed down to 10^{-9} nm. These extremely narrow bandwidths of laser beams can be used to advantage for the spectral selection of useful signals from a noise-corrupted environment.

The directivity of a laser beam is described in terms of the solid angle or corresponding plane divergence angle confining the laser radiation. The angular divergence of laser beams amounts to minutes of arc for gas lasers, several ten minutes for solid-state lasers, and units to tens of degrees of arc for semiconductor lasers. The high directivity of laser beams enables accurate spatial ranging of irradiated objects, high angular resolution in laser experiments, and intense irradiance on irradiated objects.

The radiant power or flux produced by a laser depends on a particular type. The power of continuous wave gas lasers ranges from units to tens of milliwatts, for CW semiconductor lasers it is several kilowatts, and for pulsed solid-state lasers the power can be as high as 10^{10} W. Observing that the divergence of a laser beam is in the order of a few minutes of arc, even milliwatts of laser power is capable of producing radiant intensities of up to 10^3 W ster^{-1}.

Compared with other sources of light, lasers exhibit the highest degree of coherence. This property of laser light is used in optical data transmission systems, in some standards of length, in interferometry and some other fields.

Almost all types of laser emit polarized light. By setting the output windows of a gas laser at the Brewster angle the lasing is made linearly polarized. This property of laser light is attractive for many instruments operating with polarized light beams.

The use of a laser as a light source required, for the most part, optical systems to transform the laser beam as needed for a particular purpose. The problems solvable by these optical systems include the concentration of a laser beam into a small spot (focusing), the transformation of a laser beam into one of small divergence (collimation), and the formation of a laser beam into one having the required parameters for matching with the following optical system (matching).

The aforelisted properties of laser radiation impose some specific constraints on the configuration of relevant optical systems. High power concentration, for example, produces exceedingly strong irradiances at the place of focusing where no optical elements should be located to avoid burning damage. Accordingly, the material of optical elements must possess appropriate optical resistance. To avoid disturbing the polarization of the laser beam the surfaces of reflecting and refracting optical elements should be placed so that the angle of incidence never exceeds the critical values. Interference effects which are likely to occur owing to the highly coherent radiation should be avoided by choosing suitable thicknesses of optical elements.

19.2 Relationships for Laser Beam Transformation

A laser beam leaving an arbitrarily configured laser cavity has a wavefront which is far from spherical. A spherical portion of the wavefront can be found only near the axis of the cavity. At some section the wavefront is planar. At this location — this is about the centre of the so-called confocal cavity — the beam has a *waist* of linear size $2y$ (Fig. 19.1). For a cavity of two mirrors spaced by distance L and having curvatures r_1 and r_2, the position of the waist with respect to the mirror poles can be defined in terms of the g-parameters

$$g_1 = 1 - L/r_1 \quad g_2 = 1 - L/r_2$$

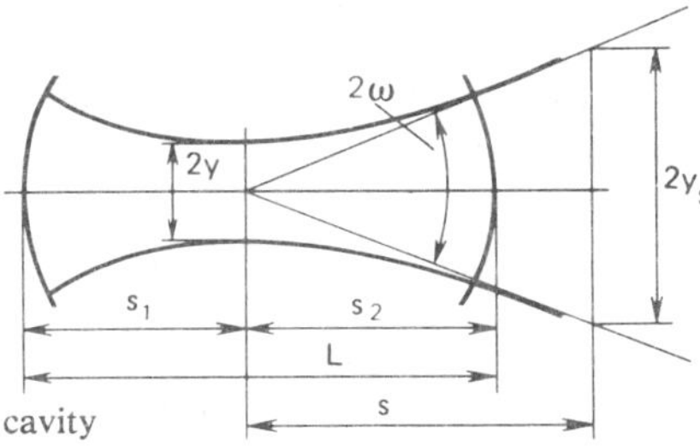

Fig. 19.1. Geometry of a laser beam inside the cavity

as [7]

$$s_1 = L \, \frac{g_2(1 - g_1)}{g_1 + g_2 - 2g_1g_2} \tag{19.1}$$

$$s_2 = L \, \frac{g_1(1 - g_2)}{g_1 + g_2 - 2g_1g_2} \tag{19.2}$$

We shall evaluate the spatial parameters of laser beams with reference to the 'equivalent confocal parameter' [7]

$$b = 2L \, \frac{\sqrt{g_1g_2(1 - g_1g_2)}}{g_1 + g_2 - 2g_1g_2} \tag{19.3}$$

If one of the cavity mirrors is plane, the waist is at the plane of this mirror, and the respective confocal parameter reduces to

$$b = 2\sqrt{(r - L)L}$$

If both mirrors of the cavity are planar, the resultant beam may be described as a train of plane waves whose divergence angle is at the diffraction limit. For this case the concepts of beam waist and equivalent confocal parameter are not necessary.

The diameter of the light spot at the waist is in general

$$2y = 2(\lambda b/2\pi)^{1/2} \tag{19.4}$$

where λ is the wavelength of the laser.

The spot diameter at a distance s from the beam waist is given as

$$2y_s = 2y\sqrt{1 + \varepsilon^2} \tag{19.5}$$

where $\varepsilon = 2s/b$ is the relative coordinate of the cross section.

At an arbitrary cross section the wavefront of a laser beam may be deemed approximately spherical with the radius

$$R = \frac{1 + \varepsilon^2}{2\varepsilon} \, b$$

The divergence of a laser beam, measured by the plane angle 2ω shown in Fig. 19.1, is a function of the beam diameter $2y_s$. For $s \gg b$ Eq. (19.5) suggests that the diameter of the beam is a linear function of the section coordinate, therefore the laser beam may be regarded as a homocentric one with the origin at the centre of the waist. The angle of divergence (in radians) is

$$2\omega = 2(2\lambda/\pi b)^{1/2} = 2\lambda/\pi y \tag{19.6}$$

We note that the solid angle corresponding to 2ω confines about 86% of the intensity for the fundamental mode which is the lowest order symmetric mode of lasing. The modes of higher orders exhibit greater divergencies.

To summarize, when we know the position of the beam waist and the equivalent confocal parameter b we are in a position to determine the parameters of the laser beam at any section.

If a laser beam meets an optical system, say a lens, the beam emergent from the system will be characterized by a new equivalent confocal parameter and a new position of the beam waist. The parameters of the emergent laser beam can be determined by the conjugate distance equation (3.7) where a and a' are replaced, respectively, with the radius of curvature R of the wavefront incident on the lens and the radius of curvature of the emergent wavefront R'.

If the beam waist of a laser is at a distance a from the thin lens of focal length f' (Fig. 19.2a), the equivalent confocal parameter of the emergent beam [7]

$$b' = \frac{b}{(1 + a/f')^2 + (b/2f')^2} \tag{19.7}$$

The beam waist for the emergent beam is located by the equation

$$1 - \frac{a'}{f'} = \frac{1 + a/f'}{(1 + a/f')^2 + (b/2f')^2} \tag{19.8}$$

The formulae (19.7) and (19.8) are also valid for an optical system of finite thickness, the distances a and a' being then measured from the principal planes.

If the beam waist is a distance z from the first focal plane, its throw to the second focal plane is given as

$$z' = \frac{-z}{(z/f')^2 + (b/2f')^2} \tag{19.9}$$

and the value of equivalent confocal parameter for the beam emergent

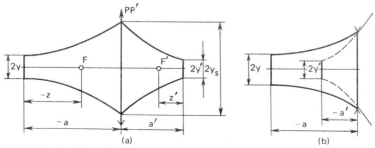

Fig. 19.2. Re-shaping a laser beam by a thin lens

from the system will be

$$b' = b \frac{4f'^2}{4z^2 + b^2} \tag{19.10}$$

From the penultimate expression it is apparent that for the case of the beam waist situating at the first focal plane ($z = 0$), the waist of the beam emergent from the system will be at the second focal plane. A laser beam surviving a negative lens has an imaginary waist and the divergence of the emergent beam increases as compared with that of the incident beam, as indicated in Fig. 19.2b.

19.3 Concentrating Laser Radiation

To produce high irradiance of a small spot the laser beam should be contracted, i.e. focused, into a small diameter. This spot can be chosen as the waist of the beam emergent from an optical system.

Equation (19.4) indicates that the smallest size $2y'$ of the transformed beam waist will be attained at the lowest equivalent confocal parameter b' of the beam emergent from the optical system. In view of Eq. (19.10), for a given laser, the parameter b' will be smaller the shorter the focal length of the optical system and the longer the distance from the laser to the first focal point of this system. The position of the waist can be found by Eq. (19.8) or (19.9). For short-focus systems, $b \gg f'$ so that by virtue of (19.9) $z' \approx 0$, that is the waist of the laser beam behind the system occurs near the second focal plane.

For efficient utilization of a laser beam, the diameter of the optical system should not be less than that of the beam section at the entrance pupil of the system.

If the optical system is a simple lens (see Fig. 19.2a), its diameter is determined subject to

$$D \geqslant 2y_s \tag{19.11}$$

where $2y_s$ is calculated by Eq. (19.5) where s is taken equal to the distance from the waist to the principal plane of the lens.

Equations (19.5) and (19.11) suggest that the laser should be placed as close to the optical system as possible to get the smallest aperture of the entrance pupil. Even with short-focal-length systems this condition will take care of the smallest possible aperture ratio D/f', thus providing favourable conditions for control of system aberrations.

Thus, if the diameter $2y'$ of the spot, in which the laser is to be concentrated, is specified, it immediately leads us to the equivalent confocal

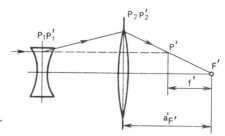

Fig. 19.3. A two-lens system of $a_F{}' > f'$

parameter b' of the emergent beam

$$b' = \frac{2\pi}{\lambda} y'^2 \qquad (19.12)$$

Having elected the appropriate type of laser and determined the equivalent confocal parameter b we select a constructionally suitable distance z from the waist to the first focal plane of the optical system. Then, by Eq. (19.10) the focal length of the optical system results as

$$f' = \frac{1}{2} \sqrt{\frac{b'}{b} (4z^2 + b^2)}$$

The clear diameter is computed by Eq. (19.5) subject to (19.11).

Short-focus systems would concentrate the laser beam short from the rear surface of the optical system which is an inconvenient configuration for most relevant applications. As a way out, a short-focus lens can be augmented by a front negative component to produce a reverse telephoto lens (Fig. 19.3) having $a_F{}' > f'$.

A two-component optical system is mandatory where the laser beam need be concentrated into a small spot at an appreciable throw [7]. The respective design can be carried out in the lines of the procedure outlined above, by successful application of Eqs. (19.9), (19.10) and (19.12) to each element of the lens.

19.4 Control of Beam Divergence

For all the high directivity of a laser beam, at long distances from the laser, divergence can expand the beam into an appreciable spot. As can be seen from Eq. (19.6) the divergence angle can be diminished by warranting greater equivalent confocal parameters. The simplest method to achieve this goal is to use small curvature mirrors in the cavity. However, this way increases the diffraction losses and renders laser stability more sensitive to misalignment.

A method of divergence control for laser beams with the help of one element, say a simple lens, is attractive owing to its simplicity. However, Eq. (19.10) suggests that to achieve a greater confocal parameter of the emergent beam, the waist of the incident beam must coincide with the first focal plane of the optical system ($z = 0$) and the system itself must be a long-focus design. This solution may be found objectionable because of considerable size of the resultant system.

A more appropriate method of divergence control seems to be the use of a two-lens configuration. The first lens of the system may be either positive or negative. A negative component means a more compact configuration. The second lens is always positive. The necessary angular magnification of the system is determined subject to (19.6) by the formula

$$\gamma = \frac{2\omega'}{2\omega} = \frac{2y}{2y'} = \sqrt{\frac{b}{b'}} \tag{19.13}$$

where 2ω and $2\omega'$ are respectively the angular divergence of the beam ahead of and behind the optical system, $2y$ and $2y'$ the waists of incident and emergent beams, and b and b' the equivalent confocal parameters of incident and emergent beams.

Let us evaluate the basic relationships of two-element system design with reference to Fig. 19.4. The position of the waist and the equivalent confocal parameter of the laser beam emergent from the first element can be determined by Eqs. (19.9) and (19.10) as

$$z_1' = \frac{z_1}{(z_1/f_1')^2 + (b_1/2f_1')^2} \tag{19.14}$$

$$b_1' = b_1 \frac{4f_1'^2}{4z_1^2 + b_1^2} \tag{19.15}$$

In order to achieve the least divergence of the laser beam after the second element of the system, the image of the waist produced by the first element must be as small as possible and located at the first focal plane of the second element ($z_2 = 0$). The first of the aforementioned conditions can be satisfied by using a short-focus element. Saying it another way, the problem solved by the first element is similar to that handled in concentrating laser radiation above. Meeting the second condition causes the second focal point of the first element to occur a distance z_1' defined by Eq. (19.14), from the first focal point of the second element. Sometimes this distance is called the 'optical interval' and denoted by Δ. For most practical applications $b_1 \gg f_1'$, therefore Δ is comparatively small. Hence, a two-element system for beam divergence control is close to the afocal system defocused by $\Delta = z_1'$.

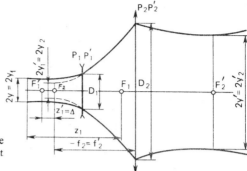

Fig. 19.4. Laser divergence control by a two-element system

With respect to the second element, the laser beam emergent from the first element can be regarded as the one in object space, that is, $2y_1' = 2y_2$ and $b_1' = b_2$ (see Fig. 19.4). Then, at $z_2 = 0$ we have from Eq. (19.10)

$$b_2' = 4f_2'^2/b_2 = 4f_2'^2/b_1'$$ (19.16)

Consequently, the angular magnification of the two-element system results from combined solution of Eqs. (19.13), (19.15) and (19.16) as

$$\gamma = \sqrt{b_1/b_2'}$$
$$= (f_1'/f_2')\sqrt{b_1^2/(4z_1^2 + b_1^2)}$$ (19.17)

At $\Delta = 0$ our system becomes afocal, and the angular magnification is given by

$$\gamma = -f_1'/f_2'$$ (19.18)

Because the radicand in Eq. (19.17) is always less than unity the comparison of Eqs. (19.17) and (19.18) suggests that the angular magnification and, hence, the divergence of the laser beam is always less for a defocused afocal system than for its afocal counterpart.

To recapitulate our evaluation, the design of a two-element system for control of laser beam divergence can be as follows. Given a laser of angular divergence 2ω, equivalent confocal parameter $b = b_1$ and waist spot diameter $2y = 2y_1$, we adopt the position of the waist, z_1, relative to the first focal point of the first component from some constructional or dimensional considerations. The diameter of the first lens, D_1, is found from Eq. (19.5) subject to the condition (19.11). The focal length of this lens, f_1', is selected such that the respective relative aperture f_1'/D_1 would not prevent the system from being freed from aberrations. Equation (19.14) yields the optical interval $\Delta = z_1'$, and Eq. (19.15) the equivalent confocal parameter of the beam emergent from the first element ($b_1' = b_2$). Since the desired angular divergence at the system output $2\omega'$ is normally specified, Eq.

(19.13) gives the angular magnification of the system, and Eq. (19.17) the focal length of the second (positive) element, namely,

$$f_2' = (f_1'/\gamma)\sqrt{b_1^2/(4z_1^2 + b_1^2)}$$

The diameter of the second lens, D_2, is determined by Eq. (19.13) subject to (19.11) and the equivalent confocal parameter b_2.

19.5 The Laser Photoelectric System

High directivity and large power of laser radiation make the laser an attractive light source for photoelectric optical systems such as optical radars. The principal configuration of an optical radar is indicated in Fig. 19.5.

Let 2ω denote the divergence of the laser beam used to irradiate distant objects. This angle is rather small, therefore the corresponding solid angle will be $\Omega = \pi\omega^2$. In the multimode operation warranting the highest power of lasing, the distribution of radiant flux over the solid angle Ω may be deemed uniform. Then the radiant intensity of the laser in the axial direction is

$$I_e = \Phi_e/\pi\omega^2$$

where Φ_e is the radiant flux of the laser.

If the distance from the laser to the irradiated object is p, and the transmission of the atmosphere at this distance is τ_a, then assuming the beam is normal to the object's surface the irradiance it produces on this surface is

$$E_e = \tau_a I_e/p^2$$

Supposing further the object's surface is Lambertian, of diffuse reflectivity ρ, we can determine the radiance of the object as a secondary source (see Eq. (7.71)) as

$$L_e = \rho E_e/\pi$$

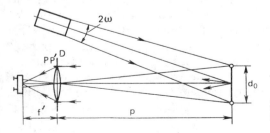

Fig. 19.5. A laser-based optical photoelectric system

The diameter of the bright spot on the object's surface irradiated by the laser is $d = 2\omega p$. This relationship is true, of course if the size of the incident spot at the distance p does not exceed that of the irradiated object.

Thus we have arrived at the radiance and the area of the secondary light source necessary to design a photoelectric optical system in the lines of the procedure outlined in Chapter 18.

We suppose that the detector of our photoelectric system is placed at the image plane of the secondary source of light. For appreciable distances p the detector is placed at the second focal plane of the optical system (see Fig. 19.5), so that the linear magnification of the optical system can be defined as $\beta = f'/p$ and the diameter of the image of the spot on the object's surface as $d_0' = d_0\beta$. If this image fits the sensitive area of the detector, the necessary aperture ratio of the optical system can be found by virtue of (18.38) as

$$\frac{D}{f'} = 2\sqrt{\frac{i_{min}}{\tau_a\tau_f\tau_s\,\pi L_e A_0'\,S(\lambda)}}$$

where A_0 is the area of the image of the irradiated spot on the object's surface, and $S(\lambda)$ the absolute spectral responsivity of the detector to the monochromatic light of the laser.

If the image of the irradiated spot overlaps the working surface of the detector, then the necessary aperture ratio of the optical system is determined by Eq. (18.37) as

$$\frac{D}{f'} = 2\sqrt{\frac{i_{min}}{\tau_a\tau_f\tau_s\,\pi L_e A_d\,S(\lambda)}}$$

The range of the system under examination can be extended by a tighter control of the divergence of the laser beam. This can be achieved by means of the two-element system we considered in Section 19.3.

19.6 Optical Systems for Holography

The recent trend in the development of holography is characteristic of the invasion of scientific and technological fields by numerous applications of this technique. The most important holographic applications include holographic interferometry, three-dimensional recording of fast processes, holographic television, high-capacity holographic memory, and pattern-recognition applications.

The process of holographic recording and subsequent reconstruction of the recorded wavefront calls for coherent sources of light. As a rule these sources are lasers offering high spatial and temporal coherence. Most

352   19 Optical Systems for Lasers

lasers, however, emit thin beams of insufficient area for holographic applications. Optical systems are useful tools to expand the laser beam to the desired cross sections.

Theoretical treatments of holographic problems assume that the recording of a hologram and the subsequent wavefront reconstruction make use of a plane monochromatic wave which from the geometric optical standpoint is a pencil of rays parallel to the optical axis. In fact, this pencil is slightly divergent, the minimum value of this divergence is controlled by diffraction.

An optimal laser-beam expander is a two-element system, close to afocal one, which is used to control the divergence of laser beams. The optics and design procedure for such a system are outlined in Section 19.4. The minimal diameter of the beam transformed by this optical system can be found by Eq. (19.13) as

$$2y' = \frac{2y}{\gamma} = \frac{1}{\gamma}\sqrt{\frac{b}{b'}}$$

The angular magnification, γ, of the afocal system is derived with Eq. (19.18). Holographic applications make use of afocal systems based on the Kepler telescope (Fig. 19.6a) or the Halilean telescope (Fig. 19.6b).

Optical data processing, transmission, and storage is a rapidly developing field of holographic application. Optical systems provide convenient means to perform Fourier transforms of coherent optical signals in optical analysis and recognition of various objects. More often than not this transformation is handled in a two-lens system (Fig. 19.7).

When the object AB being analysed (input transparency) is illuminated by a plane normally incident monochromatic wave, the first lens produces a space-frequency spectrum (Fourier image) of the object AB at the second focal plane. The second lens takes another Fourier transformation and produces an inverted image of the input transparency. By inserting various filters or patterns at the second focal plane of the first lens some portions

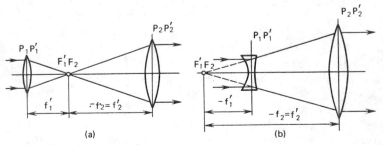

Fig. 19.6. Afocal laser-beam expanders

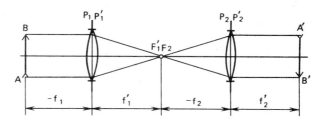

Fig. 19.7. A two-lens system for Fourier transforms

of the object's spatial spectrum can be let to pass through or blocked, i.e. modulated, thus improving the quality of the image $A'B'$. In general, such a filter performs amplitude and phase modulations of the beam passing through the system, and may serve as a coding element for a variety of data processing applications. In holographic memory systems the object matrix and filter (page composer) are holographic transparencies. Turning to the optical side of the system, we note that an accurate Fourier transformation can be achieved in this case on the condition that both elements of the system are well corrected for aberrations.

ANAMORPHIC SYSTEMS

20.1 The Anamorphic Effect — Defined

We recall that the optical systems combined of optical elements having spherical symmetry image an object plane normal to the optical axis into a geometrically similar image plane, the image scale being constant over the entire image field. Such systems have identical meridional sections.

Some applications call for systems producing images with different scale in the principal azimuths which are at right angles to each other. The systems which have a different power or magnification in one principal meridian than in the other are called *anamorphic* (or anamorphotic). Anamorphic systems are needed, for example, in the recording instruments where a narrow slit of light is required, or in wide-screen motion picture processes where the photography and projection is made with the use of the ordinary 35-mm film format. Such devices usually make use of cylinder lenses or prisms.

Anamorphic systems can transform a rectangle having *one* ratio of the sides into a rectangle with another ratio of the sides. A rectangle can be converted into a parallelogram, trapezoid, or dissimilar rectangle as shown in Fig. 20.1. Axially symmetrical systems also can effect dissimilar imagery, say for an object lying at a plane tilted with respect to the optical axis (see Figs. 3.13 and 3.14).

As a rule, anamorphic systems consist of cylindrical lenses or their combinations with ordinary spherical objective lenses. We shall refer to the principal meridian perpendicular to the axis of the lens cylinder as meridian I, and that containing the cylinder axis and oriented at 90° to the first as meridian II. An image produced by an anamorphic system can be expanded in one meridian and compressed in the other, the apparent expansion being effected by enlarging the width or by compressing the height.

We shall refer to the effect produced by an anamorphic system in terms of the image (transformed) to object width ratio, $k_a = a_t/a$, the image to *object height ratio,* $k_h = h_t/h$, and the so-called *anamorphic factor A* $= k_a/k_h$.

A slanted image results when the conditional rectangle imaged by the anamorphic system is tilted to one side. The parameters to judge the transformed image in this case are the angle of slant, ψ, and the ratios k_a and k_h.

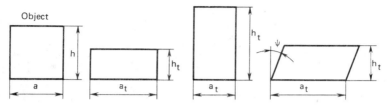

Fig. 20.1. Defining anamorphic distortions of the image

Let us look at the action of a cylinder lens. Fig. 20.2 shows a positive (planoconvex) cylinder lens illuminated by a bundle of parallel rays, say from a luminous point source at infinity. The image formed by the lens is a line of light parallel to the axis of the cylindrical surface of the lens, perpendicular to the optical axis of the lens and passing through the second focal point, F', of meridian I. The length of this light segment x_{II}' (in meridian II) is equal to the length l of the lens. If this small source of light (A in Fig. 20.3) is a distance $-a$ apart, the image produced by this lens will be a line segment perpendicular to the optical axis. The throw of the image a' from the lens can be found by the conjugate distance equation (3.7) with account of the position of the lens principal planes in meridian I.

The length x_{II} of the image (in meridian II) depends on the length of the lens l and its magnification β_I in meridian I. Assuming that the lens is indefinitely thin one we have

$$\frac{x_{II}'}{l} = \frac{a' - a}{-a} \approx -\beta_I + 1$$

Consequently, $x_{II}' \approx (1 - \beta_I)l$.

It should be noted that the residual aberrations of the lens would shape the light line into a slit concave toward the lens.

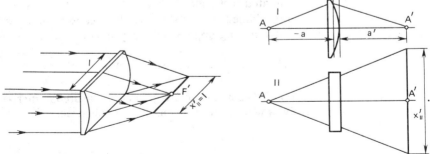

Fig. 20.2. The action of a cylinder lens in imaging an infinitely distant point object

Fig. 20.3. The action of a cylinder lens in imaging a point object at a finite working distance

23*

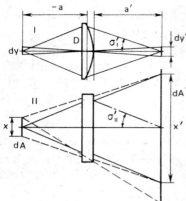

Fig. 20.4.. Distortions introduced by a cylindrical lens in imaging a plane object

If the filament of the lamp is a straight line segment x parallel to the axis of the lens cylindrical surface (Fig. 20.4), the position of the image can be found as above from the conjugate distance equation, and the length of the image will be

$$x' \approx l(1 - \beta_I) - x\beta_I$$

The height of the image in meridian I will be

$$dy' = \beta_I\, dy$$

where dy is the width of the filament.

The illumination of an image formed by a cylinder lens can be estimated with due account of the fact that the aperture angles are different in two orthogonal meridians. Let us examine this situation with reference to Fig. 20.4 where the luminous rectangle of area $dA = x \times dy$ is transformed into the illuminated rectangle of area $dA' = x' \times dy'$. Letting the luminance L of the filament be identical in all directions we may write, from Chapter 6, the following approximate expression

$$E' \approx \tau L\, d\Omega' \approx \tau L\, \frac{lD}{a'^2} \approx 4\tau L \sin \sigma_I' \sin \sigma_{II}'$$

where σ_I' and σ_{II} are the angular apertures in image space measured in meridians I and II respectively, and τ is the transmission of the system.

20.2 Cylindrical and Spherocylindrical Anamorphic Lenses

Because an anamorphic system is one which has different magnifications in the two principal meridians, for an object at infinity the anamorphic factor will be

$$A = f'_I / f'_{II}$$

where f'_I is the posterior focal length of the system in meridian I, and f'_{II} is the posterior focal length of the system in meridian II.

If the object lies within a finite distance from the optical system, the anamorphic factor is

$$A = \beta_I / \beta_{II} \tag{20.1}$$

Anamorphic lenses may be frequently encountered in many wide-screen motion picture processes, say in photography and projection of objects situated at finite distances. A specific case of such a lens is an anamorphic condenser.

Let us consider an anamorphic lens consisting of two thin elements having focal length $f'_1 = f'_I$ and $f'_2 = f_{II}$. The elements are cylinder lenses oriented at right angles to each other as indicated in Fig. 20.5. The action of the system may be regarded as combined of the actions of two subsystems with their own magnifications in meridians I and II:

$$\beta_I = (a'_2 + d)/a_1 \tag{20.2}$$

$$\beta_{II} = a'_2 / (a_1 + d)$$

where d is the spacing between the elements, a_1 the throw of the object from the first element, and a'_2 the throw of the image from the second element.

Using the conjugate distance equation in meridian I yields

$$a_1 = \frac{f'_1 (a'_2 + d)}{f'_1 - a'_2 - d} \tag{20.4}$$

$$a'_2 = \frac{f'_1 (a_1 - d) - a_1 d}{a_1 + f'_1} \tag{20.5}$$

and in meridian II yields

$$a_1 = \frac{f'_2 (a'_2 + d) - a'_2 d}{f'_2 - a'_2} \tag{20.6}$$

$$a'_2 = \frac{f'_2 (a_1 - d)}{a_1 - d + f'_2} \tag{20.7}$$

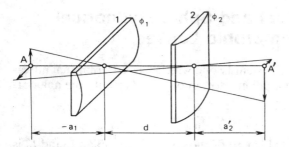

Fig. 20.5. A cylindrical anamorphic system

Equating the right-hand sides of (20.5) and (20.7) produces the equation

$$a_1^2 (f_1' - d - f_2') + a_1 d(d - 2f_1') + f_1' d^2 = 0 \qquad (20.8)$$

which can be solved with known f_1', f_2' and d, subject to either (20.4) or
(20.6) to determine a_1 and a_2' and assess the anamorphic factor by Eqs.
(20.1)-(20.3).

In the particular case of $f_1' = f_2' = f'$ Eq. (20.8) becomes

$$a_1^2 + a_1 (2f' - d) - f'd = 0 \qquad (20.9)$$

In this particular case $-a_1 = a_2'$, therefore $\beta_{II} = 1/\beta_I$, consequently, the
anamorphic factor

$$A = \beta_I/\beta_{II} = \beta_I^2 \qquad (20.10)$$

and by Eq. (20.2)

$$a_1 = d/(1 + \beta_I) \qquad (20.11)$$

Substituting this a_1 in (20.9) and rearranging leads to

$$\frac{\beta_I}{1 + \beta_I} d^2 - f'(1 - \beta_I)d = 0$$

Since we let above $d \neq 0$, the meaningful solution is

$$d = f'(1 - \beta_I^2)/\beta_I \qquad (20.12)$$

or after incorporation of (20.10)

$$d = f'(A - 1)/\sqrt{A} \qquad (20.13)$$

Example 20.1. Determine the constructional parameters of an anamorphic lens to get $A = 25$ and $f' = 100$ mm.

Solution. From Eq. (20.13) we get $d = 480$ mm, and Eq. (20.12) yields
$\beta_I = -5$, consequently, $\beta_{II} = -0.2$. By Eq. (20.11) the distance
$a_1 = -120$ mm, and $a_2' = -a_1 = 120$ mm.

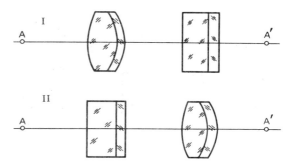

Fig. 20.6. An achromatic
process anamorphic lens

Figure 20.6 sketches the two principal meridians of an achromatic process anamorphic lens. In either meridian the system may be regarded as
consisting of two elements: lenses and plane-parallel plates in meridian I
and plane-parallel plates and lenses in meridian II. In both meridians the
distance from object to image along the axis must be the same.

A spherocylindrical anamorphic lens is a combination of cylinder and
ordinary spherical lenses. The identity of the axial object to image
distances in both principal meridians is ensured by a suitable choice of lens
thicknesses and refractive indices. In a system of two cylinder lenses and
one spherical lens, various arrangements of the lenses are possible. With
reference to Fig. 20.7 we consider a configuration with two cylinder lenses
placed on either side of the spherical element and oriented at right angles to
each other.

We shall refer to the element according to the number labels indicated
in the figure. Meridian I is where elements *2* and *3* of equivalent focal
length f_I' act, and meridian II is the plane of action of elements *1* and *2* having equivalent focal length f_{II}'. The anamorphic factor therefore is

$$A = f_I'/f_{II}' \tag{20.14}$$

From Eqs. (3.27) and (3.28) we find the focal length f_I' and f_{II}' and the
position of the equivalent focal point F' (defined by $a_{F'I}'$ and $a_{F'II}'$) as

$$f_I' = f_2'f_3'/(f_2' + f_3' - d_2)$$
$$f_{II}' = f_1'f_2'/(f_1' + f_2' - d_1)$$
$$a_{F'I}' = f_3'(f_2' - d_2)/(f_2' + f_3' - d_2)$$
$$a_{F'II}' = f_2'(f_1' - d_1)/(f_1' + f_2' - d_1)$$

With reference to Fig. 20.7, there must be

$$d_2 = a_{F'II}' - a_{F'I}'$$

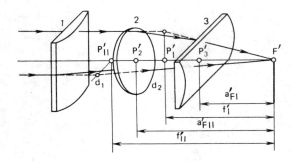

Fig. 20.7. A spherocylindrical anamorphic lens. Two cylinder lenses oriented at right angles to each other are placed on either side of a spherical element

Substituting the found f_I' and f_{II}' into (20.14) we get

$$A = \frac{f_3'}{f_1'} \frac{f_1' + f_2' - d_1}{f_2' + f_3' - d_2}$$

The constructional data usually specified in the design of such a system include the equivalent focal lengths f_I' and f_{II}', the focal length of the spherical element, f', and the distance $a_{F'II}'$ from this element to the image plane at the equivalent focal point F'.

A spherocylindrical lens can also be formed with the cylinder lenses set parallel. With such a configuration, all the three elements act in meridian I, leaving meridian II for the spherical element alone. One such arrangement designed for projection and printing processes is schematized in Fig. 20.8. Since the object in these applications is at finite distances, the anamorphic factor of such a system is

$$A = \beta_I / \beta_{II}$$

where $\beta_I = \beta_1 \beta_2 \beta_3$, and $\beta_{II} = a_{II}' / a_{II}$.

A necessary condition imposed on such a design form is that the distances between the object plane and image plane be equal in both principal meridians.

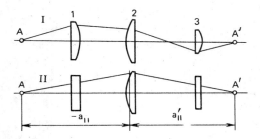

Fig. 20.8. A spherocylindrical anamorphic lens whose cylinder lenses are parallel to each other

20.3 Cylindrical Afocal Anamorphic Attachments

This is the type of system used in many wide-screen motion picture processes. The wide-angular field is used to squeeze a large horizontal field of view into an ordinary film format. The distorted picture that results is unsqueezed to normal proportions by projecting the film through a projection lens equipped with a similar attachment. The demand for wide-screen pictures in cinematographic projection has been met, therefore, without discarding standard photographic equipment. Anamorphs are configured as auxiliary telescopes set on the prime lens to minify or magnify in the horizontal direction only.

As a rule an anamorphic attachment consists of cylinder lenses set with their axes parallel. The formulation of these attachments can be an ordinary telescope (Fig. 20.9a), specifically a reversed Halilean telescope (Fig. 20.9b).

The principal optical characteristics of cylindrical anamorphic attachments include the anamorphic factor, diameter of exit pupil, angular or linear field coverage, and physical length.

In one meridian, the cylindrical afocal combination serves to change the focal length of the prime objective lens as a system of ordinary spherical lenses would do. In the other meridian, the system is equivalent to plane-parallel plates of glass and does not affect the focal length of coverage of the prime lens. Accordingly, in meridian I, the image scale varies in agreement with the magnification, while in meridian II the image scale remains intact. The anamorphic factor of the attachment is therefore

$$A = |f'_I / f'_{II}|$$

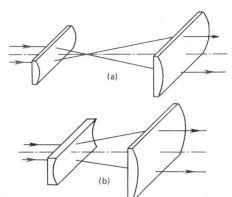

Fig. 20.9. Cylindrical afocal combinations arranged as an ordinary telescope (a) and as a reversed Halilean telescope (b)

If the element of the cylindrical attachment meeting the rays first has a shorter focal length than the second element, as is the case in modern attachments for shooting, the resultant image will be compressed ($A < 1$), while in projection, where the situation is reverse, the resultant image is expanded ($A > 1$).

Such attachments are always used with a spherical prime lens forming a real image which can be recorded on film or viewed on a screen. Observing that the prime lens is capable of transmitting ray bundles of a certain diameter, the anamorphic attachment should not be vignetting for these bundles. In other words, the diameter of the exit pupil of the attachment, D, must be equal to the diameter of the entrance pupil of the spherical prime lens. This leads us to the entrance pupil diameter of the attachment $D = AD'$.

For projection lens attachments, conversely, the diameter of the entrance pupil is the primary characteristic because it must be consistent with the diameter of the exit pupil of the projection lens (again assessing in the direction of rays). These pupil diameters determine the relative aperture of positive element in the attachments. The longer (by absolute value) the focal length of attachment lenses, the longer the radii of curvature and the lower the aberrations of the attachment as a whole. This way of aberration control, however, entails a larger size of the attachment consisting of infinitely thin elements. The length of the attachment $L = f_1' + f_2'$, therefore the focal lengths of elements are elected with due account of size limitations and aberration requirements. As a rule, these anamorphic cylindrical combinations are formulated as Halilean telescopes offering shorter lengths.

The angular field of an attachment in meridian II equals to that of the prime objective lens, while in meridian I the angular coverage depends on the anamorphic factor: $A \tan \omega_1 = \tan \omega$.

Size evaluations for an attachment are effected by tracing oblique ray bundles in both meridians. The diameter and position of the entrance pupil of the prime lens are identical for both meridians. Normally, this entrance pupil lies inside the prime lens, therefore the exit pupil of the attachment which is to coincide with the former is brought beyond the attachment. This configuration means an appreciable increase in size for the first component of the attachment. At the same time, this element must be a short-focus one which makes the control of aberrations for this component troublesome and causes the designer to formulate this element as a multilens format.

In meridian I, where the powers of cylinder lenses are active, the dimensions of the attachments are determined by tracing an oblique bundle (see Fig. 14.12). In the other meridian, the cylinder lenses are equivalent to

plane-parallel plates and the height of a ray at each surface can be found
from the equation

$$h_{k+1} = \frac{D'}{2} + \left(\sum_{k=1}^{p} \frac{d_k}{n_{k+1}} + t \right) \tan \omega$$

where D' is the diameter of the exit pupil of the attachment, d_k the spacing
between adjacent surfaces, n_{k+1} the refractive index of a medium confined
between these surfaces, t the distance from the entrance pupil to the attachment, and ω the semiangular field of the spherical prime lens.

The numbers of h, d and n increase from the prime lens, i.e. the computation is carried out in reverse raytracing.

In viewing or filming objects at finite distances from the camera both
parts of the optical system, cylindrical attachment and prime lens, have to
be focused. There exists a number of methods to bring the image of nearby
objects into focus by operating the attachment. The most popular method
of focusing consists in altering the spacings between the elements of the attachment with a simultaneous shift of the spherical prime lens.

21. THE DESIGN of OPTICAL SYSTEMS

21.1 Design Techniques: General

A design of an optical system is essentially the synthesis of the system in which the desired performance of the system or the profile of ray bundles emerging from it is given and the constructional parameters are to be determined.

In the Soviet literature on the subject, the system synthesis, called 'aberration design', is treated as the most important part of the general design which also includes layout and radiation transfer analyses. The last two parts precede aberration design and attend it.

A design of an optical system may be roughly divided into two stages. At the first stage, the designer determines the characteristics, establishes the type of the system, and decides on the number of elements required (lenses, mirrors, etc.), their shapes and arrangement. In other words, this stage roughly formulates the optical system. The final result is very sensitive to this preliminary arrangement because it determines, for a large part, whether the control of the aberrations in this system will be troublesome or not. At the second stage, the designer determines the optimal values of the constructional parameters, clear diameter of the lenses, mirrors and other optical elements, types of optical materials and their characteristics aiding in attaining the desired image quality. It is quite obvious that the entire procedure is a complicated and important problem. An efficient solution of this problem depends on both the competence of the designer and a correctly adopted method of design.

In general, the designer is to solve one of the following problems: (1) *analysis* — given the constructional parameters and characteristics of the system, trace a number of axial and oblique rays to compute the associated aberrations and judge by the values of these aberrations what the resultant performance characteristics are, (2) *synthesis* — given the amount of residual aberrations and the type of optical system, determine the constructional parameters of the system (radii of curvature, thicknesses and spacings, and coefficients in the equations of aspheric surfaces) that will ensure the desired performance.

The solution of the first problem is straightforward, as it reduces to tracing rays through the optical system and analysing the resultant aberrations. The solution of the second problem requires that the analytical relationship between the specified values of aberrations and the constructional parameters be given. This problem is especially hard to solve for original optical designs. In general form, this analytical relationship can be written for the third-order (primary) aberrations only. This explains the significance of primary aberration theory: the primary aberrations are quite representative of the system and the theory not only determines the constructional parameters of system elements but also answers the question of whether or not chosen system configuration is feasible in attaining the given performance specifications.

A formal list of design techniques would be as follows: trial and error method, algebraic method, combined method, and automatic design techniques by a computer. In practice, trial and error, combined, and automatic design techniques are the most common.

(i) The trial and error analysis is essentially the investigation (with the purpose of further use if suitable) of relationships between variations in certain system parameters of a given configuration ($r, d, n, \alpha, h, \beta, H$) and the values of aberrations arising in response.

At the outset, the designer selects from catalogues, archives of the design establishment, or patents a configuration which seems to be capable of meeting a given performance level. If necessary, this configuration is recomputed for the desired focal length or magnification. This choice will be the point where the designer begins his efforts. Now, a systematic variation of the values of certain parameters of the adopted design produces a number of versions of this design form. Then a number of axial and oblique rays are traced through each of the versions, the resultant aberrations are determined. Comparing these aberrations with those of the initial design, the designer arrives at the aforementioned relationships.

The results of the investigation are normally represented in the form of tables or graphs from which it is easy to understand the effect of parameter variations on the aberrations and other quantities characterizing the performance of the optical system. By extrapolation or interpolation of these graphical or tabular forms, the designer determines the design form which would satisfy the performance specifications.

When the variable parameters include the constructional data, such as r, d, or n, the resultant design will in addition to other aberrations receive another focal length, which is undesirable for some configurations. A more suitable choice therefore is the parameter α_i, whose variation at $h_1 = f'$ and $\alpha'_p = 1$ will leave the focal length intact. The solution of the problem simplifies if the initial design form is such that variations of different α_i entail individual effects on different aberrations.

Because the relationship between variations of parameters and alterations of the system aberrations is non-linear (the higher the order of aberrations, the stronger the non-linearity), the designer should begin with very small variations in parameters, gradually improving the performance and consistently updating the tables of parameter variation effects.

If the initial design fails to lead to the desired performance, it must be appropriately modified or discarded altogether in favour of another design configuration. It is apparent that the trial and error technique means a considerable number of ray traces through the system, i. e. the technique is quite laborious and time consuming. In general it is not the way to non-traditional patent promising systems. Whether or not a selected design form will lead to a successful solution depends on the expertise and intuition of the designer. The computer has reduced the raytracing time by several orders of magnitude and revived the technique which would remain in wide use provided that the initial system domain would be modified and developed.

(*ii*) The combined method may be thought of as consisting of two stages. At the first stage, the designer refers to the so-called algebraic method based on the analytical dependences between constructional parameters and third-order aberrations. The second stage is an accurate evaluation of ray aberrations (element of the trial and error technique).

The first stage of this design is based on the assumption that the optical system at hand suffers from the third-order aberrations only (higher-order aberrations are let to be zero).

The monochromatic transverse aberrations of the ith element are represented in the form [26]

$$\Delta y' = \sum m_i P_i + \sum n_i W_i + \sum p_i \pi_i + \sum q_i$$
$$\Delta x' = \sum m_i' P_i + \sum n_i' W_i + \sum p_i' \pi_i + \sum q_i' \tag{21.1}$$

The chromatic aberrations for the same component are as follows

$$\Delta s_{\lambda_1, \lambda_2}' = \sum r_i C_i$$
$$\Delta y_{\lambda_1, \lambda_2}' / y_{\lambda_0}' = \sum r_i' C_i \tag{21.2}$$

In these sets of equations the coefficients m_i, m_i', n_i, n_i', ..., r_i, r_i' depend on *external parameters* only (powers of the elements, spacing between elements, relative apertures, and fields). The values of these parameters are determined at the stage of system layout, i. e., assumed known.

The quantities P_i, W_i, π_i, q_i, q_i', and C_i are all related to the constructional parameters of the system elements (radii of curvature, thicknesses, refractive indices, and V-values) and also depend on the position of the object with respect to the ith component. In Section 21.5 below, it will be indicated that the parameters P_i and W_i can be replaced by the parameters $\bar{P}_i$ and $\bar{W}_i$ dependent only on the constructional parameters r, d, and n, later called the *internal parameters*.

When the external parameters are specified along with the components of aberrations, we can substitute them in Eqs. (21.1) and (21.2) to solve them for P, W, π, and C.

For the known powers of all elements ϕ_i, spacing between them d_i, and positions of the object plane, s_1, and the entrance pupil, t, the auxiliary rays are traced by the equations

$$\alpha_i' - \alpha_i = h_i\phi_i$$
$$h_{i+1} = h_i - \alpha_i' d_i$$
$$\beta_i' - \beta_i = H_i\phi_i$$
$$H_{i+1} = H_i - \beta_i' d_i$$

Now that we know the coordinates of the auxiliary rays, we can set up the equations for five monochromatic and two chromatic sums [26] as follows

$$S_{\mathrm{I}} = \sum h_i^2 \phi_i \{ h_i^2 \phi_i^2 \bar{P}_i + 4\alpha_i h_i \phi_i \bar{W}_i + \alpha_i [(4 + 2\pi_i)\alpha_i - \alpha_i'] \}$$

$$S_{\mathrm{II}} = \sum h_i \phi_i \{ H_i h_i^2 \phi_i^2 \bar{P}_i + h_i \phi_i (1 + 4\alpha_i H_i) \bar{W}_i$$
$$+ \alpha_i [1 + 2H_i \alpha_i)(2 + \pi_i) - H_i \alpha_i \alpha_i'] \}$$

$$S_{\mathrm{III}} = \sum \phi_i \{ H_i^2 h_i^2 \phi_i^2 \bar{P}_i + 2H_i h_i \phi_i (1 + 2\alpha_i H_i) \bar{W}_i \qquad (21.3)$$
$$1 + 2\alpha_i H_i (2 + \pi_i) + \alpha_i H_i^2 [\alpha_i (4 + 2\pi_i) - \alpha_i'] \}$$

$$S_{\mathrm{IV}} = \sum \phi_i \pi_i$$

$$S_{\mathrm{V}} = \sum \frac{\phi_i}{h_i} \{ H_i^3 h_i^2 \phi_i^2 \bar{P}_i + (3 + 4\alpha_i H_i) H_i^2 h_i \phi_i \bar{W}_i$$
$$+ H_i (3 + \pi_i) + 3\alpha_i H_i^2 (2 + \pi_i) + \alpha_i H_i^3 [(4 + 2\pi_i)\alpha_i - \alpha_i'] \}$$

$$S_{\mathrm{I,\,ch}} = \sum h_i^2 \phi_i \bar{C}_i$$
$$S_{\mathrm{II,\,ch}} = \sum h_i H_i \phi_i \bar{C}_i$$

where

$$\bar{C}_i = -\sum \frac{\varphi_t}{V_t} = -\frac{1}{\phi_i} \sum \frac{\varphi_t}{V_t}$$

For given external parameters and sum values, this set of equations gives the numerical values of the principal parameters. Then the principal parameters are used to determine the internal parameters (r, d, n) for the given types of elements, i. e., to determine the construction of the system.

At the second stage of the combined method, the designer is to compute the exact values of aberrations ($\Delta s'$, $\Delta y'$, $\Delta x'$) and the higher orders as the differences

$$\Delta s'_{h.o} = \Delta s' - \Delta s'_{III}$$

$$\Delta y'_{h.o} = \Delta y' - \Delta y'_{III}$$

$$\Delta x'_{h.o} = \Delta x' - \Delta x'_{III}$$

At this stage the designer evaluates the effects of the higher-order aberrations and finite thicknesses of the elements, focusing on those aberrations which should be corrected first of all by the performance specifications. For instance, the objectives of telescopes are mostly corrected for spherical aberration, coma, and axial chromatic aberrations, whereas the eyepiece designs tend to be corrected for field aberrations and chromatic aberrations because the angular fields of eyepieces are Γ_t times the angular fields of objectives. The astigmatism and distortion are not controlled for the objectives of astronomical instruments as what matters here is a high quality of axial imagery. The objectives of spectrometers are, for the most part, not corrected for chromatic aberrations, field curvature, and distortion.

21.2 Aberration Tolerances

There is no way to free an optical system from all aberrations however complicated it may be made for this purpose. Attempts to eliminate all aberrations, at least in part, result in overcomplicated formulations and prohibitively expansive designs. It would be more rational, therefore, to set a reasonable level to which aberrations should be controlled.

In real optical systems an allowance is made for residual aberrations; the list and amount of specific aberrations are assigned depending on intended application and conditions in which the system will be used. Of course, any amount of aberration degrades the image and any larger amount simply degrades it still more. Thus, it would be more accurate to call this section 'aberration allowances', in spite of the fact that just the other term is more appropriate from the optical shop standpoint. Allowable aberrations define the quality of the image produced by the system, as they are directly connected with the aberration blur spot size used to estimate the resolving power of the system. In turn, the resolution of an instrument should be consistent with the resolution of the detector sensing the image. In instruments for visual usage, for example, the eye is used as detector; the image produced by projection systems on screens or by cameras on film emulsion is also to be viewed by the eye. The allowable aberrations in these systems will not be the same because the conditions of observation are different.

The angular limit of resolution for the eye is taken to be one minute of arc within angular field 2° at the absolute contrast. At lower values of contrast, the eye resolution limit falls off at a rate depending on the brightness of the image background. For instance, the contrast of objects viewed with telescopic systems varies in the range of 0.2 to 0.8. The attendant resolution of the eye varies approximately from 2.5 to 1.5 minutes of arc. In microscopes, the contrast of objects under examination is still lower, therefore the angular resolution of the eye is taken to be 3 to 4 minutes of arc at an eye pupil diameter of 2-3 mm. Observing that in microscope work $D' \leqslant 0.5$-1 mm, the angular resolution of the eye becomes half as much, $\psi_{eye} = 6'$ to $10'$.

For simultaneous observations of axial and extra-axial points, the resolution of the eye decreases as $\psi_{eye} = 3.3'$ for the angular coverage $\pm 5°$ and $\psi_{eye} = 5'$ for the angular coverage $\pm 10°$.

Thus, the allowable values of residual aberrations should be specified in consistency with the capabilities of the eye if visual instruments are designed. On the other hand, the effect of the aberrations of optical system on the eye resolution must be also taken into account. Below we give the specific increments (in seconds of arc) added by one minute of aberration to the angle resolvable by the eye.

Chromatic aberration	3
Coma	5
Astigmatism	12
Defocusing	12

For binoculars and geodetic instruments, a residual spherical aberration of 1 to 2 minutes of arc is allowed, and with account of chromatic aberration this figure grows to 2-3 minutes of arc. The total monochromatic aberration allowable for extra-axial bundles may be 5 to 10 minutes of arc, where 2-3 minutes are due to coma. In more complicated telescopic systems such as range finders, naval periscopes, etc., the allowable spherical aberration is as high as 10 to 12 minutes of arc, and over the entire visible range even up to 20 minutes of arc.

The allowable values of curvature of field, astigmatism, and distortion are dependent on the angular fields of eyepieces. For ordinary eyepieces, the astigmatism and field curvature amount to 3-4 dioptres, for wide-angle eyepieces, these are 5-6 dioptres; the distortion allowable for ordinary eyepieces is about 3.5 to 7%, and for wide-angle pieces it should not exceed 10%.

The allowance for the transverse chromatic aberration in telescopes is up to 0.5-1%.

The aberrations behind the eyepiece are, as a rule, higher for microscopes than for telescopes. For a point on the axis, for example, the angular aberration may reach 10 to 15 minutes. The allowances for field curvature and astigmatism of microscope objectives of modest magnification (40 ×) are as follows: in achromats, 1.2-3 mm and 0.5-3 mm, respectively; and in apochromats, 2 mm and 1.5 mm. In compensated eyepieces, the allowable distortion is 1.5%, and in the Kellner eyepieces the figure is 2%.

A more accurate estimate of allowable aberrations for microscopes may be derived by evaluating the wave aberrations discussed in Section 21.19. Table 21.1 summarizes the allowable aberrations for microscope objectives.

As a rule, for photographic objectives tolerable aberrations are given in terms of the maximum allowable blur spot sizes [26]. These are 0.03-0.05 mm for photographs produced without enlargement, and 0.01-0.03 mm for photographs with subsequent enlargement.

Analysis of Soviet-made photographic objectives [11] has established an average allowable blur size of 0.01-0.02 mm for an axial point and 0.03-0.05 mm for extra-axial points.

The aforelisted values of aberration blurs contain the amounts of allowable aberrations. Specific figures in this case are rather troublesome, because the allowable aberrations vary widely depending on intended applications and characteristics of a particular design. If, for instance, a high quality of the image at the axis is thought, the spherical aberration is corrected not only for the edge of the pupil, but also for its zones. In exceedingly wide-angle objectives, the designer tends above all to correct

Table 21.1. Allowable Residual Aberrations

Microscope objective	Spherical aberration λ_e	$\lambda_{F'}$ & $\lambda_{C'}$	Lateral colour, %	Total, off-axis bundle
Achromat	0.25λ	0.5λ	⩾2	>0.5λ
Apochromat	(0.1-0.15)λ	<0.25λ		
Planobjective			<2	0.5λ

astigmatism, field curvature, distortion, and lateral colour. For normal photographic objectives, the allowable astigmatic difference is of the order of -0.15 to -0.3 mm, the average curvature of field should not exceed 0.3 mm, and the allowable distortion is 0.5% to 3% at the edge of the field.

The allowable distortion for objectives for aerial photography amounts to about 0.1%, and for extremely wide-angle designs it may be even 0.04%.

The allowable residual aberrations for objectives of projection systems are about the same as for photographic objectives. The projection aplanatic objectives have poorly corrected curvature of the image field. The allowances for distortion in motion-picture anastigmatic lenses amount to 1-2%. The constraints imposed on projection lenses were outlined in Chapter 17.

Lens condensers perform well in distributing the flux if the diameter of the smallest blur spot is under 3-10% of the source image size. In non-critical condensers, this parameter is allowed to be as high as 30%.

For tracking photoelectric systems, the allowable values of residual aberrations, defined in terms of allowable aberration blurs, are more convenient to estimate and compare if they are converted to radians. If the objective of such an instrument produces the blur spot of diameter $2y'$, and the focal length of the objective is f', the angular dimension $\Delta\sigma'$ of the spot in milliradians can be derived by the equation

$$\Delta\sigma' = \frac{2y'}{f'}1000$$

21.3 Relating the Parameters of the Auxiliary Rays

The third-order aberration equations (9.7) as will be recalled are used in solving the problem of analysis, when the aberrations of the optical systems are determined by the given constructional parameters. These equations are hard to use in the problem of synthesis, when the constructional parameters of optical elements in a chosen design form are to be determined by the given values of aberrations, because these equations contain the unknown parameters of the two auxiliary rays (α_k, h_k, β_k, H_k) and the designer has to resort to all manner of ingenious techniques to guess the values of these parameters.

The layout of a selected optical system form, especially at the stage of the thin-lens evaluation, yields the incidence heights h_k and H_k. Therefore, if we establish a relationship between the parameters of the two auxiliary rays, we obtain more convenient expressions for the sums dependent only on the parameter of one of the rays. We wish to establish this relationship below.

Figure 21.1 shows the passage of both auxiliary rays through the kth surface of the optical system separating media of refractive indices n_k and n_{k+1}.

We resort to the Lagrange invariant relating the slope of the first auxiliary ray to the image (or object) size; for axial points A_k and A_{k+1}, it is written as

$$I = n_k \alpha_k y_k = n_{k+1} \alpha_{k+1} y_{k+1} \qquad (21.4)$$

or observing that $y_k = (t_k - s_k)\beta_k$ and $y_{k+1} = (t'_k - s'_k)\beta_{k+1}$,

$$I = n_k \alpha_k (t_k - s_k)\beta_k = n_{k+1} \alpha_{k+1}(t'_k - s'_k)\beta_{k+1}$$

whence

$$\frac{I}{n_k} = \alpha_k \beta_k t_k - \alpha_k \beta_k s_k$$

$$\frac{I}{n_{k+1}} = \alpha_{k+1}\beta_{k+1} t'_k - \alpha_{k+1}\beta_{k+1} s'_k \qquad (21.5)$$

With reference to Fig. 21.1,

$$H_k = \beta_k t_k \qquad H_k = \beta_{k+1} t'_k$$
$$h_k = \alpha_k s_k \qquad h_k = \alpha_{k+1} s'_k$$

Substituting these relationships in Eqs. (21.5) and denoting $1/n_k = \mu_k$ and $1/n_{k+1} = \mu_{k+1}$ we obtain the equations

$$\alpha_k H_k - \beta_k h_k = I \mu_k$$
$$\alpha_{k+1} H_k - \beta_{k+1} h_k = I \mu_{k+1}$$

from which

$$h_k(\beta_{k+1} - \beta_k) - H_k(\alpha_{k+1} - \alpha_k) = -I\,(\mu_{k+1} - \mu_k)$$

or

$$\frac{\beta_{k+1} - \beta_k}{\alpha_{k+1} - \alpha_k} = \frac{H_k}{h_k} - I \frac{\mu_{k+1} - \mu_k}{h_k(\alpha_{k+1} - \alpha_k)}$$

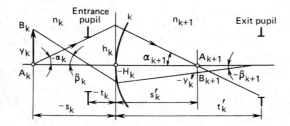

Fig. 21.1. Relationship between the auxiliary rays

which rewrites in shorthand as

$$\frac{\delta\beta_k}{\delta\alpha_k} = \frac{H_k}{h_k} - I\,\frac{\delta\mu_k}{h_k\delta\alpha_k} \tag{21.6}$$

Now we substitute this ratio for $\delta\beta_k/\delta\alpha_k$ in Eqs. (9.7).

(1) The sum S_I is independent of this ratio and remains as it stands, i. e.,

$$S_\mathrm{I} = \sum_{k=1}^{p} h_k P_k$$

(2) The expression for S_II is transformed as follows

$$S_\mathrm{II} = \sum_{k=1}^{p} h_k P_k \frac{\delta\beta_k}{\delta\alpha_k} = \sum_{k=1}^{p} h_k P_k \left(\frac{H_k}{h_k} - I\,\frac{\delta\mu_k}{h_k\delta\alpha_k} \right)$$

$$= \sum_{k=1}^{p} H_k P_k - I \sum_{k=1}^{p} P_k \frac{\delta\mu_k}{\delta\alpha_k}$$

We recall that

$$P_k = \left(\frac{\delta\alpha_k}{\delta\mu_k} \right)^2 \delta(\alpha_k\mu_k) = \frac{\delta\alpha_k}{\delta\mu_k}\, W_k$$

where

$$W_k = \frac{\delta\alpha_k}{\delta\mu_k}\, \delta(\alpha_k\mu_k)$$

consequently

$$S_\mathrm{II} = \sum_{k=1}^{p} H_k P_k - I \sum_{k=1}^{p} W_k$$

(3) The expression for S_III transforms as follows

$$S_\mathrm{III} = \sum_{k=1}^{p} h_k P_k \left(\frac{\delta\beta_k}{\delta\alpha_k} \right)^2$$

$$= \sum_{k=1}^{p} h_k P_k \left[\frac{H_k^2}{h_k^2} - 2I\,\frac{H_k\delta\mu_k}{h_k^2\delta\alpha_k} + I^2\frac{1}{h_k^2}\left(\frac{\delta\mu_k}{\delta\alpha_k} \right)^2 \right]$$

or after rearrangements

$$S_\mathrm{III} = \sum_{k=1}^{p} \frac{H_k^2}{h_k} P_k - 2I \sum_{k=1}^{p} \frac{H_k}{h_k} W_k + I^2 \sum_{k=1}^{p} \frac{\delta(\alpha_k\mu_k)}{h_k}$$

(4) The sum S_{IV} will be transformed as follows

$$S_{IV} = \sum_{k=1}^{p} \frac{\delta(\alpha_k n_k)}{h_k n_k n_{k+1}} = \sum_{k=1}^{p} \frac{1}{h_k} \Pi_k$$

where

$$\Pi_k = \frac{\delta(\alpha_k n_k)}{n_k n_{k+1}}$$

(5) The expression for S_V will be transformed as follows

$$S_V = \sum_{k=1}^{p} \left[h_k P_k \left(\frac{\delta\beta_k}{\delta\alpha_k} \right)^2 + I^2 \frac{(\delta\alpha_k n_k)}{h_k n_k n_{k+1}} \right] \frac{\delta\beta_k}{\delta\alpha_k}$$

$$= \sum_{k=1}^{p} \left[\frac{H_k^2}{h_k} P_k - 2I \frac{H_k}{h_k} W_k + I^2 \frac{\delta(\alpha_k \mu_k)}{h_k} \right.$$

$$+ I^2 \frac{\delta(\alpha_k n_k)}{h_k n_k n_{k+1}} \left] \left(\frac{H_k}{h_k} - I \frac{\delta\mu_k}{h_k \delta\alpha_k} \right) \right.$$

$$= \sum_{k=1}^{p} \left\{ \frac{H_k^3}{h_k^2} P_k - 2I \frac{H_k^2}{h_k^2} W_k + I^2 \frac{H_k}{h_k^2} \left[\delta(\alpha_k \mu_k) + \frac{\delta(\alpha_k n_k)}{n_k n_{k+1}} \right] \right.$$

$$- I \frac{H_k^2}{h_k^2} W_k + 2I^2 \frac{H_k}{h_k^2} \delta(\alpha_k \mu_k)$$

$$\left. - I^3 \frac{1}{h_k^2} \frac{\delta\mu_k}{\delta\alpha_k} \left[\delta(\alpha_k \mu_k) + \frac{\delta(\alpha_k n_k)}{n_k n_{k+1}} \right] \right\}$$

Since

$$\frac{\delta\mu_k}{\delta\alpha_k} \left[\delta(\alpha_k \mu_k) + \frac{\delta(\alpha_k n_k)}{n_k n_{k+1}} \right] = \mu_{k+1}^2 - \mu_k^2 = \delta(\mu_k)^2$$

we finally get

$$S_V = \sum_{k=1}^{p} \frac{H_k^3}{h_k^2} P_k - 3I \sum_{k=1}^{p} \frac{H_k^2}{h_k^2} W_k$$

$$+ I^2 \sum_{k=1}^{p} \frac{H_k}{h_k^2} [3\delta(\alpha_k\mu_k) + \Pi_k] - I^3 \sum_{k=1}^{p} \frac{1}{h_k^2} \delta(\mu_k)^2$$

Thus,

$$S_{\mathrm{I}} = \sum_{k=1}^{p} h_k P_k$$

$$S_{\mathrm{II}} = \sum_{k=1}^{p} H_k P_k - I \sum_{k=1}^{p} W_k$$

$$S_{\mathrm{III}} = \sum_{k=1}^{p} \frac{H_k^2}{h_k} P_k - 2I \sum_{k=1}^{p} \frac{H_k}{h_k} W_k + I^2 \sum_{k=1}^{p} \frac{\delta(\alpha_k\mu_k)}{h_k}$$

$$S_{\mathrm{IV}} = \sum_{k=1}^{p} \frac{1}{h_k} \Pi_k$$

$$S_{\mathrm{V}} = \sum_{k=1}^{p} \frac{H_k^3}{h_k^2} P_k - 3I \sum_{k=1}^{p} \frac{H_k^2}{h_k^2} W_k$$

$$+ I^2 \sum_{k=1}^{p} \frac{H_k}{h_k^2} [3\delta(\alpha_k\mu_k) + \Pi_k] - I^3 \sum_{k=1}^{p} \frac{1}{h_k^2} \delta(\mu_k)^2$$

where

$$I = -n_1\alpha_1(s_1 - t)\beta_1$$

and $I = -n_1 h_1 \beta_1$ at $s_1 = -\infty$.

These expressions are convenient in that for given values of the sums they can be solved to obtain the unknown angles (α_k) of the first auxiliary ray.

21.4 Seidel Sum Transformation for a Thin-Lens System

The thin-lens representation of an optical system halves the number of parameters controlling the aberrations of the system and, thus, reduces the efforts involved in choosing an appropriate system configuration. This approach is convenient at the initial stage of optical system synthesis, when only the general layout is known while the constructional parameters of system elements are not. It enables the designer to decide whether or not this design form is feasible already at this stage.

In the general case, the optical system can consist of a number of thin lenses. Let us consider how the third-order aberration sums for the ith thin element are transformed. We assume that the ith element consists of z indefinitely thin lenses immersed in air and having $p = 2z$ surfaces. We assume that the lens thicknesses and the spacing between them are all zero, i. e., $d_1 = d_2 = ... = d_{p-1} = d_i = 0$, where d_i is the thickness of the entire ith component. Accordingly, the incidence heights where the auxiliary rays meet the surfaces of the element lenses will be

$$h_1 = h_2 = ... = h_p$$
$$H_1 = H_2 = ... = H_p$$

Let $h_1 = h_i$ and $H_1 = H_i$ are the heights of the auxiliary rays at element i. Because these quantities are constants we factor them out from the summation operators in Eqs. (21.7).

(1) The first sum becomes

$$S_{\mathrm{I},\,i} = \sum_{k=1}^{p} h_k P_k = h_i \sum_{k=1}^{p} P_k$$

Denoting $\sum_{k=1}^{p} P_k$ by P_i we get for this sum $S_{\mathrm{I},\,i} = h_i P_i$.

(2) The second sum is

$$S_{\mathrm{II},\,i} = \sum_{k=1}^{p} H_k P_k - I \sum_{k=1}^{p} W_k = H_i \sum_{k=1}^{p} P_k - I \sum_{k=1}^{p} W_k$$

Denoting $\sum_{k=1}^{p} W_k$ by W_i we get for this sum $S_{\mathrm{II},\,i} = H_i P_i - IW_i$.

(3) The third sum is

$$S_{\text{III},\,i} = \sum_{k=1}^{p} \frac{H_k^2}{h_k} P_k - 2I \sum_{k=1}^{p} \frac{H_k}{h_k} W_k + I^2 \sum_{k=1}^{p} \frac{\delta(\alpha_k \mu_k)}{h_k}$$

$$= \frac{H_i^2}{h_i} \sum_{k=1}^{p} P_k - 2I \frac{H_i}{h_i} \sum_{k=1}^{p} W_k + I^2 \frac{1}{h_i} \sum_{k=1}^{p} \delta(\alpha_k \mu_k)$$

The last term of this sum after rearrangement becomes

$$\sum_{k=1}^{p} \delta(\alpha_k \mu_k) = \alpha_{p+1} \mu_{p+1} - \alpha_1 \mu_1$$

Since the lenses are in air, $\mu_{p+1} = \mu_1 = 1$, $\alpha_{p+1} = \alpha_i'$, $\alpha_1 = \alpha_i$, and $\alpha_i' = \alpha_i + h_i \phi_i$. Now,

$$\sum_{k=1}^{p} \delta(\alpha_k \mu_k) = \alpha_i' - \alpha_i = h_i \phi_i$$

so that we have finally

$$S_{\text{III},\,i} = \frac{H_i^2}{h_i} P_i - 2I \frac{H_i}{h_i} W_i + I^2 \phi_i$$

where $\phi_i = \sum_{t=1}^{z} \phi_t$ is the power of the entire ith element.

(4) The fourth sum will be transformed as follows

$$S_{\text{IV},\,i} = \sum_{k=1}^{p} \frac{\delta(\alpha_k n_k)}{h_k n_k n_{k+1}} = \frac{1}{h_i} \sum_{k=1}^{p} \frac{\delta(\alpha_k n_k)}{n_k n_{k+1}}$$

$$\sum_{k=1}^{p} \frac{\delta(\alpha_k n_k)}{n_k n_{k+1}} = \frac{\alpha_2 n_2 - \alpha_1 n_1}{n_2 n_1} + \frac{\alpha_3 n_3 - \alpha_2 n_2}{n_3 n_2} + \ldots$$

$$+ \frac{\alpha_p n_p - \alpha_{p-1} n_{p+1}}{n_p n_{p-1}} + \frac{\alpha_p' n_p' - \alpha_p n_p}{n_p' n_p}$$

Observing that $n_1 = n_3 = \ldots = n_{2t-1} = 1$; $\alpha_3 - \alpha_1 = h_1 \phi_1$, $\alpha_5 - \alpha_3 = h_2 \phi_2$, $\ldots$, $\alpha_p' - \alpha_p = h_z \phi_z$, and altering the subscripts of refractive indices in agreement with the numbers of the lenses we get

$$S_{\text{IV}, i} = \sum_{t=1}^{z} \frac{\phi_t}{n_t}$$

The power of the entire element is

$$\phi_i = \phi_1 + \phi_2 + \ \ldots \ + \phi_z = \sum_{t=1}^{z} \phi_t$$

The reduced power is $\varphi_t = \phi_t/\phi_i$, now

$$S_{\text{IV}, i} = \sum_{t=1}^{z} \frac{\phi_i \, \varphi_t}{n_t} = \phi_i \sum_{t=1}^{z} \frac{\phi_t}{n_t}$$

or, denoting $\displaystyle\sum_{t=1}^{z} \frac{\varphi_t}{n_t} = \pi_i$, we finally obtain $S_{\text{IV}, i} = \phi_i \, \pi_i$.

(5) The first and second terms of $S_{\text{V}, i}$ can be obtained by analogy with the transformation performed for $S_{\text{III}, i}$ and $S_{\text{IV}, i}$. The third term has the form

$$I^2 \frac{H_i}{h_i} \sum_{k=1}^{p} \left[\frac{3\delta(\alpha_k \mu_k)}{h_i} + \frac{\Pi_k}{h_i} \right] = I^2 \frac{H_i}{h_i} \left[\sum_{k=1}^{p} \frac{3\delta(\alpha_k \mu_k)}{h_i} + \sum_{k=1}^{p} \frac{\Pi_k}{h_i} \right]$$

$$= I^2 \frac{H_i}{h_i} (3\phi_i + \phi_i \, \pi_i)$$

$$= I^2 \frac{H_t}{h_i} (3 + \pi_i)\phi_i$$

The last term of $S_{\text{V}, i}$

$$I^3 \frac{1}{h_i^2} \sum_{k=1}^{p} \delta(\mu_k)^2 = 0$$

because $\mu_p' = \mu_1$.

Finally we have

$$S_{\text{V}, i} = \frac{H_i^3}{h_i^2} P_i - 3I \frac{H_i^2}{h_i^2} W_i + I^2 \frac{H_i}{h_i} (3 + \pi_i) \, \phi_i$$

Thus, our set of equations becomes

$$S_{\text{I},\, i} = h_i\, \phi_i$$

$$S_{\text{II},\, i} = H_i\, P_i - IW_i$$

$$S_{\text{III},\, i} = \frac{H_i^2}{h_i} P_i - 2I \frac{H_i}{h_i} W_i + I^2 \phi_i \qquad (21.8)$$

$$S_{\text{IV},\, i} = \phi_i\, \pi_i$$

$$S_{\text{V},\, i} = \frac{H_i^3}{h_i^2} P_i - 3I \frac{H_i^2}{h_i^2} W_i + I^2 \frac{H_i}{h_i} (3 + \pi_i)\, \phi_i$$

Analysis of these equations leads to the following conclusions on the possibilities of aberration control in a thin optical system.

(i) The spherical aberration and field curvature are independent of the position of the entrance pupil because H_i does not enter the expressions for $S_{\text{I},\, i}$ and $S_{\text{IV},\, i}$.

(ii) The field curvature defined by the sum $S_{\text{IV},\, i}$ is independent of the shapes of the lenses, as a lens figured in any manner can have a given power ϕ_i.

The parameter π_i of a thin element varies within a narrow range and cannot materially affect the value of the fourth sum. To demonstrate,

$$\pi_i = \sum_{t=1}^{t=z} \frac{\varphi_t}{n_t} = \frac{\varphi_1}{n_1} + \frac{\varphi_2}{n_2} + \ldots + \frac{\varphi_z}{n_z}$$

where all the indices vary in the range 1.5-1.7 for most common optical glasses and differ only insignificantly from the average value $n = 1.6$ so that

$$\pi_i = \frac{\varphi_1 + \varphi_2 + \ldots + \varphi_z}{n} \approx \frac{1}{n}$$

and we may safely let $S_{\text{IV},\, i} \approx \phi_i/n$, and since $\pi_{i,\, \text{av}} \approx 0.6\text{-}0.7$, the fourth sum of the thin element would remain practically unchanged, $\pi_i \approx (0.6\text{-}0.7)\phi_i$.

All the other monochromatic aberrations (safe for field curvature) can be controlled by changing the shape of the lenses.

(iii) The field aberrations, including coma, astigmatism, and distortion, defined by the sums $S_{\text{II},\, i}$, $S_{\text{III},\, i}$, and $S_{\text{V},\, i}$, respectively, depend on the position of the entrance pupil because their equations contain $H_i = t_i \beta_i$, where β_i is the angle of the second auxiliary ray, and t_i the throw of the entrance pupil.

(iv) Changing the position of the entrance pupil ($H_i = t_i$) cannot affect $(t + 1)$st aberration if the first t aberrations are corrected. Let us examine this conclusion in more detail.

If the tth aberration is corrected, i. e. $S_{I, i} = 0$, which is possible at $P_i = 0$ only, then the $(t + 1)$st aberration (in this case a coma) cannot be corrected by displacing the entrance pupil as $S_{II, i} = -IW_i$. If $W_i \neq 0$, then for any H_i the sum $S_{II, i} = -IW_i$ is a nonzero constant.

In an aplanatic thin system, astigmatism cannot be eliminated. The system is an aplanatic if $S_{I, i} = 0$ and $S_{II, i} = 0$, which is possible only for $P_i = 0$ and $W_i = 0$, then $S_{III, i} = I^2 \phi_i \neq 0$ as $I \neq 0$ and $\phi_i \neq 0$.

If the system is corrected for the first four monochromatic aberrations (spherical, coma, astigmatism, and field curvature), then displacing the entrance pupil will have no effect on the fifth aberration, distortion. Indeed, in this case all the rays of an oblique beam have a common point of intersection in the paraxial image plane (Fig. 21.2), and if the system suffers from distortion ($B'_{eye} B'_0 \neq 0$) it is immaterial which of the rays will be the principal one.

(v) An aplanatic system having its entrance pupil at the first surface ($H_i = 0$) is free from distortion.

(vi) In a thin system, the astigmatism can be corrected only when the entrance pupil is displaced from the first surface and either P_i or W_i, or both P_i and W_i are nonzero. To demonstrate,

(1) $H_i \neq 0$, $P_i \neq 0$, $W_i = 0$, then $S_{III, i} = (H_i^2/h_i)P_i + I^2 \phi_i = 0$ because we can always find a position for the entrance pupil which satisfies this equation, i. e.,

$$H_i^2 = -I^2 \phi_i h_i/P_i$$

(2) $H_1 \neq 0$, $P_i = 0$, and $W_i \neq 0$, i. e. for $S_{III, i} = -2I (H_i/h_i)W_i + I^2 \phi_i = 0$,

$$H_i = I\phi_i h_i/2W_i$$

(3) $H_i \neq 0$, $P_i \neq 0$, and $W_i \neq 0$. Now, from the condition $S_{III, i} = (H_i^2/h_i)P_i - 2I(H_i/h_i)W_i + I^2\phi_i = 0$ we get the equation $H_i^2 P_i - 2IH_i W_i + I^2 \phi_i h_i = 0$ which can be solved to determine the position of the entrance pupil in terms of H_i.

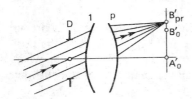

Fig. 21.2. The structure of a ray bundle at which distortion is independent of the position of the entrance pupil

Thus, the third-order monochromatic aberrations of a thin element depend on three parameters, P_i, W_i, and π_i; since the last parameter is practically constant, a thin system formally allows for correction of only two aberrations. Under favourable conditions, however, the position of entrance pupil H_i may also become a parameter of aberration control.

Consequently, an optical system where all monochromatic aberrations should be controlled must be constituted by a few air separated components. Each of these components can be a simple element, such as a single lens or mirror, or compounded, the constitution depending on the principal optical characteristics of the system and the desired performance.

21.5 The Basic Parameters of Thin Elements

From the immediately preceding sections we learned that for optical systems constituted by thin components, the sums of monochromatic aberrations depend on three parameters P_i, W_i, and π_i, which are in turn dependent on the constructional constituents of the component and position of the object. The later circumstance leads to situations where the same thin component will have different values of P_i and W_i according to the distance defining the location of the object in various designs. This is fairly inconvenient in comparing the correction potentialities of components when electing them for some or other optical formulation. At a constant power ϕ_i and index n_i of an optical element, the attendant parameter π_i is invariable.

To simplify the procedure of design, we establish a relationship between P_i and W_i as a function of object location, letting one of the object positions to be the principal one. From a multitude of object locations we select two more distinct than others, at infinity and at the twofold focal length for which the magnification of the component is -1. Of these two positions we adopt the infinite one because most optical systems are designed at this setting. The list of such systems includes objectives of astronomical and geodetical instruments, collimator lenses, lenses of binoculars, rangefinders, periscopes, camera lenses, lenses of erector systems, and eyepieces designed in reverse rays.

We shall refer to the thin-component parameters P_i and W_i derived with the object at infinity as the *basic* parameters, the respective notation being $\bar{P}_i$, $\bar{W}_i$, and $\bar{\pi}_i$. In the material that follows we derive the parameters P_i and W_i for any other position of the object in terms of the basic parameters.

Figure 21.3 schematizes a thin component (numbered i) which consists of p surfaces forming z lenses immersed in air, i.e., $n_1 = n_3 = \ldots = n_{2l-1} = \ldots = n_{p+1} = 1$ and $\mu_1 = \mu_3 = \ldots = \mu_{2l-1} = \ldots \mu_{p+1} = 1$.

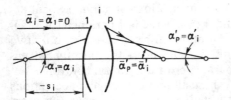

Fig. 21.3. Construction to deduce the dependences of P and W on the basic parameters $\overline{P}$ and $\overline{W}$ in a thin optical system

The ray from infinity has coordinates $\overline{\alpha}_i = \overline{\alpha}_1 = 0$, h_i, and $\overline{\alpha}_i' = \overline{\alpha}_p$ (it is often assumed that $\overline{\alpha}_p' = 1$). This ray defines the basic parameters $\overline{P}_i$ and $\overline{W}_i$. The parameters P_i and W_i can be determined by tracing a ray with the coordinates $\alpha_i = \alpha_1$, h_i, and $\alpha_p' = \alpha_i'$ for the object at a finite distance s_i.

The basic parameters ($s_i = -\infty$) for the ith component are given as follows

$$\overline{P}_i = \sum_{1}^{p} \overline{P}_k = \sum_{1}^{p} \left(\frac{\delta\overline{\alpha}_k}{\delta\mu_k}\right)^2 \delta(\overline{\alpha}_k \mu_k)$$

$$\overline{W}_i = \sum_{1}^{p} \overline{W}_k = \sum_{1}^{p} \frac{\delta\overline{\alpha}_k}{\delta\mu_k} \delta(\overline{\alpha}_k \mu_k) \qquad (21.9)$$

$$\overline{\pi}_i = \sum_{1}^{z} \frac{\varphi t}{n_t}$$

For the object at a finite throw, ($s_i \neq -\infty$) the expressions for P_i, W_i, and π_i are as follows

$$P_i = \sum_{1}^{p} P_k = \sum_{1}^{p} \left(\frac{\delta\alpha_k}{\delta\mu_k}\right)^2 \delta(\alpha_k \mu_k)$$

$$W_i = \sum_{1}^{p} P_k = \sum_{1}^{p} \frac{\delta\alpha_k}{\delta\mu_k} \delta(\alpha_k \mu_k) \qquad (21.10)$$

$$\pi_i = \overline{\pi}_i$$

In order to establish a relationship between $\overline{P}_i$, $\overline{W}_i$ and P_i, W_i, we need to relate $\delta\alpha_k/\delta\mu_k$ to $\delta\overline{\alpha}_k/\delta\mu_k$ and $\delta(\alpha_k \mu_k)$ to $\delta(\overline{\alpha}_k \mu_k)$. Whatever the object positions, the lenses in the thin component have the same radii of curvature and powers, therefore for surface k we have, in view of (4.16),

$$r_k = \frac{n_{k+1} - n_k}{\bar{\alpha}_{k+1} n_{k+1} - \bar{\alpha}_k n_k} \qquad \bar{h}_i = \frac{n_{k+1} - n_k}{\alpha_{k+1} n_{k+1} - \alpha_k n_k} h_i$$

whence

$$\alpha_{k+1} n_{k+1} - \frac{h_i}{\bar{h}_i} \bar{\alpha}_{k+1} n_{k+1} = \alpha_k n_k - \frac{h_i}{\bar{h}_i} \bar{\alpha}_k n_k = \ldots = \alpha_1 n_1 = \alpha_1 \qquad (21.11)$$

since $\bar{\alpha}_1 = 0$ and $n_1 = 1$.

We denote the ratio $h_i/\bar{h}_i = h_i \phi_i / \bar{h}_i \phi_i$ by q. Since $h_i \phi_i = \alpha_i' - \alpha_i = \alpha_p' - \alpha_1$ and $\bar{h}_i \phi_i = \bar{\alpha}_i' - \bar{\alpha}_i = \bar{\alpha}_p' - \bar{\alpha}_1 = \bar{\alpha}_p' = 1$, we have $q = \alpha_p' - \alpha_1$. Incorporating q in Eq. (21.11) we get

$$\alpha_{k+1} = q\bar{\alpha}_{k+1} + \alpha_1 \mu_{k+1} \qquad (21.12)$$

$$\alpha_k = q\bar{\alpha}_k + \alpha_1 \mu_k \qquad (21.13)$$

Subtracting the second equation from the first one yields $\delta\alpha_k$, and $\delta\alpha_k/\delta\mu_k$, $\delta\alpha_k = q\delta\bar{\alpha}_k + \alpha_1\delta\mu_k$ and

$$\frac{\delta\alpha_k}{\delta\mu_k} = q \frac{\delta\bar{\alpha}_k}{\delta\mu_k} + \alpha_1 \qquad (21.14)$$

Multiplying Eq. (21.12) by μ_{k+1} and Eq. (21.13) by μ_k and subtracting them partwise yields

$$\delta(\alpha_k \mu_k) = q\, \delta(\bar{\alpha}_k \mu_k) + \alpha_1 \delta(\mu_k)^2 \qquad (21.15)$$

Substituting Eqs. (21.14) and (21.15) into (21.10) we find by virtue of (21.9) after straightforward rearrangements

$$P_i = (\alpha_i' - \alpha_i)^3 \bar{P}_i + 4\alpha_i(\alpha_i' - \alpha_i)^2 \bar{W}_i$$
$$+ \alpha_i(\alpha_i' - \alpha_i)[2\alpha_i(2 + \pi_i) - \alpha_i' \,]$$

$$W_i = (\alpha_i' - \alpha_i)^2 \bar{W}_i + \alpha_i(\alpha_i' - \alpha_i)(2 + \pi_i) \qquad (21.16)$$

Thus, given the basic parameters $\bar{P}_i$, $\bar{W}_i$, and $\bar{\pi}_i$ one can determine the parameters P_i and W_i for any location of the object, i. e., for any magnification β and $\alpha_1 = \alpha_i = \alpha_i' \beta = \alpha_p' \beta$. Conversely, if the parameters P_i, W_i, and π_i are known, say obtained in the process of design, then Eqs. (21.16) can be resolved for the basic parameters of the element as follows

$$\bar{W}_i = \frac{W_i - \alpha_i(\alpha_i' - \alpha_i)(2 + \pi_i)}{(\alpha_i' - \alpha_i)^2}$$

$$\bar{P}_i = \frac{P_i - 4\alpha_i \, W_i + \alpha_i(\alpha_i' - \alpha_i)[2\alpha_i(2 + \pi_i) + \alpha_i']}{(\alpha_i' - \alpha_i)^3} \quad (21.17)$$

These equations have been introduced in the practice of system design by Prof. Slyusarev of the USSR.

Some designs require that the basic parameters be established for the reverse direction of rays, denoted by $\overleftarrow{P}_i$ and $\overleftarrow{W}_i$. For this purpose we assume that a ray incident on the ith thin element passes through the first focal point (Fig. 21.4a) at the inclination $\alpha_i = 1$, then $\alpha_i' = 0$ and in agreement with Eqs. (21.16) we have

$$P_i = -\bar{P}_i + 4W_i - 4 - 2\pi_i$$
$$W_i = \bar{W}_i - 2 - \pi_i$$

If now we turn this element as shown in Fig. 21.4b so that $\overleftarrow{\bar{\alpha}}_i = 0$ and $\overleftarrow{\bar{\alpha}}_i' = 1$, the parameters P_i and W_i change sign and become $\overleftarrow{\bar{P}}_i = -P_i$ and $\overleftarrow{\bar{W}}_i = -W_i$ or

$$\overleftarrow{\bar{P}}_i = \bar{P}_i - 4\bar{W}_i + 4 + 2\pi_i \quad (21.18)$$
$$\overleftarrow{\bar{W}}_i = -\bar{W}_i + 2 + \pi_i$$

By way of example, for the object at infinity, a planoconvex lens of glass with refractive index $n = 1.5$ has the following basic parameters: $\bar{P}_1 = 9$ and $\bar{W}_1 = 3$ ($H_1 = 0$, i. e. the entrance pupil coincides with the lens). The same lens but convex toward the object has $\bar{P}_2 = 2.33$ and $\bar{W}_2 = -0.33$. These values can be obtained by Eqs. (21.18) for the planoconvex lens convex toward the object when $P_2 = \overleftarrow{\bar{P}}_1$ and $\bar{W}_1 = \bar{W}_2$. A symmetric lens ($r_1 = -r_2$) has for this same case $\bar{P} = 3.33$ and $\bar{W} = 1.33$ and since it is symmetric, $\overleftarrow{\bar{P}} = \bar{P}$ and $\overleftarrow{\bar{W}} = \bar{W}$ which can be derived numerically from Eqs. (21.18).

Equations (21.18) indicate that a thin element produces the same aberrations for both directions of rays ($\bar{P}_i = \overleftarrow{\bar{P}}_i$ and $\bar{W}_i = \overleftarrow{\bar{W}}_i$) if the condition $\bar{W}_i = 1 + (\pi_i/2)$ is satisfied for any $\bar{P}_i$. Hence, coma is always significant in thin symmetric components having $H_1 = 0$.

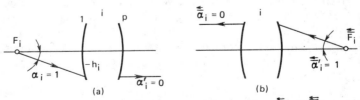

Fig. 21.4. Deduction of the formulae for the basic parameters $\overleftarrow{\bar{P}}$ and $\overleftarrow{\bar{W}}$ in reverse tracing of rays

21.6 Aberrations of Aspherics

Aspheric optical surfaces, or simply aspherics, are optical surfaces that are not spherical in form. More than one radius of curvature is assigned to the surface. The manner in which the surface varies across the curvature is generally defined by an analytic formula. The aspheric surface affords the designer additional freedom to modify the system to advantage and achieve better image quality, improved system design, smaller size, and lower weight.

To illustrate this point for reflective aspherics, a paraboloid mirror, for example, is known to produce a virtually perfect image of an infinitely distant axial point, and an ellipsoid mirror gives an errorless imagery of an axial point at a finite distance.

A simple lens with spherical surfaces cannot produce the ideal image of an axial point, whereas some asphericity added to one of the surfaces can correct the image to ideal one. We refer the reader to Sections 2.5 and 2.6 where the raytracing equations are given for aspheric surfaces specified analytically, for example, as follows

$$0 = by^2 + cx^2 + a_1z + a_2z^2 + a_3z^3 + \dots$$
$$z = B_1(y^2 + x^2) + B_2(y^2 + x^2)^2 + B_3(y^2 + x^2)^3 + \dots$$

where $a_1 = -2r_0$, $B_1 = 1/2r_0$, and r_0 is the radius of curvature at the vertex of the surface.

The computer reduced raytracing time by several orders of magnitude and a ray trace through any aspherical surface can be now performed without much trouble. Many computer programs have been written for second-order aspherics which are the most common.

The third-order aberration equations for optical systems with aspherics of second order have the same form as Eqs. (9.6), but the sums are presented in the following form

$$S_{I,a} = S_I + \Delta S_{I,a}$$
$$S_{II,a} = S_{II} + \Delta S_{II,a}$$
$$S_{III,a} = S_{III} + \Delta S_{III,a}$$
$$S_{IV,a} = S_{IV}$$
$$S_{V,a} = S_V + \Delta S_{V,a}$$

Here, S_I through S_V are the third-order aberration sums of the optical system with spherical surfaces and $\Delta S_{i,a}$ are the corrections for particular sums due to the aspherics:

$$\Delta S_{\text{I, a}} = \sum_{1}^{p} h_k b_k \frac{[\delta(\alpha_k n_k)]^3}{(\delta n_k)^2}$$

$$\Delta S_{\text{II, a}} = \sum_{1}^{p} h_k b_k \frac{[\delta(\alpha_k n_k)]^2 \delta(\beta_k n_k)}{(\delta n_k)^2} \qquad (21.19)$$

$$\Delta S_{\text{III, a}} = \sum_{1}^{p} h_k b_k \frac{\delta(\alpha_k n_k)[\delta(\beta_k n_k)]^2}{(\delta n_k)^2}$$

the correction $\Delta S_{\text{IV, a}}$ is zero because aspherics cannot correct field curvature, and

$$\Delta S_{\text{V, a}} = \sum_{1}^{p} h_k b_k \frac{[\delta(\beta_k n_k)]^3}{(\delta n_k)^2} \qquad (21.20)$$

where $b_k = -e_k^2$, and e_k is the eccentricity of the aspheric surface.

In general, an aspheric surface incorporated in a component gives the designer an additional degree of freedom for the control of third-order ray aberrations, therefore t aspherics should be incorporated if t aberrations are to be corrected [26]. As we mentioned above, one surface per lens is enough to correct the spherical aberration. For $n < n'$ this surface must be ellipsoidal, and for $n > n'$ it must be hyperboloidal.

By way of example determine the eccentricity of an ellipsoidal surface given that $s_1 = -\infty, n_1 = 1, n_2 = n' = 1.5, n_3 = 1$. The second surface must be a sphere centred on the second focal point of the lens (Fig. 21.5). To a first approximation, we assume it is a thin lens and assign an ordinary normalizing set $h_1 = 1, \alpha_1 = 0$, and $\alpha_2 = \alpha_3 = 1$. Now, for a thin glass

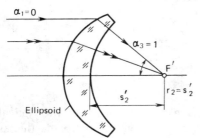

Fig. 21.5. An ellipsoid-front aberration-free lens

lens of $n_2 = n' = 1.5$ the first aberration sum $S_I = 21\alpha_2^2 - 24\alpha_2 + 9$ equals 6. The spherical aberration will be corrected if $S_{I, a} = 0$, or in view of (21.19) $\Delta S_{I, a} = -S_I$, i. e.,

$$\Delta S_{I, a} = h_1 b_1 \frac{(\alpha_2 n_2 - \alpha_1 n_1)^3}{(n_2 - n_1)^2}$$

whence $b_1 = -0.444$ and $e_i = \sqrt{0.444}$.

Consider now the procedure of eccentricity evaluation for two mirrors in reflecting systems (Fig. 21.6), for which freedom of two third-order aberrations, spherical and coma, is sought. Systems with this type of correction are called aplanatic ($S_{I, a} = 0$ and $S_{II, a} = 0$). We select as the base parameters for the system sketched at (a) the angle α_2 of the first auxiliary ray between the mirrors ($\alpha_1 = 0$, $\alpha_3 = 1$, $h_1 = 1$, $f' = 1$) and the height h_2 of this ray on the second mirror.

For a reflecting system with aspheric surfaces of second-order, the third-order aberration sums can be represented as follows

$$S_{I, a} = S_I + \frac{\alpha_2^3 e_1^2 - h_2(1 + \alpha_2)^3 e_2^2}{4}$$

$$S_{II, a} = S_{II} - \frac{(1 - h_2)(1 + \alpha_2)^3 e_2^2}{4\alpha_2}$$

$$S_{III, a} = S_{III} - \frac{(1 - h_2)^2(1 + \alpha_2)^3 e_2^2}{4\alpha_2^2 h_2^2} \qquad (21.21)$$

$$S_{IV, a} = S_{IV}$$

$$S_{V, a} = S_V - \frac{(1 - h_2)^3(1 + \alpha_2)^3 e_2^2}{4\alpha_2^3 h_2^2}$$

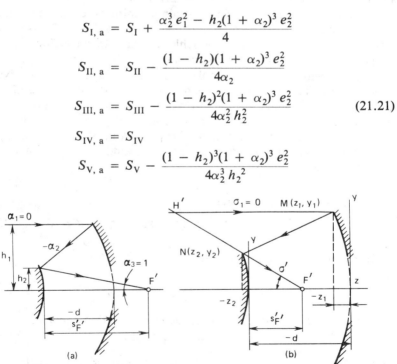

Fig. 21.6. Two-mirror systems: (a) deriving the formulae of excentricities, (b) solving in parametric form

where S_I through S_V are the respective sums of the reflecting system with spherical mirrors (see Eqs. (9.7)).

It is worth noting that these equations are written for the case of the entrance pupil coinciding with the first surface ($H_1 = 0$) which then can affect only the spherical aberration, therefore the eccentricity e_1^2 appears in the first sum $S_{I,\,a}$ alone.

Because our objective is an aplanatic design with $S_{I,\,a}$ and $S_{II,\,a}$ both zero, solving the first two equations of (21.21) for e_i^2 we obtain

$$e_2^2 = \frac{2\alpha_2 + (1 - h_2)(1 + \alpha_2)(1 - \alpha_2)^2}{(1 - h_2)(1 + \alpha_2)^3}$$
$$e_1^2 = 1 + 2h_2/(1 - h_2)\,\alpha_2^2 \tag{21.22}$$

Two other aberrations can be corrected in the general case.

The derived values of the eccentricities can be used to determine the largest deviation t_m of the aspheric from the sphere, $t_{m,i} = h_i^4\,e_i^2/32r_i^3$, and adopt some appropriate method of fabrication.

Equations (21.22) are valid in attaining third-order aplanatic designs for reflecting objectives with relative apertures less than $f/2$. Aspherics of higher orders should be used to achieve aplanatic corrections for high-speed reflecting objectives of relative apertures in the order of $f/1$.

The problem of two-mirror aplanatic system design was first solved by K. Schwarzschild and then independently by D. Maksutov and G. Chrétien. These developments remain in use thus far. In what follows we present one of the solutions in parametric form with reference to Fig. 21.6b. For a ray passing through points $M(z_1, y_1)$ and $N(z_2, y_2)$ on the first (large) and second mirrors, respectively, the condition for the image being aberration-free as follows

$$\sqrt{(z_1 - z_2 - d)^2 + (y_1 - y_2)^2} + $$
$$+ \sqrt{(s_{F'}' - z_2)^2 + y_2^2} - s_{F'}' + z_1 + d = 0 \tag{21.23}$$

Introducing $t = \tan(\sigma'/2)$ we get

$$\frac{y_2}{s_{F'}' - z_2} = \tan\sigma' = \frac{2t}{1 - t^2} \tag{21.24}$$

Given a focal length of $f' = 1$, the sine condition becomes

$$y_1 = \sin\sigma' = 2t/(1 + t^2) \tag{21.25}$$

Combined solution of Eqs. (21.23)-(21.25) produces the surface coordinates of the two-surface aplanatic system as functions of the parameter t

$$z_1 = -\frac{s'_{F'}\left(1 + \dfrac{d+1}{d}t^2\right)^{\frac{2d+1}{d+1}} - \dfrac{t^2}{d}}{(1+t^2)^2} + s'_{F'}$$

$$y_1 = \frac{2t}{1+t^2}$$

$$z_2 = \frac{d(1-t^2)}{t^2 - (d/s'_{F'})\left(1 + \dfrac{d+1}{d}t^2\right)^{\frac{1}{1+d}}} + s'_{F'}$$

$$y_2 = -\frac{2td}{t^2 - (d/s'_{F'})\left(1 + \dfrac{d+1}{d}t^2\right)^{\frac{1}{1+d}}}$$

where $t = (1 - \sqrt{1 - y_1^2})/y_1$.

Example 21.1. The surface equations derived for a high-speed ($f/1$) two-surface reflecting system of $f' = 100$ mm, $d = -60$ mm, and $s'_{F'} = 40$ mm are as follows

$$y_1^2 + x_1^2 + 399.99z_1 - 1.34694z_1^2 + 0.02825z_1^3 - 0.002042z_1^4 = 0$$
$$-z_2 + 0.356346 \times 10^{-6}(y_2^2 + x_2^2) + 0.807058 \times 10^{-6}(y_2^2 + x_2^2)^2$$
$$- 0.352283 \times 10^{-9}(y_2^2 + x_2^2)^3 + 0.117978 \times 10^{-12}(y_2^2 + x_2^2)^4 = 0$$

The aberrations are summarized in Table 21.1.

Table 21.1. Aberrations of a Two-surface Reflecting System

m	$10^2 \tan \sigma'$	s'	$\Delta s'$	$\Delta y'$	$\Delta f'$	Isoplanatic factor η
25	25.82	40.001	0.006	0.0014	0.0105	0.00
38	41.08	39.999	0.004	0.0017	0.0072	0.00
50	57.74	39.993	−0.002	0.0016	−0.0024	0.00

The system is seen to image an object point at the level of diffraction with a good accuracy for the sine condition.

Designers sometimes refer to prior art and re-scale a system to the desired focal length f' in the proportion $K = f'/f_i'$, where f_i' is the focal length of the appropriate initial system with aspheric surfaces. Prepared for upscaling or downscaling, the meridional curves of aspheric surfaces become

$$y_1^2 + Ka_1 z_1 + a_2 z_1^2 + \frac{a_3}{K} z_1^3 + \frac{a_4}{K^2} z_1^4 + \ldots = 0$$

$$z_2 = \frac{b_1}{K} y_2^2 + \frac{b_2}{K^3} y_2^4 + \frac{b_3}{K^5} y_2^6 + \frac{b_4}{K^7} y_2^8 + \ldots$$

Example 21.2. A mirror with the focal length $f_i' = 100$ mm has an aspheric surface described by the equation

$$y^2 + x^2 + 400\,z + 0.2z^2 - 0.035\,z^3 + 0.0072\,z^4 = 0$$

It is required to upscale this reflector to the one having the focal length $f' = 150$ mm.

Solution. The ratio $K = f'/f_i' = 1.5$, and the equation of the upscaled aspheric mirror becomes

$$Y^2 + X^2 + 600\,Z + 0.2\,Z^2 - 0.0233\,Z^3 + 0.0048\,Z^4 = 0$$

21.7 Minimizing the Spherical Aberration

In optical systems with a narrow field coverage in object space the performance (image quality) is decided above all by the amount of spherical aberration present in the system. Such systems include objective lenses of narrow field coverage, condensers of illumination systems, and the like systems consisting of positive lenses. The initial stage of the respective design is a layout on the basis of the thin-lens theory. Both objectives and condensers can be handled in one of the two power layouts: (1) the system is constituted by lenses of identical power each of which is designed for the spherical residuals being a minimum, (2) the system utilizes aplanatic menisci and one lens for which the spherical aberration is minimized. In the following material we examine each of these alternative versions in applications to both objectives and condensers on the basis of the third order aberration theory.

Objective constituted by lenses of identical power. A schematic diagram of such an objective is presented in Fig. 21.7. Let the number of lenses in this design be z. We assume that the thicknesses of all lenses and the spacings between them are all zero, i.e. $d_1 = d_2 = \ldots = d_{z-1} = 0$, the refractive indices of all lenses are the same, i.e. $n_2 = n_4 = \ldots = n_{2z} = n$, and the system is air separated, i.e. $n_1 = n_3 = \ldots = n_{2z+1} = 1$.

To perform our computations at unit focal length we recall the appropriate normalizing conditions (9.14) which for the object at infinity read as follows: $\alpha_1 = 0$, $\alpha_{2z+1} = 1$, and $h_1 = h_2 = \ldots = h_{2z} = 1$. The last equality refers to the indefinitely thin system.

If the powers of the individual lenses are the same and the number of lenses is z, then for the reduced system we have

$$z\,\varphi_t = 1 \tag{21.26}$$

where φ_t is the reduced power of lens t. For this lens the subscripts of the angles of the first auxiliary ray run as follows: α_{2t-1} where the ray enters the lens, α_{2t} inside the lens, and α_{2t+1} on leaving the lens.

By Eq. (3.21), $\alpha_{2t+1} = \alpha_{2t-1} + h_t\,\varphi_t$ and by virtue of (21.26) for $h_t = 1$ we get $\alpha_{2t+1} - \alpha_{2t-1} = 1/z$. Since for $t = 1$, $\alpha_{2t-1} = \alpha_1 = 0$, the last formula yields

$$\alpha_{2t-1} = (t-1)/z$$
$$\alpha_{2t+1} = t/z \tag{21.27}$$

These equations determine the odd values of angles α in an indefinitely thin objective constituted by elements of identical power.

To determine the even values of angles α, we consider the first Seidel sum for lens t,

$$S_i = \sum_{t=1}^{z} h_t P_t = \sum_{t=1}^{z} P_t$$

With the angle labels as defined above, the quantity P_t is obtained from (9.7) as

$$P_t = \frac{1}{(1-\mu_{2t})^2}\,[(2\mu_{2t}+1)(\alpha_{2t+1} - \alpha_{2t-1})\alpha_{2t}^2$$
$$-\ (2+\mu_{2t})(\alpha_{2t+1}^2 - \alpha_{2t-1}^2)\alpha_{2t}$$
$$+\ (\alpha_{2t+1}^3 - \alpha_{2t-1}^3)] \tag{21.28}$$

where $\mu_{2t} = 1/n_{2t} = 1/n$.

An optical system constituted by positive lenses will have a minimal spherical residual if each of the lenses has been designed to minimize spherical aberration. Differentiating Eq. (21.28) with respect to α_{2t} and equating the derivative to zero yields, with account of (21.27), the expres-

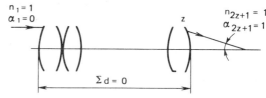

Fig. 21.7. A thin system of positive lenses

sion for α_{2t}^o corresponding to the minimal spherical aberration of each lens, namely,

$$\alpha_{2t}^o = (2n + 1)(2t - 1)/2(n + 2)z \qquad (21.29)$$

Equations (21.27) and (21.29) determine the angles of the first auxiliary ray in an indefinitely thin objective designed to keep the spherical aberration at a minimum. Having determined the angles and lens thicknesses, Eqs. (9.5) can be invoked to compute the radii of curvature for the objective of finite thickness.

The amount of coma is known to depend on the parameter W. Below we list the values of P and W of an indefinitely thin objective designed for minimal spherical residuals and for different number of lenses (all made of glass with $n = 1.5$)

Objective	P	W
Single lens	2.14	0.14
Doublet	0.44	0.15
Triplet	0.12	0.15
Four-lens	0.014	0.16

Thus, as the number of lenses in the system increases, P falls off to become practically zero at $z = 4$, whereas W is practically constant at about 0.15.

Objective with aplanatic menisci. A schematic of an indefinitely thin objective with aplanatic menisci is shown in Fig. 21.8. All the lenses except the first one are aplanatic menisci. These lenses obey the sine condition and do not contribute spherical aberration in the design. The magnification of meniscus labelled t is $\beta_t = 1/n_{2t} = 1/n$.

If the number of menisci is $z - 1$ and all of them are of the same glass, then their overall magnification is

$$\beta_{z - 1} = \frac{1}{n^{z - 1}}$$

Then subject to $\alpha_{2z + 1} = 1$ we have

$$\alpha_3 = \frac{1}{n^{z - 1}} \qquad (21.30)$$

The spherical aberration of the objective will be a minimum if its first lens is corrected for spherical aberration. By differentiating (21.28) with respect to α_2, equating the derivative to zero, and in view of (21.30) and $\alpha_{2z + 1} + 1$ we determine the value α_2^0 corresponding to the minimum spherical aberration of the entire lens, viz.,

$$\alpha_2^0 = \frac{2n + 1}{2(2 + n)n^{z-1}} \qquad (21.31)$$

The other values of α will be computed with the linear magnification of each meniscus.

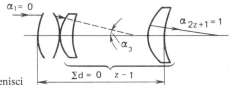

Fig. 21.8. A thin system with aplanatic menisci

The following list of P and W was computed for an indefinitely thin objective with aplanatic menisci at different number of lenses z made of glass having $n = 1.5$.

Objective	P	W
Single lens	2.14	0.14
Doublet	0.64	0.06
Triplet	0.19	0.03
Four-lens design	0.06	0.01

Comparison of these values of P and W with those for the objective constituted by lenses of identical power indicates that the objective with aplanatic menisci produces a higher amount of spherical aberration but more accurately meets the sine conditions ($W \approx 0$).

Condenser constituted by lenses of identical power. Let the magnification of the condenser consisting of indefinitely thin lenses (Fig. 21.9) be β, then in view of the normalizing conditions for the first auxiliary ray we have $\alpha_1 = \beta$, $\alpha_{2z+1} = 1$, and $h_1 = s_1\alpha_1$. If the overall power of the condenser is ϕ, then by (3.21) we get

$$\alpha_{2z+1} = \alpha_1 + h_1\phi = \beta + h_1\phi = 1 \tag{21.32}$$

Assuming the condenser being a thin design form of z lenses all having the same power, we have

$$\phi = z\phi_t \tag{21.33}$$

where ϕ_t is the power of lens labelled t.

By virtue of (21.32) and (21.33) the power of lens t is $\phi_t = (1 - \beta)/zh_t$ because for a thin system $h_1 = h_2 = \ldots = h_t$. The equation for angles (3.21) yields for each lens

$$\alpha_{2t+1} = \alpha_{2t-1} + h_t\phi_t = \alpha_{2t-1} + (1 - \beta)/z \tag{21.34}$$

thus producing all odd values of α, subject to $\alpha_1 = \beta$.

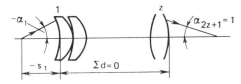

Fig. 21.9. A condenser with aplanatic menisci

To keep the spherical aberration of the condenser at a minimum each lens of the design should be controlled for the least P_t. We differentiate (21.28) with respect to α_{2t} and put the derivative to zero to obtain the expression for α_{2t}^0 corresponding to the least spherical aberration of each lens in the design, viz.,

$$\alpha_{2t}^0 = \frac{(2n + 1)(\alpha_{2t + 1} + \alpha_{2t - 1})}{2(2 + n)} \tag{21.35}$$

Condenser with aplanatic menisci. If all the lenses of the condenser, except the last one, are aplanatic menisci made of glass of index n, then for z lenses in the condenser the overall magnification of $z - 1$ menisci will be

$$\beta_m = n^{z - 1} \tag{21.36}$$

If $\beta = \alpha_1$ is the magnification of the condenser, then in view of (21.36) the parameter $\alpha_{2z - 1}$ of the first auxiliary ray at the last lens will be

$$\alpha_{2z - 1} = \alpha_1/\beta_m = \beta/n^{z - 1} \tag{21.37}$$

Because aplanatic menisci do not contribute to spherical aberration, to minimize the spherical aberration in the condenser its last lens should be designed for the least spherical aberration. This implies satisfying Eq. (21.35) for the last lens labelled z. In view of $\alpha_{2z + 1} = 1$ and Eq. (21.37) we get

$$\alpha_{2z}^0 = \frac{(2n + 1)(\alpha_{2z + 1} + \alpha_{2z - 1})}{2(2 + n)}$$

$$= \frac{(2n + 1)(1 + \beta/n^{z - 1})}{2(2 + n)}$$

21.8 A Cemented Doublet

The cemented doublet is one of the most common design forms. It may be used as an individual system or as a part of more complex formulations. In the computational procedure that follows we assume that the types of glass are specified for both lenses constituting the design.

Thin lens theory gives three degrees of freedom in the form of three radii of curvature. One of the radii is to secure the required focal length, the other two will be used in the control of aberrations. Hence, for a cemented doublet with pre-selected glass types, we are able to correct only two aberrations. From a methodological standpoint, it would be more convenient to use the parameters of the first auxiliary ray rather than the radii of curvature to control aberrations.

Figure 21.10 shows the optics of a cemented doublet in a sketch form. For the known (auxiliary) rays, we let the normalizing set (9.14) as follows $\alpha_1 = 0$, $\alpha_4 = 1$, $h_1 = h_2 = h_3 = 1$, $\beta_1 = 1$, $H_1 = t/f'$, and $I = -1$.

In agreement with the normal practice we suppose that the performance specifications stipulate the focal length f', relative aperture f'/D, angular field coverage 2ω and the residuals: the longitudinal spherical aberration at the pupil edge $\Delta s'$, and longitudinal chromatic aberration (axial colour) $\Delta s'_{\lambda_1, \lambda_2}$. These aberrations may be zero or nonzero and can be used to compensate for the residuals of the subsequent components.

At the initial design stage we assume that the lens aberrations terminate at the third order and by virtue of (9.21) and (10.1) arrive at the Seidel sums as follows

$$S_I = -2\Delta s' f' / m^2 \tag{21.38}$$

$$S_{I, ch} = \Delta s_{\lambda_1, \lambda_2}/f' \tag{21.39}$$

In the general case, the residual axial colour is selected such that to attain a necessary correction of the spherochromatism for a zonal ray of $0.7m_{cr}$.

Let the relative powers of the lens elements be φ_1 and φ_2. Then, witn account of the scaling condition and Eq. (21.3) for the chromatic aberration of a thin lens, we get

$$\varphi_1 + \varphi_2 = 1$$
$$S_{I, ch} = -(\varphi_1/V_1 + \varphi_2/V_2) \tag{21.40}$$

where V_1 and V_2 are the Abbe V-values of the spectral interval in which the lens is to be achromatized. Solving this system yields

$$\varphi_1 = V_1(1 + V_2 S_{I, ch})/(V_1 - V_2)$$
$$\varphi_2 = 1 - \varphi_1 \tag{21.41}$$

Thus, the parameter φ_1 determines the axial chromatic aberration of this cemented doublet.

The first Seidel sum is a function of α_2 and α_3. For convenience, we in-

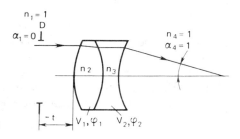

Fig. 21.10. A two-lens cemented objective

troduce a parameter Q related to α_2 and α_3 as follows

$$\alpha_2 = (1 - \mu_2)Q + \varphi_1$$
$$\alpha_3 = (1 - \mu_3)Q + \varphi_1$$

(21.42)

Since our lens is indefinitely thin, by (21.8) we have $S_I = P$, or in terms of Q

$$S_I = aQ^2 + bQ + c$$

(21.43)

where

$$a = 1 + \frac{2}{n_2} \varphi_1 + \frac{2}{n_3} \varphi_2$$

$$b = \frac{3}{n_2 - 1} \varphi_1^2 - \frac{3}{n_3 - 1} \varphi_2^2 - 2\varphi_2$$

(21.44)

$$c = \frac{n_2}{(n_2 - 1)^2} \varphi_1^3 + \frac{n_3}{(n_3 - 1)^2} \varphi_2^3 + \frac{n_3}{n_3 - 1} \varphi_2^2$$

These analytic dependences enable our design to be carried out in steps as follows.

Given the principal characteristics of the lens and the amounts of residuals, we obtain $S_I = P$ by Eq. (21.38) and $S_{I, ch}$ by Eq. (21.39), determine φ_1 and φ_2 and derive a, b, and c from (21.44). Then we solve the quadratic equation (21.43) and obtain two values of Q corresponding to the required value of S_I. The analysis of the roots is carried out on the following basis.

The parameter Q also defines the parameter W of a thin doublet

$$W = -\frac{a + 1}{2} Q + \frac{\varphi_2 - b}{3}$$

which in turn, by (21.8), affects the second Seidel sum

$$S_{II} = H_1 P + W$$

which controls coma. Therefore, of the two roots Q it would be wise to pick the one leading to S_{II} which satisfies the desired amount of coma for the lens.

If the quadratic equation (21.43) has no real-valued roots, this implies that the selected types of glass prevent the design from being simultaneously corrected for spherical aberration and axial colour, that is, another combination of glasses is needed.

Once the parameter Q has been selected, we determine α_2 and α_3 by Eq. (21.42) and compute the radii of curvature with account of the real thickness of lens elements.

Now, with the constructional data of the lens on hand we may perform a refining computation on a computer. The resultant values of spherical aberration for the edge zone and longitudinal chromatic aberration will differ from those specified owing to the effect of higher-order aberrations and lens thicknesses. If this difference exceeds the tolerable value, then a new lens design is necessary which would take into account higher-order aberrations. It should be kept in mind that the parameter φ_1 affects both the longitudinal chromatic aberration and spherical aberration, whereas the parameter Q produces an effect on the spherical aberration only. Therefore, in the correction procedure that follows one should first arrive at the desired amount of axial colour by varying φ_1, and then control the spherical aberration by varying Q.

The procedure outlined above assumes that the types of glass have been specified. In the circumstances, only two aberrations can be controlled in a two-element lens. When the selection of glasses is relegated to the designer, he obtains an additional degree of freedom to correct one more aberration. As a rule, the three aberrations to be controlled are spherical, axial colour, and coma for the edge of the field with due account of vignetting.

Slyusarev [20] has developed a procedure for cemented doublets, which involves selection of glass types. This procedure is based on an approximate relation

$$P = P_{\min} + 0.84(W - 0.15)^2 \tag{21.45}$$

where $P_{\min}$ is the minimal value of P for a combination of glasses ensuring a certain amount of longitudinal chromatic aberration. This relation holds for the case of a simple lens as well.

21.9 An Air Separated Doublet

This lens receives an additional degree of freedom as compared with the cemented doublet, namely, the fourth radius of curvature. Therefore, with due account of scaling, it lends itself for correction of three aberrations.

A schematic of a two-element airspaced lens is presented in Fig. 21.11. We assume that the lens is a thin design and incorporate the following normalizing conditions: $\alpha_1 = 0$, $\alpha_5 = 1$, $h_1 = h_2 = h_3 = h_4 = 1$, $\beta_1 = 1$, $H_1 = t/f'$, and $I = -1$.

We also assume that the performance specifications for the lens stipulate its principal characteristics, such as f', f'/D, and 2ω, and the residual aberrations: longitudinal spherical aberration for the edge of the pupil, longitudinal chromatic aberration, and coma for the edge of the field with due account of vignetting.

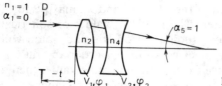

$n_1 = 1$ D
$\alpha_1 = 0$

$\alpha_5 = 1$

$-t$ V_1, φ_1 V_2, φ_2

Fig. 21.11. An airspaced doublet

With the data thus specified we compute the Seidel sums by (9.21), (9.26) and (10.1) as follows

$$S_I = -2f' \Delta s' / m^2 \tag{21.46}$$

$$S_{II} = -2f' \Delta y' / 3m^2 \omega_1 \tag{21.47}$$

$$S_{I, ch} = \Delta s'_{\lambda_1, \lambda_2} / f' \tag{21.48}$$

For a thin lens, the second Seidel sum is expressed in terms of P and W by Eq. (21.8). Having determined $P = S_I$ from (21.46) and S_{II} from (21.47) we obtain the necessary value of W at the given position of the entrance pupil (H_1) as follows

$$W = S_{II} - H_1 P$$

Thus, from a standpoint of correction of monochromatic aberrations the problem of two-element air separated lens reduces to evaluation of its constructional parameters satisfying the pre-specified values of P and W. One way to solve this problem is based on the basic parameters of thin elements.

From the normalizing conditions, as defined above, we obtain by the equation for angles $\varphi_1 = \alpha_3$. Then the condition for control of chromatic aberration in the lens is given by Eq. (21.41)

$$\alpha_3 = \varphi_1 = V_1(1 + V_2 S_{I, ch})/(V_1 - V_2)$$

The parameters P and W of the entire lens depend on the respective parameters of the elements. In the thin lens approximation,

$$P = P_1 + P_2$$
$$W = W_1 + W_2 \tag{21.49}$$

Now we express P_1 and W_1 of the first element and P_2 and W_2 of the second element in terms of their basic parameters, observing that for the first element, $\alpha_i = 0$ and $\alpha_i' = \alpha_3$; and for the second element, $\alpha_i = \alpha_3$ and $\alpha_i' = 1$. By Eqs. (21.16) we have

$$P_1 = \alpha_3^3 \bar{P}_1$$
$$W_1 = \alpha_3^2 \bar{W}_1 \tag{21.50}$$

$$P_2 = (1 - \alpha_3)^3 \bar{P}_2 + 4\alpha_3(1 - \alpha_3)^2 \bar{W}_2$$
$$+ \alpha_3(1 - \alpha_3)[2\alpha_3(2 + \pi) - 1]$$
$$W_2 = (1 - \alpha_3)^2 \bar{W}_2 + \alpha_3(1 - \alpha_3)(2 + \pi)$$

(21.50)

where $\bar{P}$ and $\bar{W}$ are the basic parameters of a thin lens, and $\pi_i \approx 0.7$.

The basic parameters of a thin lens are related by the approximate expression (21.45) as follows

$$\bar{P}_1 = \bar{P}_{\min, 1} + 0.85(\bar{W} - 0.15)^2$$
$$\bar{P}_2 = \bar{P}_{\min, 2} + 0.85(\bar{W} - 0.15)^2$$

(21.51)

The quantity $\bar{P}_{\min}$ for each element can be determined by

$$\bar{P}_{\min, 1} = \frac{(4n_2 - 1)n_2}{4(2 + n_2)(n_2 - 1)^2}$$
$$\bar{P}_{\min, 2} = \frac{(4n_4 - 1)n_4}{4(2 + n_4)(n_4 - 1)^2}$$

(21.52)

Since the types of glass for this lens have been specified, Eqs. (21.52) can be used to determine $\bar{P}_{\min}$ for each element. Substituting (21.50) into (21.49) and observing (21.51) we arrive at two equations, quadratic and linear, in the unknown basic parameters $\bar{W}_1$ and $\bar{W}_2$ of the thin elements of the lens. If the quadratic equation has real-valued roots, then the one should be selected which corresponds to lower absolute values of $\bar{W}_1$ and $\bar{W}_2$.

Having determined $\bar{W}$ for each element, we can obtain the respective values of the parameters

$$\bar{\alpha}_2 = \frac{n_2}{n_2 + 1} - \frac{n_2 - 1}{n_2 + 1} \bar{W}_1$$
$$\bar{\alpha}_4 = \frac{n_4}{n_4 + 1} - \frac{n_4 - 1}{n_4 + 1} \bar{W}_2$$

The transition from $\bar{\alpha}$ computed for each element to the parameters α_2 and α_4 of the airspaced doublet is effected by Eq. (21.12).

For the first element, $\alpha_1 = 0$, $\bar{\alpha}_3 = 1$, $q = \alpha_3$, $\alpha_2 = q\,\bar{\alpha}_2$.

For the second element, $\alpha_5 = 1$, $\bar{\alpha}_5 = 1$, $q = 1 - \alpha_3$, $\alpha_4 = q\,\bar{\alpha}_4 + \alpha_3/n_4$.

Having determined all the values of α for a thin lens and found the thicknesses of the elements and spacings between them, we derive the parameters of the initial design by Eq. (9.5). This follows by a check-out computation of residual aberrations on a computer; the results of this computation form a basis for a subsequent correction of the design.

21.10 A High-Speed Separable Objective

One of the simplest design forms having a high relative aperture ($f/2$ to $f/1.5$) and a modest angular coverage ($2\omega \leqslant 20°$) is a lens consisting of two positive components spaced some distance from each other. These lenses, called separable or Petzval projection lenses, find their use as high-speed objectives for motion-picture projection, night-vision instruments, illumination of photocathodes, etc.

A meridional section through the separable objective is sketched in Fig. 21.12. If φ_I and φ_{II} are the relative powers of the components, and d is the reduced spacing between them (thin lens), then Eq. (3.27) for a two-component system yields

$$\varphi = \varphi_I + \varphi_{II} - \varphi_I\varphi_{II}\, d = 1$$
$$h_{II} = 1 - \varphi_I d$$

The normalizing set for the first auxiliary ray will be this: $\alpha_I = 0$, $h_I = 1$, and $\alpha_{III} = 1$. Then Eqs. (3.21) and (3.22) give

$$\varphi_I = \alpha_{II}$$
$$\varphi_{II} = (1 - \alpha_{II})/h_{II} \qquad (21.53)$$
$$d = (1 - h_{II})/\alpha_{II}$$

These formulae define the external parameters of the lens which have an effect on both chromatic aberrations, $S_{I,\,ch}$ and $S_{II,\,ch}$, and on the Petzval curvature, S_{IV}. We recall here that the *Petzval curvature* is a sort of basic curvature associated with every optical system; it is a function of the index of refraction of the lens elements and their surface curvatures. When there is no astigmatism, the sagittal and meridional (tangential) image surfaces coincide with each other and lie on the Petzval surface. When there is primary astigmatism present, the tangential image surface lies three times as far from the Petzval surface as the sagittal image, both on the same side of the Petzval surface (see, e.g. Fig. 9.12).

When the desired form must be simple in construction and have a high

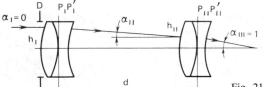

Fig. 21.12. A high-speed two-element lens

relative aperture, both components of the lens must be positive. If the system is constituted by only positive thin components, the associated Petzval curvature cannot be eliminated in principle and the arrangement under consideration has the Petzval sum, S_{IV}, as high as 0.8 to 1.2. This prevents the use of a greater angular coverage than 15° to 20°.

In this design, the correction of spherical aberration, coma, and both cromatic aberrations should be given a due consideration. This type of system is noted for its strongly inward curving field ($S_{IV} \approx 1$). The field is artificially flattened by negative (overcorrected) astigmatism ($S_{III} < 0$) which is introduced at the cemented surface of the rear doublet. Observing the high relative aperture of the lens, major attention should be placed on the adjustment of the spherical aberration of higher orders.

Let us assume that the entrance pupil of the system is the rim of the front doublet ($t = 0$). Then the parameters of the second auxiliary ray will be as follows: $H_I = 0$, $\beta_I = \beta_{II} = 1$, and $H_{II} = -d$.

Under the said normalizing conditions, in view of (21.3), the system will be free from axial colour if

$$S_{I,\,ch} = \varphi_I \bar{C}_I + h_{II}^2 \, \varphi_{II} \, \bar{C}_{II} = 0 \qquad (21.54)$$

To free the system from the lateral colour we need, by (21.3),

$$S_{II,\,ch} = H_{II} h_{II} \varphi_{II} \bar{C}_{II} = 0 \qquad (21.55)$$

This and penultimate equations suggest that both chromatic abberations of the lens can be eliminated if each component of the lens is achromatized. This can be achieved by using cemented or airspaced (broken) doublets.

With account of the values of the parameters of auxiliary rays, Eqs. (21.53) and (21.8) lead to the following Seidel sums defining the monochromatic aberration of the lens

$$
\begin{aligned}
S_I &= P_I + h_{II} P_{II} \\
S_{II} &= W_I + H_{II} P_{II} + W_{II} \\
S_{III} &= \alpha_{II} + \frac{H_{II}^2}{h_{II}} P_{II} + 2 \frac{H_{II}}{h_{II}} W_{II} + \frac{1 - \alpha_{II}}{h_{II}}
\end{aligned}
\qquad (21.56)
$$

If we substitute the basic parameters of thin lens components for P and W in these expressions, we obtain three equations in four unknowns: $\bar{P}_I$, $\bar{W}_I$, P_{II}, and W_{II}. To extract the appropriate roots, the following notions are useful. The high relative aperture of the lens and positive spherical aberrations of higher orders dictate the first Seidel sum equal to 0.2 or 0.3. The second sum may be let equal zero. To partially compensate for (flatten) the field curvature, the third Seidel sum should be a small negative value in the range from about -0.05 to -0.10.

Thus, to correct three aberrations, it will be sufficient to have three free parameters (degrees of freedom) in the set (21.56). For the remaining fourth parameter, it would be appropriate to impose a constraint requiring that the higher-order spherical aberration be a minimum.

The higher-order spherical aberration is mainly affected by the construction of the front doublet where the height of the extreme ray grazing the rim of the entrance pupil is 2 or 3 times that on the rear component. Therefore, the basic parameters $\bar{P}_I$ and $\bar{W}_I$ of the front component should be chosen so as to minimize the higher-order spherical aberration contributed by this component. Experimental evidence indicates that a cemented doublet will have a minimal higher-order spherical aberration if its basic parameter $\bar{P}_I$ is positive and $\bar{W}_I$ is close to zero. An appropriate computational procedure for this type of lens would be as follows.

Investigations into the correcting potentialities of this design have indicated that the region of most suitable solutions is covered by the parameters of the first auxiliary ray chosen as $\alpha_{II} = 0.5\text{-}0.7$ and $h_{II} = 0.5\text{-}0.3$. This will be the point of departure for our design. With α_{II} and h_{II} selected in these intervals we find the external parameters of the lens (φ_I, φ_{II}, and d) by Eqs. (21.53).

The parameters P_I, W_I, P_{II}, and W_{II} should be expressed in terms of the basic parameters by Eqs. (21.16), subject to $\alpha_i = 0$ and $\alpha_i' = \alpha_{II}$ for the front component, and $\alpha_i = \alpha_{II}$ and $\alpha_i' = 1$ for the rear component. In order to minimize the higher-order spherical aberration, we put the basic parameter $\bar{W}_I$ equal to zero. Now we equate the Seidel sums to the values as defined above and substitute $\bar{P}_I$, $\bar{P}_{II}$, and $\bar{W}_{II}$ in (21.56). This leads us to a set of three equations in three unknowns solvable for $\bar{P}$ and $\bar{W}$ of each component.

The conditions (21.54) and (21.55) for correction of chromatic aberrations lead us to obtain the chromatic parameters $\bar{C}_I$ and $\bar{C}_{II}$ of the lens components. In solving Eq. (21.54) one should keep in mind that the chromatic aberration should evolve a bit undercorrected in order to adjust the spherical zonal for $m_z = 0.7\, m_{cr}$.

Having determined the basic parameters $\bar{P}$, $\bar{W}$, and $\bar{C}$ of each component (we assume that they are cemented doublets), we invoke Slyusarev's technique [20] to find the structural parameters of these doublets, for example, by looking up in the tables presented in the book of this author.

The radii of curvature of the components will be obtained with the focal lengths of each doublet determined in terms of the relative powers derived by (21.53), namely,

$$f_I' = f'/\varphi_I$$
$$f_{II}' = f'/\varphi_{II}$$

where f' stands for the focal length of the lens.

The thickness of lens elements in each doublet depends on their diameters. If the entrance pupil of the lens is at the front doublet, its diameter equals the diameter of the entrance pupil. For the rear doublet, the diameter can be determined from the condition of the passage of an oblique bundle corresponding to the edge of the field with account of vignetting.

21.11 An Airspaced Triplet

Airspaced anastigmats utilize a large separation between positive and negative components to correct the Petzval sum. A simplest form of this type is the *Cooke triplet* consisting of three single lenses spaced at a finite distance from each other. This objective relates to the group of universal lenses covering a rather wide total field in excess of 50°-60° at a speed of $f/2.8$.

A more useful design of the triplet has a negative lens between two positive elements (Fig. 21.13a). Another possible arrangement with a positive lens between two negative elements has an obvious setback in the form of an exceedingly large power requirement for the positive element to secure a positive focal length. Other deviations from the symmetry in the power layout lead to considerable efforts required to adjust the distortion.

To put the name of the triplet in a correct historical perspective we note that this form was developed by H. Teiler of the United Kingdom in 1894 and has been mass-produced by many optical suppliers throughout the world. An elaboration of this system is the Tessar which may be regarded as a triplet with the rear positive element compounded for two glasses.

Owing to comparative simplicity of this arrangement, the triplet lends itself well to analysis and design on the basis of third-order aberration theory. Assuming the elements of the triplet to be infinitely thin, we may select for them parameters in terms of which most aberrations of the lens become linear functions. The computational procedures used in this coun-

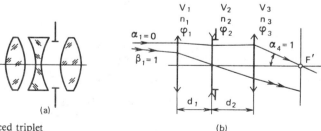

Fig. 21.13. An airspaced triplet

try are based on the methods developed by G. Slyusarev [20] and D. Volosov [2]. It will be noted that all the methods involve the separation of parameters in the external, i. e. independent of the lens shapes, and the internal which define the construction of the lenses.

The triplet design is essentially the solution of nine equations for the correction of five third-order aberrations, two chromatic aberrations, and for two layout specifications. To meet all these conditions, the triplet lends five external parameters (three powers of the elements and two spacings between them), three internal parameters (element shapes) and six optical constants of the glasses (refractive indices and V-values). It is worth noting that from a mathematical standpoint the optical constants of glasses are not full-value parameters, for they can assume only discrete values within a limited range.

The optics of a Cooke triplet is schematized in Fig. 21.13b. The focal length of the lens is assumed to be unity. The normalizing set for the first auxiliary ray is as follows: $\alpha_1 = 0$, $\alpha_4 = 1$, $h_1 = 1$; for the second ray: $\beta_1 = 1$ and $I = -1$.

Let us examine first the analytical formulations which define the scaling conditions and the correction of aberrations dependent on the external parameters. It would be appropriate to turn to the conditions for the control of spherical aberration, coma, and astigmatism after the evaluation of the external parameters, because the adjustment of these aberrations is achievable by means of altering the internal parameters of the elements, i. e. by adjusting the radii of curvature of the refracting surfaces.

Because the aperture stop is usually placed inside the lens, we assume that this stop coincides with the second element, i. e. $H_2 = 0$, for the initial design.

Thus, the choice of external parameters of a triplet must satisfy the six conditions as follows.

(1) For a given focal length (scaling condition)

$$\varphi_1 + h_2 \varphi_2 + h_3 \varphi_3 = 1 \qquad (21.57)$$

where φ_1, φ_2, and φ_3 are the relative powers of the triplet elements.

(2) For a given back focal length

$$h_3 = s_F'. \qquad (21.58)$$

This condition is not always mandatory.

(3) For Petzval curvature

$$S_{IV} = \varphi_1/n_1 + \varphi_2/n_2 + \varphi_3/n_3 \qquad (21.59)$$

(4) For the longitudinal chromatic aberration

$$S_{\text{I, ch}} = -\left(\frac{\varphi_1}{V_1} + \frac{h_2^2 \varphi_2}{V_2} + \frac{h_3^2 \varphi_3}{V_3} \right) \qquad (21.60)$$

(5) For the transverse chromatic aberration

$$S_{\text{II, ch}} = -\left(\frac{H_1 \varphi_1}{V_1} + \frac{H_3 h_3 \varphi_3}{V_3} \right) \qquad (21.61)$$

(6) For the distortion

$$S_V = 3.65 H_1 \varphi_1 + 3.65 \, (H_3/h_3)\varphi_3 \qquad (21.62)$$

The fifth Seidel sum which defines the amount of distortion is express-ed through P and W by Eq. (21.8). Since for most triplet formulations the values of d_1 and d_2 range between 0.1 and 0.2, the heights of the second auxiliary ray on the first and third elements have about the same values. Therefore, the terms containing H_1 and H_3 raised to the third and second power may be safely omitted from Eqs. (21.8), and by letting $\pi_1 = \pi_3 \approx 0.65$ we arrive at the approximate expression (21.62) giving the amount of residual distortion.

The relations (21.61) and (21.62) can be simplified still further by eliminating the parameters H_1 and H_3 of the second auxiliary ray. For $\beta_1 = 1$ and $H_2 = 0$ we have from the simple-lens formulae $\beta_2 = 1 + H_1\varphi_1$ and $H_1 = d_1\beta_2$. Consequently, $\beta_2 = 1/(1 - \varphi_1 d) = 1/h_2$, i. e.,

$$\begin{aligned} H_1 &= d_1/h_2 \\ H_3 &= -d_2/h_2 \end{aligned} \qquad (21.63)$$

From the relation between d and h in the equation for angles we have

$$\begin{aligned} h_2 &= 1 - \varphi_1 d_1 \\ h_3 &= 1 - \varphi_1 d_1 - d_2(\varphi_1 + \varphi_2 - \varphi_1\varphi_2 d_1) \end{aligned} \qquad (21.64)$$

When the condition (21.58) is not mandatory, we may substitute Eqs. (21.63) and (21.64) into Eqs. (21.57)-(21.62) to arrive at five equations in five unknowns, φ_1, φ_2, φ_3, d_1, and d_2. This system is rather hard to solve because the equations are nonlinear in the unknowns. In addition, a straightforward mathematical approach may lead to unfeasible solutions dictating, say, prohibitively large powers of the elements, large spacing be-tween the elements, and the like. Accordingly, the steps in the control of aberrations for a triplet should be as follows.

We assign to φ_1 a number of values between 1 and 2, to φ_2 a number of values between -3 and -4, and evaluate φ_3 by Eq. (21.59) subject to the selected types of glass. Then we derive the heights h_2 and h_3 subject to the conditions for scaling (21.57) and axial colour (21.60). It is desirable that the condition (21.58) be met. Then we use Eq. (21.64) to compute d_1 and d_2, and Eqs. (21.61) and (21.62) to compute $S_{\text{II, ch}}$ and S_{V}. These computations are to be carried out for different combinations of optical glasses to see which set of the external parameters is optimal.

The correction for the other monochromatic aberrations is effected by choosing a suitable set of parameters for the first auxiliary ray inside each element with the help of the curvature radii of the refracting surfaces. This stage of design is appropriate to carry over from the thin-lens design to the one involving lens thicknesses. Having one degree of freedom for each element of the triplet we are in a position to adjust three aberrations, namely, spherical, coma, and astigmatism. By Eq. (21.8) we arrive at the three first Seidel sums for the triplet

$$S_{\text{I}} = P_1 + h_2 P_2 + h_3 P_3$$
$$S_{\text{II}} = H_1 P_1 + W_1 + W_2 + H_3 P_3 + W_3 \qquad (21.65)$$
$$S_{\text{III}} = H_1^2 P_1 + 2 H_1 W_1 + \varphi_1 + \varphi_2 + \frac{H_3^2}{h_3} P_3 + 2 \frac{H_3}{h_3} W_3 + \varphi_3$$

where the external parameters h, H, and φ were determined at the preceding stage of the procedure, and the parameters P and W refer to each element and depend on the angles α inside the elements.

Equations (21.65) are rather complex functions of α, and the angles proper call for a great deal of research effort for their evaluation.

Because this design form lends itself well to correction work, the triplet residuals can be brought down to a level at which the limiting resolution is about 30 mm^{-1} at the axis and about 10-15 mm^{-1} over the field. A way for further improvement of the triplet performance is being paved by the development of new types of glass, specifically extradense crowns.

21.12 Reflecting Systems

The recent decades have seen an increasing use of optical systems in the non-visual regions of the spectrum, that is in the ultraviolet and infrared regions. This has resulted in a corresponding increase in the use of reflecting optics. Fortunately, most applications permit the use of relatively unsophisticated reflecting systems free from chromatic aberrations and from the need for procuring satisfactory refractive materials for the optical regions of interest. In such systems, refractive elements are of comparatively low power and play the role of correctors, the attendant chromatic contributions being usually negligible. Another advantage of reflecting and catadioptric systems over lens designs is due to their small size. The disadvantages include comparatively narrow angular fields, vignetting of the central portion of the entrance pupil, and sensitivity to misalignment.

Consider the simplest reflecting systems which obviously include reflecting surfaces both spherical and aspherical.

Aberrations of a spherical reflector. More often than not a spherical reflector is employed either to image an infinitely distant object, or to produce an image at infinity. The aberrations of a spherical mirror will be determined by the third-order aberration equations. The normalizing set for the parameters of auxiliary rays will be as follows (Fig. 21.14): $\alpha_1 = 0$, $\alpha_2 = 1$, $h_1 = -1$, $\beta_1 = 1$, $H_1 = t/f'$, and $I = 1$.

Because for a reflecting surface $n_1 = -n_2 = 1$, Eq. (21.10) yields $P = -1/4$ and $W = 1/2$. Substituting these values of P and W into Eq. (21.8) gives the third-order aberration coefficients as follows

$$S_I = -1/4$$

$$S_{II} = -\frac{1}{4}H_1 + \frac{1}{2}$$

$$S_{III} = -\frac{1}{4}H_1^2 + H_1 - 1 \qquad (21.66)$$

$$S_{IV} = 1$$

$$S_V = -\frac{1}{4}H_1^3 + \frac{3}{2}H_1^2 - 2H_1$$

Now, for a meridional ray fan ($M = 0$) at $n_p' = n_2 = -1$ the third-order aberration equations for a spherical mirror will be defined by Eq. (9.17) and Eqs. (21.66) as follows:
the transverse spherical aberration

$$\Delta y_{III}' = -m^3/8f'^2 \qquad (21.67)$$

the longitudinal spherical aberration

$$\Delta s_{III}' = -m^2/8f' \qquad (21.68)$$

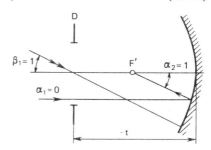

Fig. 21.14. A spherical reflector

the meridional coma

$$\Delta y_{III} = \frac{3m^2\omega_1}{8f'}(-H_1 + 2) \qquad (21.69)$$

the astigmatism

$$\Delta z_a' = z_s' - z_m' = -f'\omega_1^2(-\frac{1}{4}H_1^2 + H_1 - 1) \qquad (21.70)$$

the distortion

$$\Delta y_{III}' = \frac{f'}{2}\omega_1^3(-\frac{1}{4}H_1^3 + \frac{3}{2}H_1^2 - 2H_1) \qquad (21.71)$$

Comparing the spherical aberration of a single lens ($S_1 = 2.14$ at $n = 1.5$) with that of a single reflector ($S_1 = -1/4$) we see that the contribution of the lens is 8.5 times that of the reflector. Equations (21.69)-(21.71) make it clear that if the centre of the entrance pupil coincides with the vertex of the reflector ($H_1 = 0$), the meridional coma, field curvature, and astigmatism cannot be eliminated. The distortion, however, is absent in this case.

If we let in Eqs. (21.66) $H_1 = 2$ (at $f' = 1$), i. e. make the centre of the entrance pupil coincide with centre of curvature of the reflector, then $S_{II} = S_{III} = S_V = 0$. This implies that such a location of the entrance pupil frees the reflecting system of all aberrations except spherical and Petzval curvature.

For a paraboloidal mirror, $S_1 = P = 0$, that is, no spherical aberration is present. Equations (21.66) indicate for this case that $S_{II} = 1/2$ for any H_1, i. e., the coma of a paraboloidal mirror is independent of location of the entrance pupil. At $H_1 = 0$, which holds when the entrance pupil coincides with the rim of the mirror, the field aberrations of a parabolic mirror are the same as that of a spherical mirror.

Aberrations of a two-mirror objective (Fig. 21.15). The distance c from the vertex of the primary (large) mirror to the image plane, normally given in performance specifications on a system design, depends on the intended

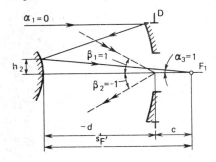

Fig. 21.15. A two-mirror objective

application of the objective. We assume that the centre of the entrance pupil coincides with the vertex of the primary mirror, and let the normalizing set for the auxiliary rays be as follows: $\alpha_1 = 0$, $h_1 = 1$, $\alpha_3 = 1$, $\beta_1 = 1$, $H_1 = 0$, $\beta_2 = -1$, and $I = -1$. Referring to Fig. 21.15 we note that the distances $-d$, c, and s'_F are reduced to the unit focal length.

The two-mirror objective has two degrees of freedom (parameters α_2 and d) which will be selected subject to the given dimensional constraints. With reference to Fig. 21.15 we see that at $\alpha_3 = 1$

$$-d + c = h_2 = s_{F'}$$

and the equation for incidence heights gives at $h_1 = 1$

$$h_2 = 1 - d\,\alpha_2$$

Combining the two last expressions yields the relation

$$d = (c - 1)/(1 - \alpha_2) \tag{21.72}$$

connecting α_2 and d for a given c. In addition, d and c define the height h_2 of the first auxiliary ray on the secondary mirror. Since this mirror obscures the central portion of the system's entrance pupil, it is desirable that $h_2 \leqslant 0.5$.

Let us derive the expressions for P and W subject to the aforementioned normalizing conditions. Observing that $n_1 = n_3 = 1$ and $n_2 = -1$, we have from (21.10)

$$\begin{aligned}
P_1 &= -\alpha_2^3/4 \\
W_1 &= \alpha_2^2/2 \\
P_2 &= (1 - \alpha_2)^2(1 + \alpha_2)/4 \\
W_2 &= (1 - \alpha_2^2)/2
\end{aligned} \tag{21.73}$$

The Seidel sums, defined by Eqs. (21.8), become

$$\begin{aligned}
S_I &= P_1 + h_2 P_2 \\
S_{II} &= W_1 + H_2 P_2 + W_2 \\
S_{III} &= \varphi_1 + \frac{H_2^2}{h_2} P_2 + 2\frac{H_2}{h_2} W_2 + \varphi_2 \\
S_{IV} &= \alpha_2 - \frac{1 + \alpha_2}{h_2} \\
S_V &= \frac{H_2^3}{h_2^2} P_2 + 3\frac{H_2^2}{h_2^2} W_2 + 2\frac{H_2}{h_2} \varphi_2
\end{aligned} \tag{21.74}$$

where $h_2 = 1 - d\alpha_2$, $H_2 = d$, $\varphi_1 = \alpha_2$, and $\varphi_2 = (1 + \alpha_2)/h_2$. Relating d and α_2 by (21.72) for a given c, one may investigate the controllability of the aberrations in this system as a function of α_2.

Additional correcting capabilities are provided for the system by lens correctors or asphericity introduced into the spherical reflecting surface. Asphericity is equivalent to an additional degree of freedom used for correction. If an asphericity is introduced at the surface coinciding with the aperture stop, the height of the second auxiliary ray will be zero on this surface ($H_i = 0$). Therefore, by (21.19), the asphericity will affect the spherical aberration alone, without having any effect on the third order field aberrations.

21.13 Catadioptric Systems

A variety of compensating lenses and variously configured plates, called correctors for short, are used to compensate for axial and field aberrations in reflecting systems. The correctors are located ahead of the mirror, i. e. in parallel rays if the object is at infinity, and inside the system to work in convergent bundles of rays. A corrector is shaped so that its chromatic contribution is a minimum. Below we wish to analyze some correcting lenses.

Meniscus corrector for parallel rays. One of the simplest catadioptric systems consists of a compensating lens and a spherical mirror, Fig. 21.16. For its analysis, we take the following normalizing set assuming that the object is at infinity: $\alpha_1 = 0$, $\alpha_4 = 1$, $h_1 = f'$, $\beta_1 = 1$, and $I = -f'$.

In order that the correcting lens of finite thickness may be free of longitudinal chromatic aberration, it has to obey the condition (10.1), namely,

$$S_{\mathrm{I, ch}} = \sum_{k=1}^{2} h_k C_k = 0$$

After incorporation of the normalizing conditions this expression becomes

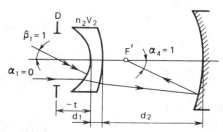

Fig. 21.16. A spherical reflector with a meniscus corrector

$$S_{\text{I, ch}} = h_1 C_1 + h_2 C_2 = -\frac{1}{V_2}\left[(h_1 - h_2)\alpha_2 + h_2\alpha_2\right] = 0$$

Since from the equation for heights we have $h_2 = h_1 - d_1\alpha_2$, this equation rewrites as

$$S_{\text{I, ch}} = -\frac{1}{V_2}(d_1\alpha_2^2 - d_1\alpha_2\alpha_3 + h_1\alpha_3) = 0$$

The solutions to this quadratic equation establish a relation between the angles α_2 and α_3 as follows

$$\alpha_2 = \frac{\alpha_3}{2}\left(1 \pm \sqrt{1 - \frac{4h_1}{d_1\alpha_3}}\,\right) \tag{21.75}$$

Observing that $h_1/\alpha_3 = f_c'$, the focal length of the corrector, we finally have

$$\alpha_2 = \frac{\alpha_3}{2}\left(1 \pm \sqrt{1 - \frac{4f_c'}{d_1}}\,\right) \tag{21.76}$$

It will be noted that the condition (21.76) does not contain the constringence V_2 of the lens glass. This implies that the lens is corrected for axial colour at any wavelength, that is, the corrector is a complete apochromat for any type of glass.

For real-valued roots in (21.76), it must be $d_1 \geqslant 4f_c'$. It is apparent that for a positive correcting lens ($f_c' > 0$) this condition would lead to practically unfeasible construction of the lens.

The condition (21.76) is always satisfied for a negative member ($f_c' < 0$). Such a corrector is called *Maksutov's meniscus* in deference to D. Maksutov of the Soviet Union who devised this corrector in 1941. Taking the root in (21.76) with a minus sign means a meniscus convex toward the mirror, the plus sign means a meniscus concave to the mirror.

The major objective of Maksutov's meniscus is to compensate for the spherical aberration of the spherical mirror or mirror system without aspheric surfaces. The Seidel sum for the system under examination has the form

$$S_{\text{I}} = \sum_{k=1}^{3} h_k P_k = h_1 P_1 + h_2 P_2 + h_3 P_3 \tag{21.77}$$

where

$$P_1 = \frac{\alpha_2^3 \mu_2}{(1 - \mu_2)^2}$$

$$P_2 = \left(\frac{\alpha_3 - \alpha_2}{\mu_2 - 1}\right)^2 (\alpha_3 - \alpha_2 \mu_2)$$

$$P_3 = -\frac{(1 - \alpha_3)^2}{4}(1 + \alpha_3) \qquad (21.78)$$

$$h_1 = f', \quad h_2 = f' - d_1 \alpha_2$$

Basing on Eqs. (21.77) and (21.78), a recommended correction procedure for this system may be as follows. For a given focal length we choose the thickness of the meniscus in proportion to its diameter D as $d_1 = (0.08 \text{ to } 0.12) D$. Next we assign a value for α_3 and determine the respective α_2 by Eq. (21.75), while the surface coefficients P_1, P_2, P_3, and the height h_2 by Eq. (21.78). Then by virtue of (21.77) we find subject to $S_I = 0$

$$h_3 = -\frac{h_1 P_1 + h_2 P_2}{P_3}$$

and after that the distance between the meniscus and the mirror

$$d_2 = \frac{h_2 - h_3}{\alpha_3}$$

The above computations should be performed for a number of combinations of α_3 and d_1.

In order to estimate the field aberrations we calculate the parameters W_1, W_2, and W_3 to determine then S_{II} and S_{III} by Eq. (21.8). Having completed the computational runs for all the initial sets of data, we are in a position to elect a more suitable design form.

The meniscus corrector can be used to advantage in reflecting systems of relative apertures up to $f/2$-$f/3$ because at larger apertures the image quality will markedly deteriorate owing to strong contributions of spherochromatism.

Afocal compensating lens for convergent beams. Higher requirements imposed on the corrector performance call for a more elaborate design. Let us examine a corrector consisting of two thin lenses placed in convergent rays. It is quite obvious that the design should be free from chromatic aberrations. For a thin-lens system this requirement boils down to satisfying the condition

$$\frac{\varphi_1}{V_1} + \frac{\varphi_2}{V_2} = 0 \qquad (21.79)$$

where φ_1 and φ_2 are the powers of corrector elements.

The power of the corrector will be

$$\varphi = \varphi_1 + \varphi_2 \qquad (21.80)$$

To eliminate the secondary spectrum in an indefinitely thin system, the relative partial dispersions of the optical materials must be the same. This condition is satisfied if both elements of the corrector are made of the same optical glass, i.e., when $V_1 = V_2$. Then, by Eqs. (21.79) and (21.80) we get $\varphi_1 = -\varphi_2$ and $\varphi = 0$.

Hence, a thin-lens broken contact corrector is a complete apochromat if its lenses are of the same glass and form an afocal system. This type of corrector can be used, for instance, in a Cassegrain objective. To use a smaller corrector, it should be placed behind the secondary mirror in convergent rays. A respective arrangement is sketched in Fig. 21.17.

First of all, we assume that the entrance pupil coincides with the rim of the primary mirror, therefore the normalizing conditions for the auxiliary rays are as follows: $\alpha_1 = 0$, $h_1 = 1$, $\alpha_3 = \alpha_7 = 1$, $\beta_1 = 1$, $\beta_2 = -1$, $H_1 = 0$, and $I = -1$.

At the layout stage, the designer is to secure (1) the desired throw of the image plane from the vertex of the primary mirror (distance c), (2) the shortest feasible length of the system $(-d_1 + c)$, and (3) an admissible amount of obscuring of the central portion of the entrance pupil ($h_2 \leqslant 0.5h_1$).

Requiring that the corrector should be afocal means the equality of the angles $\alpha_3 = \alpha_7 = 1$. This fact allows an independent investigation of the correcting capabilities of the reflecting system and the corrector. The value of angle α_5 between the lenses depends on the powers of the corrector lenses, therefore the corrector provides two degrees of freedom (α_4 and α_6) to adjust two aberrations, spherical and coma. In deciding on the constructional parameters of the corrector the designer should not overlook the fact

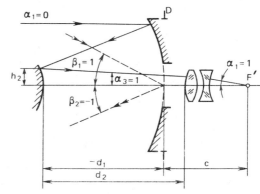

Fig. 21.17. A two-mirror objective with an afocal corrector in converging bundles of rays

that the correcting power of this lens depends on its position with respect to the vertex of the secondary mirror. If the corrector is placed closer to the secondary mirror, its effect on the spherical aberration of the system will be stronger, conversely, if the corrector is placed closer to the second focal plane of the system, its effect on the field aberrations will be increased.

This type of corrector lens was introduced in optical engineering by V.Churilovsky, a Soviet researcher, in 1934. The design performs well in outlined embodiments with relative apertures up to $f/5$.

Afocal corrector for parallel rays. The performance of reflecting systems with relative aperture of $f/1.5$-$f/1$ is substantially improved by an afocal corrector placed in collimated beams, i.e. ahead of the primary mirror. This type of corrector was devised by D. Volosov, D. Galperin, and Sh. Pechatnikova in 1942. The corrector has the same diameter as the primary mirror. A respective arrangement is sketched in Fig. 21.18.

Assuming that the plane of the entrance pupil coincides with the thin-lens corrector, we arrive at the normalizing conditions for the auxiliary rays as follows: $\alpha_1 = \alpha_5 = 0$, $\alpha_7 = 1$, $h_1 = h_2 = h_3 = h_4 = h_5 = 1$, $\beta_1 = \beta_5 = 1$, $H_1 = H_2 = H_3 = H_4 = 0$, and $I = -1$.

As a rule the constructional parameters of a reflecting system are evaluated so as to ensure the required length of the entire system, throw of the image plane, the amount of obscuring for the central portion of the entrance pupil, and such. Once the aberrations of the reflecting system are known, the designer is to develop a corrector which will do the minimizing work in an optimal manner.

Because this type of corrector is afocal ($\alpha_5 = 0$), and the parameter α_3 is defined by the power of the lenses, we have two degrees of freedom, α_2 and α_4. A very compact arrangement can be achieved by using the last surface of the corrector as a secondary mirror. We assume that this option is taken, and since the radius of the secondary mirror is evaluated in the layout of the reflecting system, the parameter α_4 is determined by choosing the radius r_6 of the secondary as

$$\alpha_4 = \frac{n-1}{n}\frac{h_4}{r_6}$$

where n is the refractive index of the corrector glass.

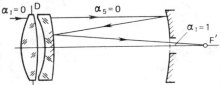

$\alpha_1 = 0$ $\alpha_5 = 0$ $\alpha_1 = 1$ F'

Fig. 21.18. A two-mirror objective with an afocal corrector in parallel bundles of rays

To close this section we note that catadioptric objectives make use of both afocal and focal correctors placed in parallel and convergent beams. An effective means of improving the correcting capabilities of such systems is the use of aspheric surfaces. This and other relevant topics are treated in depth in the book of D. Volosov [2].

21.14 Automatic Correction by Computer

The correction of an optical system is a process of introducing certain modifications into the system parameters so that to arrive finally at a set of parameters which suits the functions selected by the designer in the best possible way. The correction parameters p_i may be the radii of curvature, lens thickness, airspacings, coefficients of aspheric surface equations, parameters of optical materials, etc. The functions to be controlled (denoted by Φ_i) may be ray aberrations, third order aberration coefficients, monochromatic and polychromatic modulation transfer functions, paraxial quantities (focal lengths, back focal lengths) and the like.

In automatic correction, a computer modifies the selected parameters according to a program to find out a solution for the given system configuration and to make the selected functions equal to the desired values.

The existent programs are run on an interactive basis and the designer is to decide which system configuration will be used as an initial trial, what modifications in the configuration are necessary to arrive at the desired solution in shortest time, what functions will be corrected, and what parameters will be modified.

The methods of automatic correction may be divided into specialized techniques and those using common software. Specialized techniques are developed for certain types of optical system and involve equations and procedures usable for manual evaluations, such as, say, third order aberration equations. Many computer programs for the design of doublets are currently existent. These programs require less interactive effort of the designer, because for certain types of systems there exists an exact analytical relation between constructional parameters and aberrations. The resultant solution, however, is accurate to the third order of aberrations, and in most applications the system needs a "touching-up" improvement by trial and error.

Ordinary methods of correction are based on iterative techniques, i.e. they use systematic variation of parameters to effect successive approximations to the solution of a set of nonlinear equations.

Most popular numerical methods of solving nonlinear systems include

Newton's technique, the least squares technique, and gradient techniques [18].

The Newton technique is useful when the initial optical system is close to the desirable. It consists essentially in solving the set of equations

$$\sum_{i=1}^{k} \frac{\partial \Phi_k}{\partial p_i} \Delta p_i^{(1)} = \bar{\Phi}_k - \Phi_k^0$$

to find the modifications of the correction parameters $\Delta p_i^{(1)}$ which modify the functions of the initial system to $\bar{\Phi}_k$ which are to fall within the given intervals from the desired values, Φ_k^0 being the values of these functions in the initial system. The partial derivatives $\partial \Phi_k / \partial p_i$ are determined either by exact expressions, or by the finite difference technique.

The least squares technique is suitable when the number of functions is two or three times the number of correction parameters. An important constraint in this case concerns the requirement that not all functions be iterated to the given values. In working with this technique, the system of inconsistent equations (the number of unknowns is less than the number of equations) is transformed to a system of t simultaneous equations in t unknowns for which a higher (as compared with Newton's technique) accuracy of solution is required.

The convergence of iterations to a solution is controlled by the function

$$L = \sum_{i=1}^{k} (\Delta p_i^{(1)})^2$$

When the iterative process is converging, the ever smaller values of L occur at each next iteration step.

When the number of correction parameters equals the number of functions, the least squares fit becomes Newton's technique. The least squares fit imposes rather modest requirements on the skill of the designer and may be recommended for automatic design at initial stages.

When it is known beforehand that the given values of functions cannot be reached simultaneously, the problem is solved by minimizing the *merit*, or *performance, function*

$$F = \sum_{j=1}^{k} a_j (\Phi_j - \bar{\Phi}_j)^2$$

where Φ_j are the current values of the functions to be corrected (specifically aberrations), $\bar{\Phi}_j$ are the given values of these functions, and a_j are weighting coefficients.

Gradient techniques (steepest descent, say) iterate in the direction of the fastest descent of the merit function, which is exactly opposite to the direction of the gradient dF/dP, where

$$dF = \sum_{i=1}^{t} \frac{\partial F}{\partial p_i} \, dp_i$$

is the differential of the merit function, and

$$dP = \left[\sum_{i=1}^{t} (dp_i)^2 \right]^{1/2}$$

is the differential of the 'distance' in the parameter space measured between an initial point P_i^0 and a point with coordinates $P_i^0 + \Delta p_i$, i.e.,

$$P = \left[\sum_{i=1}^{t} (\Delta p_i)^2 \right]^{1/2}$$

Thus, the gradient technique of automatic correction relates to the techniques minimizing a merit function which is viewed as a total performance of a lens being designed.

A most common approach is to represent the merit function as a weighted sum of system aberrations squared

$$F = \sum_{i=1}^{k} a_i (\Delta y_i')^2$$

where $\Delta y_i'$ stand for the transverse aberrations of the optical system, and a_i are positive weighting coefficients representing the effect of each particular aberration on the system performance.

It is required therefore to find values of correcting parameters such that minimize the merit function. Different methods of automatic design use for this purpose different numerical techniques. The least squares fit has proved to be very efficient in the development of the relevant programs.

In automatic design by common numerical methods (Newton's technique and the least squares) the control of process convergence to a local solution must be effected by some (better automatic) means. In order to prevent divergence of the iterative process, the above techniques have been modified so that the computer can determine the steepest direction and appropriate step of iteration. Two respective modifications are known for Newton's technique and two modifications of the least squares. One of the modified Newton techniques selects a step of iteration such that the difference $\Phi_j^{(s)} - \bar{\Phi}_j$ between the value after step s and the given value of function j is consistently reducing in absolute value. None of the aberrations being adjusted become worse in this modification, but the iteration rate falls off sharply.

The other modification of Newton's technique determines the step of iteration by minimizing the merit function, thus accelerating the convergence process; however, some deterioration for certain aberrations may occur at intermediate stages of the design.

The modifications of the least squares technique confine the variations of correction parameters at each iterative step.

It should be noted that automatic design cannot produce good results unless there exist initial systems close to those desired. Ordinarily the designer has to look up for the close initial trial in books, periodicals, patent and designer own files, therefore the process of automatic design will be facilitated by the development of data banks and retrieval systems which would assist in selecting a suitable initial design form.

21.15 Summation of Aberrations

Suppose that we wish to evaluate the image quality for a sophisticated optical system constituted by optical members designed on an individual basis or assembled of mass-produced elements with known aberrations. The problem will be solved by summing up the aberrations of individual components.

Consider first the summation of longitudinal aberrations with reference to Fig. 21.19 showing arbitrary component k of a sophisticated system. The linear magnification of this component for the conjugate planes is β_k. Point A_k is the object point for this component, or it may be viewed as the aberration-free image A'_{k-1} produced by the preceding portion of the system. Point A'_{0k} is the ideal image of point A_k. Point A'_k is the intersection of the optical axis by a real ray which crosses the axis at A_k at an angle

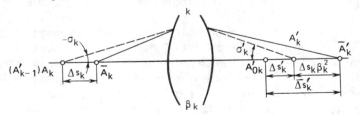

Fig. 21.19. Summing up the longitudinal aberrations

σ_k. As before, $\Delta s_k'$ denotes the longitudinal aberrations of component k for a ray emerging from this component at an angle $\overline{\sigma_k'}$.

If the object point A_k is displaced into $\overline{A}_k$, then the segment Δs_k may be regarded as the object aberration for component k, or as the aberration of the image space for the preceding portion of the system. Since Δs_k is small, we may assume that the displacement of point $\overline{A}_k'$ in image space $A_k' \overline{A}_k'$ is $\Delta s_k \beta_k^2$. Then the overall aberration $\overline{\Delta s_k'}$ after component k will be

$$\overline{\Delta s_k'} = \Delta s_k \beta_k^2 + \Delta s_k'$$

Extending the last relation to a system of p components we get

$$\overline{\Delta s_p'} = \Delta s_1' \prod_2^p \beta^2 + \Delta s_2' \prod_3^p \beta^2 + \dots + \Delta s_{p-1}' \beta_p^2 + \Delta s_p' \qquad (21.81)$$

where $\Pi_2^p \beta^2$ is the product of the squared magnifications of all components from 2-nd to p-th.

Consider now the transverse aberrations with reference to Fig. 21.20. Similar to above, B_k is an object point or the aberration-free image B_{k-1}' produced by the preceding portion of the system. Point B_{0k}' is the perfect image of point B_k. Point B_k' is the intersection with the image plane of a real ray passing through B_k, and $\Delta y_k'$ is the transverse aberration of component k for this ray.

Let us shift B_k into $\overline{B}_k$. This introduces the object aberration Δy_k which is the aberration in the image space of the preceding part of the system. If β_k is the magnification of component k, then to the object aberration Δy_k there corresponds the displacement $\Delta y_k \beta_k$ in the image plane. Consequently, the total transverse aberration is

$$\overline{\Delta y_k'} = \Delta y_k \beta_k + \Delta y_k'$$

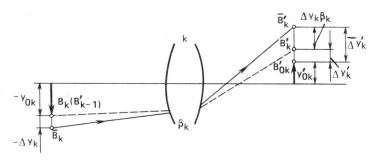

Fig. 21.20. Summing up the transverse aberrations

27*

For a system of p components we get

$$\overline{\Delta y_p'} = \Delta y_1' \prod_2^p \beta + \Delta y_2' \prod_3^p \beta + \ldots + \Delta y_{p-1}' \beta_p + \Delta y_p'$$ (21.82)

According to Eqs. (21.81) and (21.82) the summation of aberrations must be effected by tracing one ray through the entire system. It should be kept in mind that the design of individual components of the system is carried out so that the images of the points for which the aberrations are evaluated occur above the axis, i.e. $y' > 0$. In a particular system arrangement this condition can be violated resulting in $y' < 0$, which implies sign reversal for the transverse aberration, too. Therefore care should be exercised to observe the sign of the magnification in Eq. (21.82).

Individual components constituting optical systems may be designed at any methodological convenience. This allows for the reverse direction of rays to that used in the intended application. For example, all eyepieces are traced from the side of the eye, i.e. from the side of an infinitely distant object. Projection lenses of large magnification are traced from the screen. In the circumstances, the residual aberrations must be evaluated in reverse rays. In agreement with the convention adopted in this text, we denote the reverse ray aberrations by an overhead arrow pointing in this direction. If β is the magnification for a given pair of conjugate planes, and $\Delta y'$ and $\Delta s'$ are the transverse and longitudinal aberrations for the normal ray direction, then the aberrations for reverse rays will be $\Delta \bar{y}' = \Delta y'/\beta$ and $\Delta \bar{s}' = \Delta s'/\beta^2$.

Some sophisticated system arrangements can present to the designer situations with the linear magnification of some component being zero. This situation is typical, for example, of erecting systems with parallel rays between the components of the erector. The optics of such a system is schematized in Fig. 21.21. Here, $\Delta s_1'$ is the longitudinal aberration of the first component evaluated for a collimated beam, and $\Delta s_2'$ is the longitudinal aberration of the second component. If the first component is aberration-free, then a real ray from A_1 comes to A_2'. In the presence of aberration due to the first component, this ray meets the axis at a point $\overline{A_2'}$ displaced from A_2' to a distance defined through the magnification for the conjugate planes of A_1 and A_{02}'. This magnification is the ratio of the component focal lengths, $\beta_{1,2} = -f_2'/f_1'$. Consequently, the total longitudinal aberration of the two-component system will be

$$\Delta \bar{s}_2' = (f_2'/f_1')^2 \Delta s_1' + \Delta s_2'$$

Accordingly, for the total transverse aberration $\Delta \bar{y}_2'$ we have

$$\Delta \bar{y}_2' = -(f_2'/f_1')\Delta y_1' + \Delta y_2'$$

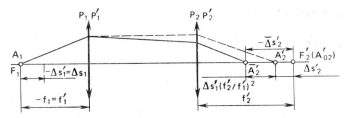

Fig. 21.21. Summing up the aberrations in a two-element system with a parallel course of rays

It is apparent that in a symmetric erector system ($f_1' = f_2'$) coma and distortion automatically cancel out.

For an optical system producing an image at infinity, there is no possibility of expressing its aberrations in linear units. A more appropriate measure for such systems is angular units characterizing the deviation of rays from a parallel bundle, or dioptres characterizing the convergence or divergence of a bundle. This type of characterization relates to the aberrations of telescopes, microscopes, afocal systems reducing divergence of laser beams, and for some other systems.

For telescopes, the aberrations are summed in the first focal plane of the eyepiece. If $\Delta s_1'$ and $\Delta y_1'$ are respectively the longitudinal and transverse aberrations of the system minus the eyepiece, and $\Delta \bar{s}_e'$ and $\Delta \bar{y}_e'$ are the longitudinal and transverse aberrations of the eyepiece derived for the reverse direction of rays (from the eye), then the overall aberrations in the first focal plane of the eyepiece will be as follows

$$\Delta \bar{s}' = \Delta s_1' + \Delta \bar{s}_e'$$
$$\Delta \bar{y}' = \Delta y_1' - \Delta \bar{y}_e'$$

Then the angular aberration (in radians) is given as

$$\Delta \sigma' = \Delta y'/f_e'$$
$$= (\Delta y_1' - \Delta \bar{y}_e')/f_e'$$

The longitudinal aberrations independent of aperture, such as astigmatism, field curvature, axial chromatic aberration, are conventionally measured in dioptres as follows

$$L = -\frac{\Delta s_1' + \Delta \bar{s}_e}{f_e'^2} 1000$$

where f_e' is in millimetres.

To close this section we note that the summing expressions derived in this section are only approximate ones and are commonly used in evaluating interim designs of a system under development. After the final correction of aberration for individual components of a system and its final arrangement, the residual aberrations of the system should be calculated by Eqs. (8.2).

21.16 Optical Specifications and Tolerances

In addition to keeping close track of the factors ensuring feasibility of designs, optical engineers must take extreme care to place their designs within the realm of the available manufacturing techniques by means of accurate specifications and tolerances. A particular design form or system must be technologically feasible and lend itself to the subsequent assembly, alignment and adjustment. A complete account of all these fabrication and mounting requirements will lead to a trade-off of the desirable system performance against the unavoidable production variations and misalignment in the mounting. This trade-off is issued in the optical shop in the form of reasonably large tolerances (for convenience and cost efficiency of manufacturing) on the geometrical features of the design.

How large tolerances on the constructional parameters and locations in the system (including centring) may be allowed is dictated by the values of permissible residual aberrations. The residuals in turn depend on the intended application of the system (see Section 21.2). The smaller the allowable residuals, the more narrow the respective manufacturing tolerances. In optical systems of almost perfect imagery requirements, for example, the allowable deviation of the wavefront from the perfect form must not exceed a quarter wavelength (Rayleigh's criterion).

If we denote the permissible wave aberration (to be discussed in the last section of the text) by l_N, then for visual optical instruments at the principal wavelength $\lambda_e = 546.07$ nm, it amounts to $l_N = 135.5$ nm. About two thirds of this allowance (≈ 100 nm) is assigned to manufacturing and mounting tolerances.

The total allowance established for an optical system is alotted to individual components and their features such as surfaces, thicknesses, angles, etc., in proportion to the effect of variations in this element on the system performance and the costs associated with more accurate fabrication and mounting.

In tolerancing the engineer should observe the position of an optical surface or element about the optical axis and image plane. For example, in tolerancing the optical surfaces and elements perpendicular to the axis of a ray bundle one may use the expression

$$\Delta_t = g l_N$$

where g is a coefficient relating the tolerable surface roughness Δ_t to the allowable wave aberration l_N at this surface.

For a refracting surface separating two media of refractive indices n_1 and $n_2, g = -1/(n_1 - n_2)$.

For a refracting surface in air $(n_1 = 1$ and $n_2 = n), g = -1/(1 - n)$.

For reflecting surfaces $(n_1 = -n_2 = n), g_r = -1/2n$.

For a reflecting surface in air $(n = 1), g_m = -1/2$.

By way of example, at $n_1 = 1.5$, we have $g = 2, g_r = -1/3$, $g_m = -1/2$, and for a surface separating two media of $n_1 = 1.5$ and $n_2 = 1.6, g \approx 10$.

Thus, if we take conventionally the requirements for the surface quality of a refracting surface in air (g) to be unity, then the manufacturing accuracy requirements for a first surface mirror (g_m) are four times as strong, and the requirements for the second surface mirror (g_r) are six times as strong as the first ones, whereas the requirements for a surface separating media of $n_1 = 1.5$ and $n_2 = 1.6$ are only 20% of the former.

The tolerances assigned depend on the place in the working bundle where the element or member is located. For instance, the larger the area of beam section, the smaller the cylindricity tolerance and the tolerances for local defects of the surfaces and centring, and the stronger the constraints imposed on the optical material constants, optical homogeneity, and birefringence. For smaller sections of the working beams, more severe tolerances are assigned with respect to such defects as air bubbles, stones, scratches, flaking, and manufacturing surface roughness.

The Soviet standard GOST 3514-76 established the tolerancing parameters in terms of categories and classes as follows.

For the limiting deviation Δn_e of refractive index, it stipulates five categories and four classes depending on the largest difference in a lot of optical blanks.

For the limiting deviation of mean dispersion $\Delta(n_{F'} - n_{C'})$, it stipulates five categories and two classes depending on the largest difference of mean dispersions in a lot of optical glass blanks.

For optical homogeneity of optical blanks under 250 mm in size (diameter or side) it establishes five categories which are characterized by the resolving power at $\lambda = 0.55 \mu$m. The resolving power of a glass is defined by the ratio of the angular resolution of the diffractometer, which has the blank in its collimated beam, to the angular resolution of the instrument without the blank.

For birefringence, it establishes five categories characterized by the optical path difference in the direction of the largest size of the blank.

For the attenuation factor measured as the inverse distance in cm at which the flux from the A type source is attenuated owing to absorption and scattering to 0.1 its initial intensity, the standard establishes eight categories.

For striation content, the standard establishes two categories limiting the frequency and size of admissible striations, and two classes defining the number of orthogonal directions in which the blank should conform to the given category.

For bubble frequency, depending on the diameter of the largest bubble admissible in the blank the standard establishes eleven categories and seven classes which are characterized by the average number of bubbles of 0.03 mm diameter per one kg of raw glass.

In drawings of optical elements, the indicated technological requirements for manufacture are as follows.

(1) The deviation N from the given curvature which indicates the maximum allowable deviation of the sag z (Fig. 21.22) of the surface of member 1 from the sag z_{test} of the surface of the test glass 2. This parameter is measured in interference fringes. The same test and indication are used to tolerance sphericity of a plane surface. One interference fringe equals $\lambda/2$. The sag difference $z - z_{\text{test}} = \Delta z = N\lambda/2 \approx -D^2\Delta r/8r^2$.

For lenses of shallow curvature and small diameter, the number of interference fringes observed is directly proportional to the deviation of the curvature ($\Delta c = -\Delta r/r^2$) and the squared diameter of the circle containing the interference fringes, viz.,

$$N = \Delta c D^2/4\lambda$$
$$= -D^2\Delta r/4\lambda r^2$$

Assuming $\lambda = 0.555\ \mu$m and omitting the sign we get

$$N = 450 D^2 \Delta r/r^2$$

In estimating the allowance N to fit the test glass subject to the admissible variation in the focal length $\Delta f'$ at surface k, one may use the expression

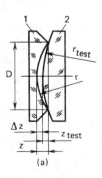 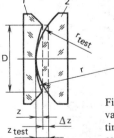

Fig. 21.22. Testing the curvature by a test glass contacting at the edge (a) and at the centre (b)

$$N_k = \frac{450D_k^2\Delta f' h_1}{\sqrt{p}f'^2(n_k' - n_k)h_k}$$

where $\Delta f' \leqslant$ (1/3 to 1/5) of the entire allowance for the focal length variation, p the number of surfaces, h_1 and h_k the ray heights on surfaces 1 and k, respectively, and n_k and n_k' are the refractive indices of media separated by surface k.

The variation of curvature of various surfaces that causes the same variation in the focal length, inflicts different variations of the aberrations, therefore a more accurate approach to the tolerancing of various surfaces would be on the basis of the allowable variations in the aberrations.

(2) The deviation ΔN from a correct spherical or plane shape which causes a cylindricity which results in astigmatism even for axial object points. If we denote the allowable astigmatism $\Delta z_a'$, the limiting deviation ΔN for surface k of an optical system having focal length f' may be determined by the formula

$$\Delta N_k \leqslant \frac{D_k^2\Delta z_a' h_1}{4\sqrt{p}f'^2\lambda(n_k' - n_k)h_k} \tag{21.83}$$

At $\lambda = 0.555$ μm this formula takes the form
(i) for photographic objectives

$$\Delta N_k \leqslant \frac{450D_k^2\Delta z_a' h_1}{\sqrt{p}f'^2(n_k' - n_k)h_k}$$

(ii) for telescopes

$$\Delta N_k \leqslant \frac{0.450D_k^2\Delta z_{a,D}' h_1}{\sqrt{p}(n_k' - n_k)\Gamma_t^2 h_k}$$

where $\Delta z_{a,D}' = 1000\Delta z_a'/f_e'^2$ in dioptres, f_e is the focal length of the eyepiece, and Γ_t is the magnifying power of the telescope.

If the allowance ΔN_1 for the first surface of the system is known and the deviation of any surface from the sphere causes the same amount of astigmatism, then for surface k

$$\Delta N_k = \Delta N_1 \left(\frac{D_k}{D_1}\right)^2 \frac{(n_1' - n_1)h_1}{(n_k' - n_k)h_k}$$

To ensure reliable control of asphericity ΔN which is measured as the difference of interference fringes in the principal sections of the surface, the total deviation N should not exceed the evaluated quantity ΔN three- to fivefold, to be tenfold the figure in the limiting case.

(3) The decentration, c, defined by the displacement c_1 of the centre O_2

of the lens second surface (Fig. 21.23), the displacement c_2 of the optical axis 2 in the second principal plane of the lens relative to the axis of rotation 1, the greatest difference in thickness over the lens, c_3, and the angle of rotation γ of the second surface about the vertex with respect to the first surface. All the listed quantities are related by the following expressions

$$c_2 = c_1(n - 1)f'/r_2$$
$$= \sigma(n - 1)f'$$
$$c_3 \approx c_1 D/r_2 = \sigma D$$
$$\sigma = c_1/r_2$$

The admissible inclinations of the surfaces for the most part lie within 5″ to 30″.

The tolerances for centring of optical elements should be assigned subject to the method of mounting.

(4) The surface roughness of optical surfaces.

(5) The smallest tolerable focal length f'_{min} of plates and prisms due to the sphericity of their surfaces (in m or mm).

(6) The limiting nonparallelism of plates, measured as the principal angle of the prism, θ, causing a deviation through an angle γ, and the angular chromatic aberration $\Delta\gamma_{\lambda_1, \lambda_2} = \theta(n_{\lambda_0} - 1)/V$.

For practical evaluations, another formula seems to be more convenient:

$$\theta = \Delta\gamma_{\lambda_1, \lambda_2} V D'_{eff}/(n_{\lambda_0} - 1)D_k$$

where D'_{eff} is the effective diameter of the exit pupil, D_k the diameter of the working bundle at the location of the element, and V the Abbe V-value.

The tolerance for wedge-shape prism unfolding is expressed in terms of two components θ_c and θ_π due to the wedge-shape property in the principal section of the prism and due to the pyramid-shape property of prism unfolding (in the section orthogonal to the principal plane), $\theta = \sqrt{\theta_c^2 + \theta_\pi^2}$.

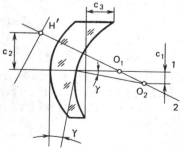

Fig. 21.23. A decentred lens

(7) The limiting nonparallelism of the prism edges expressed in an angular measure [21].

(8) The limiting difference of equal angles of the prism. In the 90° roof prism design, for example, the tolerance may be $\delta_{45} = 1'$ to control the nonparallelism and chromatic aberrations.

(9) The class of accuracy of the test glass or the limiting deviations of surface radius in per cent. In order that the error of the test glass should not exceed the manufacturing tolerances on the surfaces to be fabricated, the design must obey the following relation

$$\frac{\Delta r}{r} = \left(\frac{1}{3} \text{ to } \frac{1}{5}\right) \frac{400 r N_l \lambda}{D_l^2}$$

where $\Delta r / r$ is in per cent, D_l is the lens diameter, and N_l is the deviation from the given curvature.

21.17 Image Evaluation

Any design of an optical system terminates by producing a list of residual aberrations. Along with other data such as focal length or magnification, relative aperture, numerical aperture or entrance pupil diameter, and angular coverage, this list enters the system specifications. This document also gives the wavelength at which the monochromatic aberrations were controlled, and the wavelength interval in which the system was achromatized.

Specifications contain the structural parameters of the system (r, d, n, V), clear diameters of optical surfaces, and sags from clear diameter planes. In addition, designers indicate the position of the aperture stop, distance from the first surface to the entrance pupil, and from the last surface to the exit pupil. The diameters of the pupils and aperture stop are also indicated.

For objects at finite working distances (close conjugate work as with microscope objectives, process lenses, etc.), the distances from the first surface to the object, from the last surface to the image, and between the object and image planes are also indicated.

The numerical values of residual aberrations are summarized in the respective tables or represented in graphical form. First of all, the table of aberration is compiled for an axial point (spherical aberration, isoplanatism, longitudinal chromatic aberration, and spherochromatism). Then the table of aberrations is arrayed for the principal rays and thin astigmatic bundles for various points of the field: meridional and sagittal field curvature, astigmatism, distortion, and transverse chromatic aberration. These tables are followed by the arrays of ray aberrations for oblique meridional and sagittal ray fans. These aberrations may be quoted for the principal wavelength and for wavelengths at which the system must be achromatized. In some cases, for example, at large relative apertures and wide fields, the tables of aberrations for skew rays may also be added. The specifications of high performance optical systems, say, microscope objectives, give the tables of wave aberrations.

The data of these tables are used to plot the curves of system aberrations. The graphs of transverse aberrations are usually represented to the same scale.

For objectives of modest relative apertures, the spherical aberration, spherochromatism, and isoplanatism are evaluated for two rays, marginal ($m_{cr} = D/2$) and zonal ($m_z = m_{cr}/\sqrt{2}$). At high relative apertures the number of rays in the axial bundle is increased to three or four ($m_z = m_{cr}/2$, $m_z = \sqrt{3}\,m_{cr}/2$). The secondary spectrum is normally plotted for the rays incident on the entrance pupil at a height of $m_z = m_{cr}/\sqrt{2}$.

The aberrations of the principal rays and oblique bundles are evaluated for two or three inclinations for objectives with an angular coverage up to 60°, and for four or five inclinations for wide-angle objectives. The aberrations of oblique bundles are determined by tracing four or five rays in the meridional plane and two or three rays in the sagittal plane. At high relative apertures the number of rays in the bundle is doubled.

Photographic objectives are normally vignetted for a portion of the oblique bundle. The vignetting diaphragms are lens mounts or special stops. Aberration curves normally indicate what surface or stop is vignetting.

The correction of aberration in optical systems is for the most part effected relative to the Gaussian plane. However, residual spherical aberration and astigmatism cause a displacement of the plane with the favorable distribution of rays over the blur spot from the Gaussian plane. To locate this plane, the graphs of transverse aberrations plotted in the coordinates $\Delta y'$ against $\Delta \tan \sigma'$ have a line *aa* passing through the origin; the deviation of this line from the curve is minimal (see *Appendix 3*).

A preliminary evaluation of image quality for a system is based on the blur spot size formed by the family of rays for various points of the field. If δ' is the effective blur spot diameter, determined with account of the entire spectral interval detected by the system, then the resolving power of the system will be defined by a quantity inverse to the blur spot size.

In conclusion we note that other methods of image evaluation are based on the determination of wave aberrations, and modulation transfer function of the system. The criteria derived in this way provide an image quality evaluation prior to manufacturing a trial lens.

21.18 Wave Aberration

Wave aberrations define the deterioration of homocentricity of the ray bundles emergent from the optical system. This factor causes a redistribution of illuminance on the image of a self-luminous point and, hence, affects the quality of the image. To produce an aberration-free image of a point, the emergent wavefront must be spherical. Real optical systems deform the wavefront. The wave aberration is a measure of this deformation.

For an axial object point, the perfect optical system produces a wavefront concentric with the image of the point, called the reference point. The distance from this point to the vertex of the last surface gives the radius of the reference sphere. Deviations of a wavefront, emanating from the reference point, from the reference sphere, measured along the radii, are known as *wave aberrations*.

In a real optical system, the reference point should be placed at the axial point of the film plane spaced a distance ξ from the Gaussian plane.

Figure 21.24 shows a meridional section through a wavefront I and the respective reference sphere, II. The reference sphere of radius R is centred on $\overline{A}'$ in the film plane $\overline{Q}'$ shifted a distance ξ from the Gaussian plane. The wave aberration (also called optical path difference) for radius I is $l_1 = M_1 N_1$, and along radius 2 is

$$l_2 = M_2 N_2 = KN_2 + M_2 K = M_1 N_1 + M_2 K$$
$$= l_1 + dl$$

The longitudinal spherical aberration $\Delta s' = A_0' A'$ corresponding to the wave aberration l_1 is small in a corrected optical system, hence $\overline{\sigma}' \approx \sigma'$.

Let us derive the relation between the longitudinal spherical aberration and wave aberration. With reference to Fig. 21.24, $\overline{A}'T = (\Delta s' - \xi)\sin\sigma'$. Since the longitudinal spherical aberration $\Delta s'$ is small, the angle

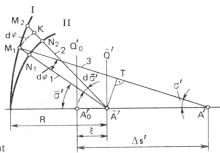

Fig. 21.24. Meridional section of a wavefront

$d\varphi$ between rays 1 and 3 is also small, therefore

$$M_1 \bar{A}' = R + l = \bar{A}' T / d\varphi$$

$$= \frac{\Delta s' - \xi}{d\varphi} \sin \sigma'$$

To the increment $dl = M_2 K$ there corresponds a small angle $d\bar{\sigma}'$ between the adjacent rays 1 and 2, therefore

$$M_1 K = M_1 \bar{A}' d\sigma'$$

$$= \frac{\Delta s' - \xi}{d\varphi} \sin \sigma' \, d\sigma'$$

($d\sigma' \approx d\bar{\sigma}'$ as $\sigma' \approx \bar{\sigma}'$).

At small $d\sigma'$ and $d\varphi$ the increment of wave aberration is

$$M_2 K = dl = M_1 K \, d\varphi$$

$$= (\Delta s' - \xi) \sin \sigma' \, d\sigma'$$

The desired relation between the wave and longitudinal spherical aberration will be as follows

$$l = \int_0^{\sigma'} (\Delta s' - \xi) \sin \sigma' \, d\sigma'$$

or

$$l = -\xi(1 - \cos \sigma') + \int_0^{\sigma'} \Delta s' \sin \sigma' \, d\sigma' \tag{21.84}$$

If the film plane coincides with the plane of ideal image, then $\xi = 0$ and

$$l = \int_0^{\sigma'} \Delta s' \sin \sigma' \, d\sigma'$$

Wave aberrations can be evaluated by graphical integration. By substituting the new variable $p = 1 - \cos \sigma'$ in (21.84) we obtain

$$l = -p\xi + \int_0^p \Delta s' \, dp = -l_0 + \Delta l \tag{21.85}$$

If a graph of on-axis spherical aberration is available for some system, $\Delta s' = f(\sigma')$, we can use it to plot the curve $\Delta s' = \varphi(p)$ shown in Fig. 21.25a. The shadowed area in this graph is proportional to

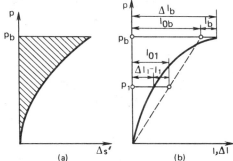

Fig. 21.25. Construction to determine the amount of wavefront aberration and to locate the film plane

$$\Delta l = \int_0^{p_b} \Delta s' \, dp$$

where p_b corresponds to the exit aperture angle. The variation of this area with p is indicated in Fig. 21.25b.

The first term in (21.85) is the equation of a straight line, $l_0 = p\xi$, shown by broken line at (b). This line is traced so that for some values p_1 and p_b the wave aberrations l_1 and l_b are equal in absolute value, which corresponds to the least optical path differences within the entire aperture angle.

It is apparent from the figure that the largest aberrations will be

$$l_1 = -l_{01} + \Delta l_1$$
$$l_b = -l_{0,b} + \Delta l_b$$

The position of the film plane, defined by ξ, can be obtained, for example, from the equation

$$\xi = -l_{01}/p_1$$

If a wave aberration $l/\lambda \leqslant 0.25$ wavelength, the image of the point is considered as satisfying Rayleigh's criterion.

Reflecting Prisms

Name and sketch	Geometry	Function and properties
Dove prism	$a = 2\xi D/(\xi - 1)$ $\xi = \sqrt{2n^2 - 1}$ n = index of glass $b = D, h = D$ Optical path length in the prism $d = 2nD/(\xi - 1)$	Operates in a collimated beam. Produces a mirror image which at prism rotation along the horizontal axis rotates at twice the speed of the prism.
Right-angle prism	$a = D$ $b = D$ $d = D$	Substitutes a plane mirror. Placed ahead of an objective to redirect the line of sight by rotating the prism around the axis.
Right-angle roof prism	$a = D$ $b = 0.366D$ $c = 1.414D$ $h = 1.732D$ $d = 1.732D$	The simplest prism producing a full inversion and rotating the beam through 90°.
Constant deviation 90° prism	$a = 2D$ $b = D$ $d = 2D$	Regardless of the angle at which a ray enters the prism, the emergent ray will be parallel and displaced a distance D.

Name and sketch	Geometry	Function and properties
Rhomboid prism	$a = 2D$ $b = D$ $h = D$ $d = 2D$	Displaces the optical axis without deviation or reorientation of the image.
Semi-penta prism	$a = 1.082D$ $b = D$ $c = 1.707D$ $d = 1.707D$	Deviates a ray by 45°. Produces an erect image. Used for convenience of sighting.
60°—90°—30° prism	$a = D$ $b = D$ $c = 2D$ $d = 1.732D$	Deviates a ray by 60°. Produces an erect image. Used for convenience of sighting.
Penta prism	$a = D$ $b = D$ $c = 1.082D$ $d = 3.414D$	Deviates a ray by 90°. A constant deviation prism: it deviates the line of sight through the same angle regardless of its orientation to the line of sight. Produces an erect image.

Name and sketch	Geometry	Function and properties
Rangefinder prism	$a = D$ $b = 1.414D$ $c = 1.155D$ $d = 2.823D$	Deviates the line of sight in two orthogonal planes: through 60° in one and through 90° in the other.
90° roof prism	$a = 2.225D$ $b = D$ $c = 1.414D$ $c_1 = 0.366D$ $d = 2.957D$	Produces a complete inversion top to bottom and right to left and deviates a ray by 180°.
Roof penta prism	$a = 1.237D$ $b = D$ $c = 1.082D$ $d = 4.223D$	Deviates the line of sight by 90°. The angle of deviation is independent of the angle of incidence. Produces a mirror image.
Triangle roof prism	$a = D$ $b = D$ $c = 2.618D$ $h = 1.618D$ $d = 2.802D$	Deviates the axial ray through 60°. Produces a mirror image.

Name and sketch	Geometry	Function and properties
Leman prism	$a = b = D$ $c = 2.5D$ $c_1 = 2D$ $d = 4.33D$	The axis is displaced by a value of c but not deviated.
Schmidt prism	$a = 1.082D$ $b = D$ $c = 1.414D$ $d = 2.414D$	Deviates the axial ray by 45°. Used in optical instruments with offset optical axes.
Leman prism	$a = b = D$ $c = 2D$ $c_1 = 2.618D$ $d = 4.535D$	The axis is displaced but not deviated. The image is inverted and reversed left to right. Has considerable optical path length and is used to shorten the size of instruments.
Schmidt prism	$a = 1.781D$ $b = D$ $c = 1.363D$ $d = 3.04D$	Deviates the axial ray by 45°. The image is inverted and reversed left to right. Has appreciable optical path length in the prism.

28*

Name and sketch	Geometry	Function and properties

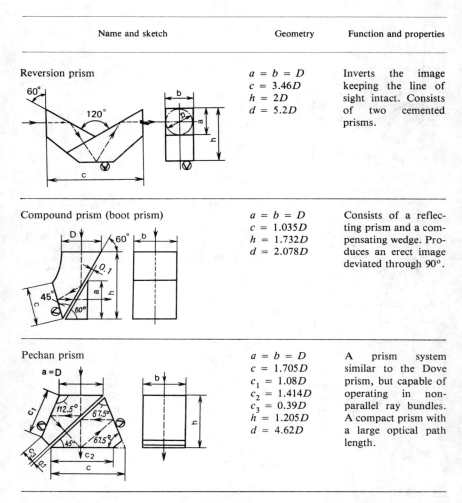

Reversion prism

$a = b = D$
$c = 3.46D$
$h = 2D$
$d = 5.2D$

Inverts the image keeping the line of sight intact. Consists of two cemented prisms.

Compound prism (boot prism)

$a = b = D$
$c = 1.035D$
$h = 1.732D$
$d = 2.078D$

Consists of a reflecting prism and a compensating wedge. Produces an erect image deviated through 90°.

Pechan prism

$a = b = D$
$c = 1.705D$
$c_1 = 1.08D$
$c_2 = 1.414D$
$c_3 = 0.39D$
$h = 1.205D$
$d = 4.62D$

A prism system similar to the Dove prism, but capable of operating in non-parallel ray bundles. A compact prism with a large optical path length.

Note. The encircled angle symbol indicates that the surface is silvered.

Appendix 2

Soviet-made Optical Materials and Products

Table A1. Selected Soviet-made Glass Types

Type	Refractive index		Mean dispersion	
	n_e	n_D	$n_{F'} - n_{C'}$	$n_F - n_C$
LK6	1.4721	1.4704	0.00708	0.00704
LK7	1.4846	1.4828	0.00732	0.00728
K8, K108	1.5183	1.5163	1.00812	0.00806
K100	1.5237	1.5215	0.00882	0.00875
BK6, BK106	1.5421	1.5399	0.00913	0.00905
BK8, BK108	1.5489	1.5467	0.00877	0.00871
BK10, BK110	1.5713	1.5688	0.01024	0.01015
TK2, TK102	1.5749	1.5724	0.01005	0.00996
TK14, TK114	1.6155	1.6130	0.01020	0.01012
TK16, TK116	1.6152	1.6126	0.01059	0.01050
TK20, TK120	1.6247	1.6220	0.01107	0.01097
TK21, TK121	1.6600	1.6568	0.01299	0.01285
TK23	1.5915	1.5891	0.00970	0.00962
KF4	1.5203	1.5181	0.00886	0.00879
BF12, BF112	1.6298	1.6259	0.01622	0.01601
BF16	1.6744	1.6709	0.01435	0.01419
BF24	1.6386	1.6344	0.01750	0.01726
LF5, LF105	1.5783	1.5749	0.01409	0.01392
F1	1.6169	1.6128	0.01681	0.01659
F101	1.6179	1.6138	0.01681	0.01659
F104	1.6290	1.6247	0.01762	0.01738
F6	1.6170	1.6031	0.01611	0.01590
TF1, TF101	1.6522	1.6475	0.01940	0.01912
TF3	1.7232	1.7172	0.02469	0.02431
TF5, TF105	1.7617	1.7550	0.02788	0.02743

Table A2. Physical Characteristics of Optical Ceramic Glasses

	CO 115 M	CO 156	CO 21
Refractive index	$1.535 \pm 5 \times 10^{-3}$	$1.545 \pm 2 \times 10^{-3}$	$1.553 \pm 2 \times 10^{-3}$
Mean dispersion	$0.0102 \pm 3 \times 10^{-4}$	$0.0104 \pm 2 \times 10^{-4}$	$0.0108 \pm 2 \times 10^{-4}$
Temperature range of linear expansion coefficient equal to $0 \pm 1.5 \times 10^{-7}$ per °C	25 ± 20	25 ± 20	220 ± 50
Heat resistance, °C	500 ± 50	350 ± 50	650 ± 50

Table A3. Resolving Power of Soviet-made Films N_{ph}, mm^{-1}

	Roll film			Cine film					
foto 32	foto 65	foto 130	foto 250	TsO-2	TsO-3	KN-1	KN-2	KN-3	TsO-5
116	92	75	70	45	45	135	100	78	70

Note. The 'foto' and KN types are black and white negative films, TsO stands for 'colour reversible'.

Table A4. Soviet-made Photographic Objectives

Group and name of lens	f', mm	F	2	N_{lf} mm^{-1} CF	EF	NL	NG
Universal							
Triplet	40 -135	2.8 -6.8	50-60	40-24	28-10		
Triplet T-43	40	4	56	37	17	3	2
Hindustar	24 -300	2.8 -4.5	40-58	50-25	30-12		
Hindustar-61	52		45	46	30	4	3
Vega	35 -100	2.8	39-50	50-30	30-15		
Vega-1	52		44.5	47	22	5	3
High-speed							
Jupiter	12 -180	1.5 -2.8	14-45	45-30	24-12		
Jupiter-8	52		45	39	24	6	5
Jupiter-9	85	2	29	32	23	7	6
Helios	28-85	1.5 -2.8	28-56	50-32	35-16	—	—
Helios-44	58	2	40	46	22	6	3
Helios-40	85	1.5	28	48	34		5

Group and name of lens	f', mm	F	2	N_{lf} mm^{-1} CF	EF	NL	NG
Wide-angle							
Jupiter-12	35	2.8	63	60	23	6	5
Mir	20-66	2.8-3.5	62-94	55-40	35-14	—	—
Mir-1	37	2.8	62	55	35	6	5
Mir-10	28	3.5	75	40		8	4
MR-2	20	5.6	95	35	20	6	5
Orion-15	28	6	75	45	18	4	2
Telephoto							
Tair	135	2.8	8-18	45-24	30-18		
	-300	-4.5					
Tair-11	135	2.8	18	44	24	4	3
Tair-3	300	4.5	8	45	38	3	
Telemar-22	200	5.6	12	52	31	4	2
MTO-500	500	8	5	38	18	3 + 2M	3
TOZ-500M		3.5		40	25	2 + 2M	2
Zoom lens							
Rubin-1	37-80	2.8	60-30	35-30	15-12	14	11

F, f-number; N_{lf}, 'photographic' resolution at the centre of the field (CF) and at the edge of the field (EF); NL, number of lenses; NG, number of glasses; M, mirrors.

Table A5. Soviet-made Camera Tubes

	Image orthicons			Vidicons		Dissector
	LI-216	LI-221	LI-407	LI-418	LI-420	LI-601
Photocathode $h \times w$, mm	24 × 32	24 × 32	4.5 × 6	15 × 20	9.5 × 12.7	24 × 24
Number of lines per frame	625	625	350	625	625	625
Number of active lines, z_a	577	577	320	577	577	
Specific resolution, mm^{-1}	24	24	71	39	61	
Line height, mm	0.041	0.041	0.014	0,026	0,0165	
λ_m, nm	570	580	500	575	580	500
$\lambda_1 - \lambda_2$, nm	375	380	350	400	420	440
	-750	-750	-700	-750	-780	-580
SNR	40	40	40	39	40	50
E_{min}, lx	0.01	0.01	5	0.1		
Working illuminance of photocathode, lx	2	1.8		5	1.5-10	100

Table A6. Soviet-made Picture Tubes

Type	Screen			Image contrast	Persistance	No. of lines	Application
	$h \times w$, mm	colour	cd/m^2				
59LK3Ts	375 × 475		110				Projection TV sets
6LK1B	34 × 38		4000	35			
10LK2B	54 × 72		3000			625	
13LK1B	70 × 93	white	32	30	moderate		TV camera viewfinders
13LK2B	85 × 85		35	25			Displays
18LK8Zh	90 × 124	cyan	300	35	very short		Flying spot for B/W and colour TV
18LK12B	100 × 100	white	300	35		1000	Image photography
23LK5B	160 × 160		32	45	moderate	625	Displays
23LK6I	124 × 170	green	700	40		1000	Image recording on cine film

Soviet-made picture tubes bear type designations which read as follows :
type 47LK2B is a 47-cm-diagonal (or diameter) cathode-ray picture tube, Model 2, white phosphor ;
type 59LK3Ts is a 59-cm-diagonal cathode-ray picture tube, Model 3, colour.

Table A7. Soviet-made TV Camera Lenses

Type	f, mm	F	2ω	τ	T_N at $N' = 13$ mm^{-1}	
					centre	edge
Mir-10-T	27	$f/3.5$	71°	0.75		0.55
Mir-1-T	37	$f/2.8$	58°		0.85	0.35
Helios-95A-T	50	$f/2$	44°			0.65
Era-4-T	85	$f/1.5$	27°	0.8	0.80	0.55
Era-2-T	100	$f/2$	23°		0.85	0.53
Tair-51-T	135	$f/3$	17°	0.75		0.64
Tair-45-T	180	$f/2.8$	13°			0.70
Tair-48-T	210	$f/3.5$	10°	0.8	0.80	
Tair-44-T	300	$f/4.5$	7.5°	0.75		0.60
Tair-47-T	400	$f/4.5$	6°	0.70		0.65

Type	f, mm	F	2ω	τ	T_N at $N' = 13$ mm^{-1}	
					centre	edge
Tair-46-T	500	$f/5.6$	5°	0.75		
Tair-50-T	750	$f/6.3$	3°		0.70	0.60
Tair-52-T	1000	$f/8$	2.3°	0.65	0.65	

Note. The spatial frequency $N' = 13$ mm^{-1} is adopted on the following grounds. The lens does not impart the definition of a TV image if it ensures the modulation transfer factor, T_N equal to at least 0.75 both at the centre and at the edge of the tube target at the following frequencies

$$N' = z_n(1 - q)/2h_{ph}$$

where q is the relative vertical flyback time (≈ 0.07). For $z_n = 625$ and $h_{ph} = 24$ mm, $N' = 12$ mm^{-1}.

Performance Specification of a Photographic Objective

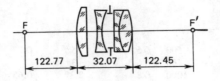

The $f/4$ Hindustar-55 lens having $f' = 140$ and $2\omega = 56$ °; corrected for $\lambda_D = 589.3$ nm and achromatized for $\lambda_{G'} = 434.05$ nm and $\lambda_C = 656.28$ nm.

The aperture stop is between the second and third elements at a distance of 5.23 mm from the second lens. The diameter of the stop is 27 mm. The distance from the front surface to the entrance pupil is $t_1 = 21.409$ mm. The distance from the last surface to the exit pupil is $t' = -9.822$ mm. The diameter of the exit pupil is $D' = 29.4$ mm.

Constructional data

	n_D	V	Glass	D_c	Sag in D_c
$r_1 = 38.475$	$n_1 = 1$			32.2	3.53
	$d_1 = 5.6$ $n_2 = 1.6126$	58.3	TK16		
$r_2 = \infty$				32.2	0
	$d_2 = 6.35$ $n_3 = 1$				
$r_3 = -86.068$				27.6	1.11
	$d_3 = 2.5$ $n_4 = 1.5749$	41.3	LF5		
$r_4 = 35.29$				26.3	2.54
	$d_4 = 8.12$ $n_5 = 1$				
$r_5 = -271.48$				27.1	0.34
	$d_5 = 2.0$ $n_6 = 1.5294$	51.8	OF1		
$r_6 = 36.149$				27.5	2.72
	$d_6 = 7.5$ $n_7 = 1.6227$	56.9	TK10		
$r_7 = -55.65$				27.7	1.75
	$n_8 = 1$				
$f' = 140.47$					
$s_F = -122.77$					
$s'_{F'} = 122.45$					

Aberration tables

On-axis point

h_1	$\tan \sigma'$	D					G'			C			$s'_G - s'_C$
		s'	$\Delta s'$	$\Delta f'$	$\eta, \%$	$\Delta y'$	s'	$\Delta s'$	$\Delta y'$	s'	$\Delta s'$	$\Delta y'$	
0	0	122.450	0	0	0	0	122.285	−0.165	0	122.600	0.150	0	−0.315
$15.6\sqrt{1/2}$	0.079	121.855	−0.595	−0.579	0.038	−0.047	121.977	−0.473	−0.037	121.954	−0.496	−0.039	0.023
15.6	0.112	122.502	0.052	0.057	0.002	0.006	123.045	0.595	0.067	122.516	0.066	0.007	0.529

Off-axis point

σ_1	t_1	t'	z'_s	z'_m	$z'_s - z'_m$	y'	$\Delta y'$		$y'_G - y'_C$
							mm	%	
−19° 57.4′	22.386	−9.779	−0.899	−0.149	−0.750	51.100	0.091	0.18	0.043
−25° 00.1′	23.192	−9.774	−0.136	0.100	−0.236	65.686	0.179	0.27	0.078

Aberrations of oblique meridional ray fans

$\sigma_1 = -19°57,4'$

m	D			G'		C		$y'_{G'} - y'_C$	D		
	$\tan \sigma'$	$\Delta \tan \sigma'$	y'	y'	$\Delta y'$	y_1	Δy		h_1	h_D	h_7
13,1	−0,295	0,092	50,992	51,039	−0,108	50,993	−0,107	0,046	5,1	10,5	14,4
13,1√1/2	−0,321	0,066	51,069	51,099	−0,031	51,067	−0,033	0,032	1,1	7,5	11,4
0	−0,387	0	51,100	51,134	0,034	51,091	−0,009	0,043	−7,8	0	3,7
−13,1√1/2	−0,456	−0,069	51,033	51,057	−0,067	51,016	−0,084	0,041	−16,1	−8,1	−4,9

$\sigma_1 = -25°00,1'$

m	D			G'		C		$y'_{G'} - y'_C$	D		
	$\tan \sigma'$	$\Delta \tan \sigma'$	y'	y'	$\Delta y'$	y'	$\Delta y'$		h_1	h_D	h_7
11,2	−0,417	0,082	65,505	65,551	−0,181	65,503	−0,135	0,048	0,4	8,9	13,8
11,2√1/2	−0,440	0,057	65,622	65,665	−0,064	65,616	−0,021	0,049	−2,9	6,3	11,2
0	−0,497	0	65,686	65,746	0	65,668	0,060	0,078	−10,2	0	4,8
−11,2√1/2	−0,556	−0,059	65,523	65,606	−0,163	65,493	−0,080	0,113	−16,9	−6,8	−2,6

Aberrations of oblique sagittal ray fans

M	$\sigma_1 = -19°57.4'$			M	$\sigma_1 = -25°00.1'$		
	$\tan \psi'$	$\Delta y'$	$\Delta x'$		$\tan \psi'$	$\Delta y'$	$\Delta x'$
$13.1\sqrt{1/2}$	0,062	0,009	−0,067	$11.2\sqrt{1/2}$	0,051	−0,001	−0,002
13.1	0,088	−0,008	−0,041	11,2	0,072	−0,022	0,035

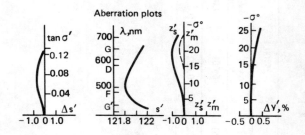

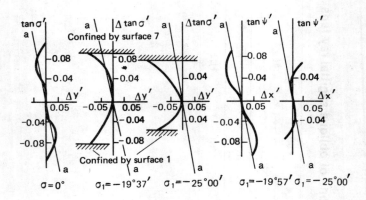

Note. The line *aa* corresponds to the film plane displaced by − 0.5 mm wi respect to the Gaussian plane.

References

In Russian

1. M.I. Apenko, A.S. Dubovik. *Applied Optics*. Nauka, Moscow 1982.
2. D.S. Volosov. *Photographic Optics. Theory, Fundamentals of Design, Optical Characteristics*. Iskusstvo, Moscow 1978.
3. M.M. Rusinov (ed.) *Computational Optics Handbook*. Mashinostroyeniye, Leningrad 1984.
4. M.M. Gurevich. *Photometry (Theory, Methods, Instrumentation)*. Energoatomizdat, Leningrad 1983.
5. N.P. Zakaznov, V.V. Gorelik. *Manufacturing of Aspherics*. Mashinostroyeniye, Moscow 1978.
6. N.P. Zakaznov. *Applied Geometrical Optics*. Mashinostroyeniye, Moscow 1984.
7. Yu.M. Klilkov. *Applied Laser Optics*. Mashinostroyeniye, Moscow 1985.
8. G.V. Kreopalova, D.T. Puryayev. *Investigation and Control of Optical Systems*. Mashinostroyeniye, Moscow 1978.
9. G.V. Kreopalova, N.L. Lazareva, D.T. Puryayev. *Optical Measurements*. Mashinostroyeniye, Moscow 1984.
10. L.Z. Kriksunov, *A Handbook of Elementary Infrared Technology*. Sovetskoye Radio, Moscow 1978.
11. C.V. Kulagin, E.M. Aparin. *Photo and Cine Instrumentation Design*. Mashinostroyeniye, Moscow 1986.
12. L.P. Lazarev. *Flying Vehicle Guiding Systems*. Mashinostroyeniye, Moscow 1984.
13. D.D. Maksutov. *Astronomical Optics*. Nauka. Leningrad 1979.
14. *International Dictionary of Photometry*. Russky Yazyk, Moscow 1979.
15. N.N. Mikhelson. *Optical Telescopes. Theory and Design*. Nauka, Moscow 1976.
16. V.A. Panov, V.N. Andreyev. *Optics of Microscopes*. Mashinostroyeniye, Leningrad 1976.
17. S.A. Rodionov. *Computer Aided Design of Optical Systems*. Mashinostroyeniye, Leningrad 1982.
18. M.M. Rusinov. *Technical Optics*. Mashinostroyeniye, Leningrad 1979.
19. L.N. Andreyev, A.P. Grammatin, S.I. Kiryushin, V.I. Kuzichev. *Problems in Optical System Theory*. Mashinostroyeniye, Moscow 1987.
20. G.G. Slyusarev. *Optical System Design*. Mashinostroyeniye, Leningrad 1975.
21. V.A. Panov (ed.). *A Handbook of Optico-Mechanical Instrumentation Designer*. Mashinostroyeniye, Leningrad 1980.
22. S.M. Kuznetsov and M.A. Okatov (eds.). *A Handbook of Optical Technologist*. Mashinostroyeniye, Leningrad 1983.
23. Yu.G. Yakushenko. *Theory and Design of Optoelectronic Instrumentation*. Sovetskoye Radio, Moscow 1980.

In English

24. R. Shannon and J. Wyant (eds.) *Applied Optical Engineering*. Vol. VIII. Academic Press, New York 1980.
25. *Optical Encyclopaedia and Dictionary for Engineers and Specifiers*. Optical Spectra Magazine, Vol. 2, 1979.
26. G.G. Slyusarev. *Aberration and Optical Design Theory*, Adam Hilger, Bristol 1984.
27. R. Feder. Optical calculations with automatic computing machinery. *J. Opt. Soc. Amer.* **41**, 630-635, 1951.
--- Automatic Optical Design. In: *Applied Optics*, Vol. II, 1209-1226, 1963.

INDEX

A

Abbe number, 61
Abbe sine condition, 136
Abbe-Young formula, 146
Aberrations
    allowable residual, 370
    chromatic, 176
    monochromatic, 154
    of aspherics, 385
    primary, 156
    ray, 154
    secondary, 156
    spherical, 162
    spherical reflector, 407
    summation of, 418
    third-order, 156
    tolerances, 368
    two-mirror objective, 408
    wave, 429
Absorptance, 129
Absorption filters, 134
Afocal anamorphic attachments, 361
Afocal corrector, 413
Airy disc, 333
Anamorphic factor, 354
Anamorphic systems, 354
Angular apertures, 101
Aperture
    diffraction effects of, 217
    stop, 101
Aperture angles, 101
Aperture ratio, 138
Aplanatic axial points, 169
Apochromat, 180
Apparent field, 230
Astigmatic ray pencils, 145
Astigmatism, 145, 170, 173
Automatic correction, 415
Auxiliary ray parameters, 371
Axial chromatic aberration, 176
Axikon, 94
Axis of vision, 187

B

Back field of view, 107
Back focal length, 57
Binocular parallax, 191
Bouguer law, 132

C

Candela, 115
Cardinal points, 33
Cassegrain reflecting system, 285
Catadioptric surface, 52
Catadioptric system, 410
Cauchy dispersion formula, 60
Celestial-body recording system, 331
Cemented doublet, 394
Chromatic difference of magnification, 181
Chromatic variation of spherical aberration, 182
Collimator, 195
Compensator, 90
Condenser, 205
    aplanatic menisci, 394
    identical-power-lens, 393
Conjugates, 31
Conjugate distance equation, 35
Constringence, 61
Continuous spectrum, 112
Critical angle, 18
Cylindrical afocal attachment, 361

D

Darc adaptation, 189
Depth of field, 215, 275
Diascopic projection, 308
Dioptre, 67
Dioptric surface, 52
Directional screen, 100
Dispersion, 87
Dispersive power, 61

Distortion, 174
   barrel, 175
   pin-cushion, 175
Doublet, 49
   air separated, 397
   cemented, 394

E

Emissivity, 325
Enlarger, 315
Entrance port, 108
Entrance pupil, 103
Episcopes, 310
Episcopic projection, 308
Exit port, 108
Exit pupil, 103
Eye, 185
   defects, 192
   relief, 228
Eyepieces, 221, 232

F

Fermat's principle, 14
Field aperture, 101
Field curvature, 170
Field lens, 236
Field of view, 107
Field stop, 101
Film format applications, 269
Filters
   absorption, 134
   interference, 135
Flying-spot projection system, 304
F-number, 138
Focal length, 32
Focal plane, 32
Focal ratio, 138
Fraunhofer lines, 112

G

Gaussian rays, 53
Geometrical optics, 12
Ghosting, 78

H

Homocentric ray bundle, 20
Hyperfocal distance, 277

I

Illumination, 117
Illumination systems, 222
Image evaluation, 427
Image illuminance, 135
Image space, 10
Isoplanatism, 168

L

Lagrange invariant, 155
Lagrange law, 55
Lambertian diffuser, 123
Lambert's cosine law, 123, 132
Laser beam,
   divergence control, 347
   expanders, 352
   properties, 342
   transformation, 343
   waist, 343
Lateral colour, 181
Law of reflection, 16
Law of refraction, 15
Lens
   anamorphic, 357
   biconcave, 69
   cemented doublet, 394
   concentric spherical, 71
   condenser, 205
   conoid, 98
   decentred, 426
   field, 221
   focon, 93
   Fresnel, 94
   high-speed two-element, 400
   objective, 265
   photographic, 280
   planoconcave, 70
   planoconvex, 69
   reverse telephoto, 289
   single, 64
   Soviet TV-camera, 440
   spherocylindrical, 360
   spheroelliptic, 73
   telescopic, 70
   types of, 65
   wide-angle, 290
   with aspheric surfaces, 72
   with inverted principal planes, 72

zoom, 255, 287
Light exposure, 118, 278
Lightguides, 91
    aperture angle of, 92
Linear coverage, 107
Line spectrum, 112
Longitudinal magnification, 38
Lumen, 117
Luminance, 118
Luminous efficiency, 122
Luminous emittance, 118
Luminous flux, 115
Luminous intensity, 115
Lux, 118

M

Magnification, 31
    angular, 36
    between pupils, 305
    defined, 31
    lateral, 31, 36, 105
    longitudinal, 38
    microscope, 208
    telescope, 225, 227
    transverse, 31, 36, 107
    TV-channel, 299
    useful, 229
Mangin mirror, 200
Maksutov meniscus, 411
Mean dispersion, 61
Meniscus, 70
    corrector, 410
Meridional coma, 167
Microscope, 208
    compound, 211
    resolution of, 214
Modulation (contrast) transfer factor, 273,
        302
Modulation transfer function (MTF), 271,
    302
Mirrors
    ellipsoidal, 201
    Mangin, 200
    plane, 76
    spherical, 77, 199, 204

N

Newton's equation, 35

Nodal points, 37
Numerical raytracing, 140, 149

O

Object space, 10
Objectives, 218, 230
    air separated doublet, 397
    high-speed separable, 400
    concentric, 291
    identical-power-lens, 390
    Maksutov, 220
    microscope, 220
    photographic, 265
    TV-camera, 299
    two-lens cemented, 394
    two-mirror, 408, 414
    with aplanatic menisci, 392
Ocular, 226
Opacity, 132
Optical axis, 19
Optical density, 132
Optical detectors, 319
    integral sensitivity of, 320
    optical arrangement for, 329, 338
    sensitivity threshold, 321
    spectral responsivity, 319
Optical glass manufacturers, 63
Optical invariant, 55
Optical path, 14
Optical specifications, 422
Optical system
    anamorphic, 354
    Cassegrain reflection, 285
    catadioptric, 201
    celestial-body recording, 331
    centred, 19, 52
    collimator, 196
    components, 60
    concept, 10
    condenser, 205
    cylindrical anamorphic, 358
    design, 364
    energy flow in, 112, 129
    eye as, 185
    holography, 351
    illumination, 195, 222
    image illuminance, 135
    laser, 342

linear coverage of, 107
materials, 60
multielement, 47
numerical raytracing for, 140
pancratic, 255
perfect, 31
photoelectric, 319
power of, 46
projection, 304, 308
pupils of, 103
radiative transfer, 123
raytracing in, 39
reduced, 161
single lens, 330
telescopic, 49
thin lens, 376
TV, 292
two-lens, 333, 353
zoom, 251
Optical raster, 99
Optical transfer function, 273
Optics of TV system, 292

P

Pancratic erector system, 256
Paraxial approximation, 53
Paraxial rays, 33, 53
Partial dispersions, 61
Performance function, 416
Petzval curvature, 400
Phase transfer function, 274
Photometry, 114
Photopic vision, 121
Picture tubes, 297
Planck's law, 324
Plane-parallel plates, 74
Power of optical system, 46, 49
Principal focus, 32
Principal planes, 32
Principal points, 32
Principal ray, 101
Prisms
        Amici, 81
        angular dispersion of, 87
        deviation of, 85
        direct vision, 88
        Dove, 83, 432
        erecting, 241
        half-speed, 83
        Leman (springer), 245, 435

Malafeyev, 245
Pechan, 436
penta, 82, 433
Porro, 82
power of, 85
principal section of, 80, 85
rangefinder, 434
reflecting, 79, 432
refracting, 85, 88
refracting angle of, 80
refracting edge of, 85
roof, 81, 432
Rutherford, 87
Schmidt, 245, 435
Projection lantern, 312
Projection systems, 308
Pupil
        effective aperture of, 109
        entrance, 103
        exit, 103
Purkinje shift, 190

Q

Quarter-wavelength coating, 130

R

Radiometry, 114
Rayleigh criterion, 229
Raytracing
        for perfect system, 46
        for tilted object planes, 43
        in optical system, 39
Rear field, 230
Reflectance, 129
Reflecting systems, 406
Reflection from
        aspheric surface, 23
        convex spherical surface, 26
        plane surface, 23
        second-order surface, 26
        spherical surface, 25
Reflection law, 16
Refracting and reflecting surfaces, 18
Refraction at
        aspheric surface, 27
        second-order surface, 21
        spherical surface, 21
Refractive index, 13, 14
        of liquids, 64
        variation of, 60

Index   453

Relative aperture, 138
Resolving power, 229, 269
     of camera tubes, 296
Responsivity function, 121
Retro-reflector, 78
Reversibility principle, 16

S

Screen phosphors, 298
Searchlight, 196
Secondary spectrum, 179
Seidel sums, 158
     transformation of, 376
Sign convention, 14
Single lens, 64
Smith-Helmholtz equation, 55
Snell's law, 15
Soviet glass designations, 63
Spectral characteristics, 323
Spectral luminous efficacy, 121
Spectral responsivity, 319
Spherical aberration, 162
     minimization of, 390
Spherochromatism, 182
Spread function, 272
Stefan-Boltzmann law, 324
Stereoscopy, 191
Stop
     aperture, 101
     field, 101
     telecentric, 102
Straubel invariant, 128

T

Telecentric stop, 102
Telescopes, 225
     Galilean, 242
     Kepler, 239
     lens erecting, 248

     pancratic, 255
     stereoscopic, 259
     terrestrial, 248
Telescopic system, 49
Thin element, 381
Threshold of vision, 190
Tolerances, 422
Transverse chromatic aberration, 180
Triplet
     airspaced, 403
     Cooke, 404
Telecentric ray bundle, 102
Telescopic lens, 70
Television episcope, 146
Threshold modulation curve (TM), 274
Total internal reflection, 18
Transmittance, 129
TV camera and picture tubes, 292
TV systems, 393
     MTF of, 302
     resolution of, 301

V

Vignetting, 108
Virtual image, 23
Visual sensitivity, 190
V-value, 61

W

Wedge, 89
Wien's displacement law, 324
Window
     entrance, 108
     exit, 108

Z

'Zero-ray' concept, 56
Zoom systems, 251

TO THE READER

Mir Publishers welcome your comments on the content, translation, and design of the book.

We would also be pleased to receive any suggestions you care to make about our future publications.

Our address is:

USSR, 129820, Moscow, I-110, GSP, Pervy Rizhsky Pereulok, 2, Mir Publishers.

N. Rykalin, A. Uglov, I. Zuev, A. Kokora
LASER AND ELECTRON BEAM MATERIAL PROCESSING HANDBOOK
MIR PUBLISHERS, MOSCOW, 1988

The handbook contains information on the basic thermophysical and hydrodynamic processes involved in the treatment of materials by laser and electron beams and also describes the methods of calculation of thermal processes. The coverage includes heat treatment, machining to dimensions, film and coat forming, metal welding and cutting, and the equipment used for the purpose. Gives information on the control of electron-beam heating. The handbook is intended for workers in mechanical engineering.

L. Sharupich, N. Tugov

OPTOELECTRONICS

MIR PUBLISHERS, MOSCOW, 1987

The book aims at introducing the reader to the elements of most important physical phenomena used in semiconductor optoelectronics and also the physical properties, parameters, and characteristics of basic optoelectronic, primarily semiconductor, devices.

In describing each class of devices, the text covers the physical processes, principles of action and structures of the devices, their parameters, characteristics, fields of application, and basic types. Considers main manufacturing methods and production cycles in technology of fabrication of optoelectronic devices.